A Comparison of the Dynamical Evolution of Planetary Systems

A Comparison of the Dynamical Evolution of Planetary Systems

Proceedings of the Sixth Alexander von Humboldt Colloquium on Celestial Mechanics Bad Hofgastein (Austria), 21–27 March 2004

Edited by
Rudolf Dvorak and Sylvio Ferraz-Mello

Reprinted from *Celestial Mechanics and Dynamical Astronomy*, Volume 92(1–3), 2005

Springer

A C.I.P. catalogue record for this book is available from the Library of Congress

ISBN 1-4020-4218-3

Published by Springer,
P.O. Box 17, 3300 AA Dordrecht, The Netherlands

Sold and distributed in North, Central and South America
By Springer,
101 Philip Drive, Norwell, MA 02061, USA

In all other countries, sold and distributed
By Springer,
P.O. Box 322, 3300 AH Dordrecht, The Netherlands

Printed on acid-free paper

Printed in the Netherlands

Cover photograph: Alexander von Humboldt und Aimé Bonpland in der Urwaldhütte (*Alexander von Humboldt and Aimé Bonpland in the Forest Cottage*). ***Picture by:*** Eduard Ender, ca. 1850. ***Supplied courtesy:*** Berlin-Brandenburgische Akademie der Wissenschaften, Berlin, Germany.

TABLE OF CONTENTS

Celestial Mechanics and Dynamical Astronomy (2005) 92:vii–viii
DOI 10.1007/s10569-005-3582-2

PREFACE

The traditional Alexander von Humboldt Colloquium on Celestial Mechanics, the sixth in the series after the first one 20 years ago, was held in Bad Hofgastein, in the Hotel Winkler, on 21–27 March, 2004. Like in the former colloquia, some 45 scientists from all over the world gathered in the Austrian Alps to present and discuss their newest results in different fields of our science.

We had a wide spectrum of talks covering the most recent developments in our area from the theoretical point of nonlinear dynamical systems to the application to 'real' astrodynamical problems. We had interesting talks and discussions on the formation of planetary systems, their stability and also the problem of habitable zones in extrasolar planetary systems. A special topic was the stability of Trojans in our planetary system, where more and more realistic dynamical models are used to explain their complex motions: besides the important contribution from the theoretical point of view, the results of several numerical experiments revealed the structure of the stable zone around the libration points. Even in extrasolar planetary systems, such stable orbits for terrestrial like planets may exist and be stable. The *Heinrich Eichhorn lecture*, given by Zoran Knežević, from Beograd, in honor of this Austrian astronomer, was devoted to another very important topic in Celestial Mechanics: the determination of orbits of Near Earth Asteroids.

All authors were encouraged to write papers of a length that they considered suitable for the presentation of their results. The editorial board of *Celestial Mechanics and Dynamical Astronomy* arranged for competent and fast refereeing so that all papers could be reviewed and, when necessary, revised before publication.

Each preface in the five preceding colloquia showed that as time is going on it is more and more difficult to find sponsors for our scientific meetings, a fact from which many of us are suffering. Nevertheless we managed to find the necessary funds also for the 2004 meeting. Special thanks have to go to the director of the IAP in Potsdam, Prof. Klaus Strassmeier, without whom, our efforts would have been unsuccessfully. His interest on the subject of extrasolar planetary systems, a hot topic in Astronomy, and his financial support saved our meeting. We also have to thank the Austrian government, which supported us with a notable amount of money, furthermore the Salzburger Landesregierung, the Österreichische Forschungsgemeinschaft and the Austrian Space Agency. Thanks are also due to the Marktgemeinde

Bad Hofgastein who made the very successful Salzburger Abend with indigenous music from Salzburg possible. Special thanks also to the former director of the Institute of Astronomy in Vienna, Prof. Paul Jackson for his generous private donation. We should not forget our hosts Mr. and Mrs. Winkler and their employees from the hotel who made the stay quite enjoyable. None of us will forget the very last evening, when the staff of kitchen under the leadership of the cook himself came to offer us as farewell the famous Salzburger Nockerln, a traditional Austrian dessert. Everyone got a lot of scientific input during the lectures and the discussions and, to summarize, we all had a splendid week in Salzburg in the Hotel Winkler. We all hope to come again in 2008 to discuss new results and new perspectives on a high level scientific standard in the Gasteinertal.

Rudolf Dvorak and Sylvio Ferraz-Mello

Celestial Mechanics and Dynamical Astronomy (2005) 92:1–18
DOI 10.1007/s10569-005-3314-7

FROM ASTROMETRY TO CELESTIAL MECHANICS: ORBIT DETERMINATION WITH VERY SHORT ARCS

(Heinrich K. Eichhorn Memorial Lecture)

ANDREA MILANI[1] and ZORAN KNEŽEVIĆ[2]

[1]*Department of Mathematics, University of Pisa, via Buonarroti 2, 56127 Pisa, Italy, e-mail: milani@dm.unipi.it*

[2]*Astronomical Observatory, Volgina 7, 11160 Belgrade 74, Serbia and Montenegro, e-mail: zoran@aob.bg.ac.yu*

(Received: 1 October 2004; revised: 15 February 2005; accepted: 7 March 2005)

Abstract. Contemporary surveys provide a huge number of detections of small solar system bodies, mostly asteroids. Typically, the reported astrometry is not enough to compute an orbit and/or perform an identification with an already discovered object. The classical methods for preliminary orbit determination fail in such cases: a new approach is necessary. When the observations are not enough to compute an orbit we represent the data with an attributable (two angles and their time derivatives). The undetermined variables range and range rate span an *admissible region* of solar system orbits, which can be sampled by a set of *Virtual Asteroids* (VAs) selected by an optimal triangulation. The attributable results from a fit and has an uncertainty represented by a covariance matrix, thus the predictions of future observations can be described by a quasi-product structure (admissible region times confidence ellipsoid), which can be approximated by a triangulation with each node surrounded by a confidence ellipsoid. The problem of identifying two independent short arcs of observations has been solved. For each VA in the admissible region of the first arc we consider prediction at the time of the second arc and the corresponding covariance matrix, and we compare them with the attributable of the second arc with its own covariance. By using the penalty (increase in the sum of squares, as in the algorithms for identification) we select the VAs which can fit together both arcs and compute a preliminary orbit. Even two attributables may not be enough to compute an orbit with a convergent differential corrections algorithm. The preliminary orbits are used as first guess for constrained differential corrections, providing solutions along the *Line Of Variations* (LOV) which can be used as second generation VAs to further predict the observations at the time of a third arc. In general the identification with a third arc will ensure a least squares orbit, with uncertainty described by the covariance matrix.

Key words: asteroid recovery, ephemerides, orbit determination

1. Introduction

The astrometric observations of a small body by themselves do not provide an orbit for the observed body, thus do not provide information on the nature of the object (asteroid, comet, satellite, Transneptunian). The first complete mathematical method to convert astrometry into orbits had been established by Gauss (1809): he devised an algorithm to compute a preliminary orbit satisfying three given observations in different nights. When additional observations became available, Gauss proposed to correct the preliminary orbit by solving a least squares problem. This method is now called *differential corrections*, and this sequence, preliminary orbit followed by least squares,[1] is now the algorithm almost universally used and considered classic. At the time when the asteroids were detected as "intruders" not found in star charts, the observations were indeed typically only one per night, and the algorithms found by Gauss were the optimal solution of the orbit determination problem. The fact is, the circumstances of the observations of asteroids (and other small bodies) are now deeply changed: the historical discovery procedure should not be conditioning our way of thinking about orbit determination to be performed with modern data.

The number of asteroid observations has in recent years increased dramatically, mostly because of the automated surveys like LINEAR, LONEOS, Catalina, Spacewatch, NEAT. The procedures of operation of these surveys are basically the same, although they can differ in details. A number of images of the same area on the celestial sphere are taken over a short time span, typically within a single night. The images are then digitally blinked and all the changes from one to another logged. If an object moves from image to image, at a constant rate and along a straight line, this is probably the detection of a real body. The series of observations, usually consisting of 3–5 positions over a time span 1–2 hours, are reported as a sequence of individual observations of the same object (note that this initial identification is done by the observer); we shall refer to this sequence as a *very short arc*.

This method of work is optimal for the discovery of asteroids and comets, but it is not suited for the determination of their orbits. In Gauss' method for preliminary orbit determination the curvature of the arc on the celestial sphere appears as divisor already in the first iteration. The smaller the curvature, the less accurate the resulting orbit: taking also into account the observational errors, often the standard algorithm fails to provide the solution. Either the preliminary orbit cannot be determined at all, or it can be computed, but the differential corrections do not converge. In such a case, we speak of a *Too Short Arc* (TSA), by which we mean too short for

[1]In the modern orbit determination, the least squares solution is computed by solving numerically the perturbed N-body problem, while Gauss was using the analytical solution of the 2-body problem, but the spirit is not changed.

orbit determination. The reported sequence of observations can be considered the detection of a moving object, rather than a discovery.

This situation is unsatisfactory, given the wealth of information contained in these observations, which remains unused due to the failure of the orbit determination procedure. Without an orbit, on the other hand, we cannot determine what kind of body we are looking at, we can neither compute an ephemeris for the later follow up observations, nor identify the observed object with any other associated to a known orbit. We need to establish a new paradigm for the process leading from astrometry to orbit determination, working efficiently and reliably under the prevailing observing conditions of today.

Our goal is to develop the procedure which would allow to extract all the existing information from the TSA, and to combine it with some plausible assumptions about the nature of the motion of the detected body, in order to get preliminary orbits. Our research plan consists of several steps, of which three are completed (Milani et al., 2004, 2005a,b); these are described in the present paper. The basic idea is as follows: a TSA comprises a number of observed positions with deviations from alignment compatible with a random observational error. We can fit a straight line to the data and compute two average angular coordinates and their corresponding average angular rates, assigning the results to the reference epoch (simple mean of the observing times). We shall call such a set of data an *attributable*.

Note that an attributable does not provide any information on the geocentric distance (range) of the body and its radial velocity (range rate) at the reference time. However, the range and range rate are constrained if we assume that the body belongs to the solar system, but that it is not a satellite of the Earth. Hence we introduce a concept of *admissible region*, which in our algorithm replaces the conventional confidence region as defined in the classical orbit determination procedure. Such a region can be sampled by virtual asteroids, and we can compute a sort of generalized ephemerides which allow identification in the sense of attribution (Milani et al., 2001), linking of two TSA's and computation of preliminary orbit, detection of virtual impactors, etc.

Our method represents an extension and important upgrade of the existing methods developed by Virtanen et al. (2001), by Tholen and Whiteley (2003) and by Goldader and Alcock (2003). We introduced several significant improvements making our method more efficient and reliable, but we agree with these authors' main conclusion that ephemerides prediction is often possible, with an accuracy good enough e.g. for recovery planning, even when the orbit cannot be computed in the usual way.

2. Definition of the Admissible Region

We assume that at time t an asteroid $\mathscr{A}$ with heliocentric position P is observed from the Earth, which is at the same time in $P_\oplus$. Let (r, α, δ) be spherical coordinates for the geocentric position $P - P_\oplus$.

DEFINITION 1. *We shall call* **attributable** *a 4-dimensional vector* $A = (\alpha, \delta, \dot{\alpha}, \dot{\delta})$, *observed at a time* t.

Here t has to be interpreted as the mean of the observation times. The angles (α, δ) can be specified as necessary: usually the geocentric equatorial coordinates, right ascension and declination for the standard epoch J2000, are used. Also, with the data contained in the observations we optionally can have another component of the attributable – an average apparent magnitude h. Note that range and range rate $(r, \dot{r})$ are left completely undetermined by this definition.

The conditions to constrain $(r, \dot{r})$ make use of the following well–known quantities:

Heliocentric two-body energy

$$\mathscr{E}_\odot(r, \dot{r}) = \frac{1}{2}\|\dot{P}\|^2 - k^2 \frac{1}{\|P\|}, \tag{1}$$

where $k = 0.01720209895$ is Gauss' constant;
Geocentric two-body energy

$$\mathscr{E}_\oplus(r, \dot{r}) = \frac{1}{2}\|\dot{P} - \dot{P}_\oplus\|^2 - k^2 \mu_\oplus \frac{1}{\|P - P_\oplus\|}, \tag{2}$$

where $\mu_\oplus$ is the ratio (mass of the Earth)/(mass of the Sun);

Radius of the sphere of influence of the Earth

$$R_{\mathrm{SI}} = a_\oplus \left(\frac{\mu_\oplus}{3}\right)^{1/3} = 0.010044\,\mathrm{AU},$$

that is the distance from the Earth to the collinear Lagrangian point L_2, apart from terms of order $\mu_\oplus^{2/3}$. Here $a_\oplus$ is the semimajor axis of the orbit of the Earth;

Physical radius of the Earth

$$R_\oplus \simeq 4.2 \times 10^{-5}\,\mathrm{AU}.$$

The following four conditions make now obvious physical sense:

(A) $\mathscr{D}_1 = \{(r, \dot{r}) : \mathscr{E}_\oplus \geqslant 0\}$ ($\mathscr{A}$ is not a satellite of the Earth);
(B) $\mathscr{D}_2 = \{(r, \dot{r}) : r \geqslant R_{\mathrm{SI}}\}$ (the orbit of $\mathscr{A}$ is not controlled by the Earth);
(C) $\mathscr{D}_3 = \{(r, \dot{r}) : \mathscr{E}_\odot \leqslant 0\}$ ($\mathscr{A}$ belongs to the solar system);
(D) $\mathscr{D}_4 = \{(r, \dot{r}) : r \geqslant R_\oplus\}$ ($\mathscr{A}$ is outside the Earth).

DEFINITION 2. *Given an attributable A, we define as* **admissible region** *the domain*

$$\mathscr{D} = \{\mathscr{D}_1 \cup \mathscr{D}_2\} \cap \mathscr{D}_3 \cap \mathscr{D}_4 .$$

3. Borders of the Admissible Region

The multi-line border of the admissible region can be mathematically described in a rigorous way. The procedures and all the results are described in full detail in Milani et al. (2004).

The admissible region cannot have more than two connected components. More precisely, the degree six polynomial resulting from condition (C) cannot have more than three real positive roots: when there is only one such root, the admissible region has only one component, when there are three, it has two components (see Figures 1 and 2 in Milani et al., 2004).

The boundary of the admissible region consists of

1. part of the algebraic curve $\mathscr{E}_\odot = 0$. If the degree six polynomial has three positive roots there is another component, a simple closed curve, at larger values of r: this includes the case when this curve reduces to a single point, if there is a double positive root;
2. two segments of the straight line $r = R_\oplus$;
3. two portions of the curve $\dot{r}^2 = G(r)$ (corresponding to $\mathscr{E}_\oplus = 0$) and one segment of the straight line $r = R_{\mathrm{SI}}$ if $R_{\mathrm{SI}} < r_0$; if $R_{\mathrm{SI}} \geqslant r_0$ the two portions of the $\dot{r}^2 = G(r)$ are joined at $r = r_0$.

Note that $G(r)$ derives from condition (A):

$$\dot{r}^2 \geqslant \frac{2k^2 \mu_\oplus}{r} - \eta^2 r^2 := G(r),$$

where η is the proper motion and $G(r) > 0$ for

$$0 < r < r_0 = \sqrt[3]{\frac{2k^2 \mu_\oplus}{\eta^2}} .$$

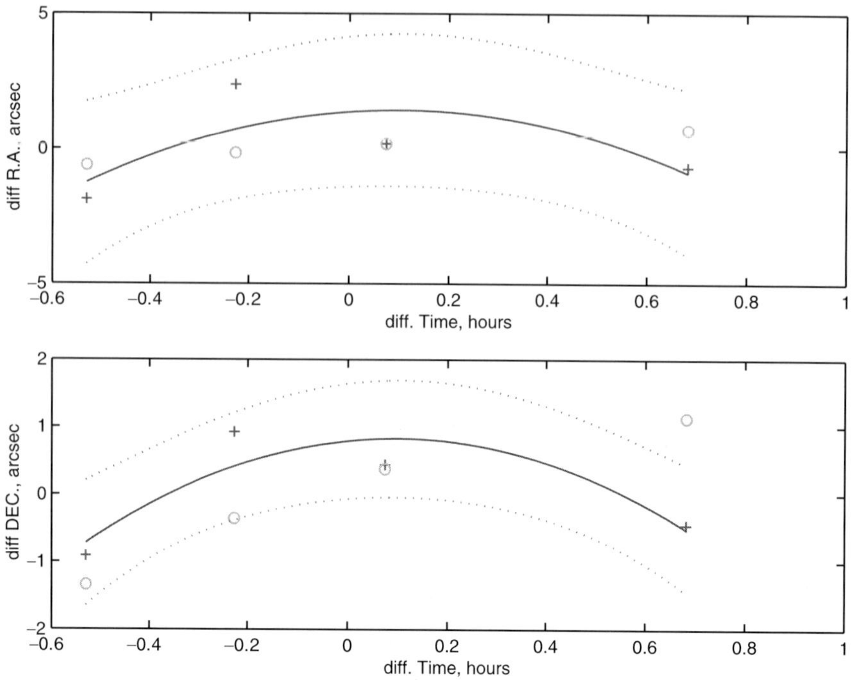

Figure 1. Residuals in right ascension (above) and declination (below) with respect to the fit of the observations of the asteroid 2004 SA_1 taken by LINEAR on 14 January 2004. The crosses represent the observations as reported, the continuous line is the best fitting parabola, the dotted lines the confidence boundaries for the parabolic fit, the circles are the observations "without astrometric error", computed from a very well determined orbit known a posteriori.

This result provides full analytical and topological description of the admissible region. From the metric point of view, however, the definition of the region is not entirely satisfactory, since the inner boundary might be too close to the observer. Instead of condition (D), therefore, one can set an upper limit for the absolute magnitude of the body to exclude very small and very close objects of minor importance:

(E) the absolute magnitude H of the object is $\leqslant H_{\max}$.

The region defined by condition (E) is a half plane $r \geqslant r_H$ (the difference between the absolute and the observed apparent magnitude does not depend on the range rate). As an example, if we set $H_{\max}=25$, corresponding to a body which most likely would not result in a very significant damage on the ground in case of impact, and the measured apparent magnitude is $\simeq 20$, the resulting $r_H \simeq 0.01$ AU.[2]

[2]For an accurate computation of r_H the phase effect has to be taken into account.

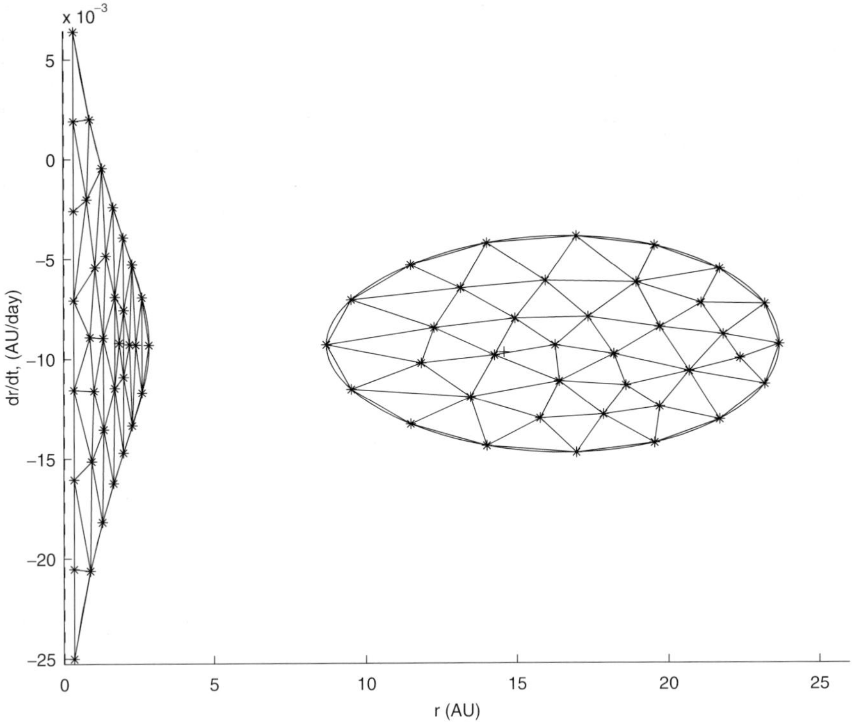

Figure 2. Triangulation of the modified admissible region for the prediscovery observations of 2000 EC_{98}. The cross marks the actual position, in the range, range rate plane, of the Centaur, as determined a posteriori with additional observations.

Sometimes it is also desirable to limit the upper boundary of the admissible region to exclude from consideration the long periodic comets with large orbital semimajor axes $a > a_{\max} = 100\,\mathrm{AU}$; condition (C) in such a case becomes (C_{100}) $\mathscr{E}_\odot \leqslant -k^2/(2a_{\max})$. If we apply these alternative conditions (E) and (C_{100}) in place of (C) and (D), we speak of the *modified admissible region*.

4. Sampling the Admissible Region

Given a very short arc of observations and the corresponding attributable $A_0 = (\alpha, \delta, \dot\alpha, \dot\delta)$ at the mean observation time t, we need to represent the uncertainty of the orbit, hence of the ephemerides, in a different way with respect to the traditional method of covariance matrices and confidence ellipsoids. This because a least square solution for the orbital elements X, with its normal matrix C and covariance matrix Γ, is in general not computable.

If the arc is very short, to the point that there is no significant information on the curvature of the path on the celestial sphere, the values of $(r, \dot{r})$ are not constrained by the observations. If, to the contrary, there is significant curvature, this allows to constrain both the range and the range rate (Danby, 1989, Ch. 6). Figure 1 shows a typical example of a set of four observations spanning only one hour, for which the curvature is not significant, that is less than the uncertainty resulting from the known performance of the station (Carpino et al., 2003.)

Even without curvature information, if we assume the object belongs to the solar system we can limit the uncertainty in the $(r, \dot{r})$ plane to the admissible region. The (modified) admissible region is a compact subset of the $r > 0$ half plane. However, this is still an infinite set, with a complex shape, thus there are infinite possible orbits and we need to find efficient ways to sample them with a finite number of points in the initial conditions space, the *Virtual Asteroids* (VAs) sharing the reality of the object, in the sense that the orbit through one of them is a good approximation of the orbit of the real object, but we do not know which one.

This requires to sample the admissible region $\mathcal{D}$ with a number of points. The most natural and geometrically significant way to sample a 2-dimensional region is a *triangulation*, with nodes and sides joining them. Regions with simple boundaries (such as the $r > 0$ half plane) can be triangulated by equilateral triangles, but the admissible region has a complicated boundary. A *Delaunay triangulation* has a number of optimal properties, e.g., it is the triangulation with the largest minimum angle (among all the triangles).

There is an efficient algorithm to compute a Delaunay triangulation, starting from a finite sampling of the boundary. Thanks to the explicit analytic description of the admissible region we can sample the boundary with approximately uniform distances and compute a Delaunay triangulation with the given nodes on the boundary. Then the nodes are selected as points $(r_i, \dot{r}_i), i = 1, N$ sampling the admissible region, with the sides and the triangles providing an additional geometric structure.[3]

As an example, from the prediscovery observations of the Centaur 2000 EC_{98} taken on February 5, 2000 (one month before the official "discovery") we have computed the attributable, the modified admissible region and its

[3]The properties defining the Delaunay triangulation are metric ones, thus the nodes selected depend upon the choice of a metric. Different metrics can be used to enhance the resolution in some portions of the admissible region, e.g., when the main goal is to search for either NEO, or main belt asteroids, or TNO.

Delaunay triangulation, shown in Figure 2. Note that in this case, as it is common for Centaurs and Transneptunians, the admissible region has two connected components and the real object is in the one farther from the Earth.

The set of VAs selected is then

$$X_i = \left(\alpha, \delta, \dot{\alpha}, \dot{\delta}, r_i, \dot{r}_i\right), \quad i = 1, N.$$

The set of six coordinates forming the vector $X_i = (\alpha, \delta, \dot{\alpha}, \dot{\delta}, r_i, \dot{r}_i)$ can be considered as a set of orbital elements.[4] They can be converted into Cartesian position and velocity (topocentric, then heliocentric since the observer's position is well known), as well as into other types of elements, e.g., Keplerian.

To assess how representative are the set of orbital element vectors $\{X_i\}$ with respect to the full set of possible orbits we need to take into account that even the measured part of the X_i vectors, that is the 4-dimensional vector A_0, has some uncertainty. The values of the angles and their rates are computed by least square fit to a set of observations, thus their uncertainty can be represented in the conventional way with a 4×4 covariance matrix Γ_A. If we assume the value of two variables $(r_i, \dot{r}_i)$, without uncertainty, then the uncertainty of the 6-vector X_i is represented by the *conditional covariance matrix*, a 6×6 matrix Γ_X

$$\Gamma_X = \begin{bmatrix} \Gamma_A & \underline{0} \\ \underline{0} & \underline{0} \end{bmatrix}$$

with $\underline{0}$ suitable matrices with null coefficients. This matrix is obviously not positive-definite, but has the $(r, \dot{r})$ subspace as kernel (null space).

A geometric description of the confidence region for the orbits compatible with the given attributable with uncertainty (A_0, Γ_A) can be given as follows: for each point A of a confidence ellipsoid centered in A_0 (in the 4-dimensional attributable space) we need to compute the admissible region $\mathcal{D}(A)$. This gives as confidence region a subset of the 6-dimensional elements space with a positive 6-dimensional volume. However, if the attributable A_0 is well determined, the confidence region is flattened, with a two dimensional "base plate" $\{A_0\} \times \mathcal{D}(A_0)$. This set is not exactly a product (confidence ellipsoid $\times$ admissible region), because the confidence region changes with the attributable; however, it is not very different from a product and the set of VAs $X_i, i = 1, N$ selected along $\{A_0\} \times \mathcal{D}(A_0)$ is quite representative.

[4]The epoch corresponding to the initial conditions is $t_j = t - r_j/c$, with c the speed of light, to take into account the light travel time.

5. Propagating to Another Epoch

Given a triangulation $B_i, i = 1, N$, each (X_i, Γ_X) is an orbital element set with uncertainty.

If these elements are converted into other coordinates Y, such as Keplerian elements, and propagated (nonlinearly) to some later epoch t_1, the covariance matrix can be propagated (linearly) to a new one $\Gamma_Y(t_1)$. The fact that the covariance matrix Γ_X is not invertible does not matter, but of course $\Gamma_Y(t_1)$ will also have a 2-dimensional kernel. Then it is possible to compute the predicted attributable (for observation time t_1) with uncertainty (A^i, Γ_{A^i}) for each node. The ephemerides (with confidence ellipse) computed in the conventional case of a well determined orbit are thus replaced by the union of N confidence ellipsoids, one for each VA, in the 4-dimensional space of the attributables A' at the new epoch.

If another very short arc of observations is available at the second epoch, its attributable with uncertainty (A_1, Γ_{A_1}) can be compared with the predictions for each VA. As an example, for the same asteroid of Figure 2, we have computed the attributable A_1 corresponding to the official discovery observations of March 3–4, 2000: Figures 3 and 4 show the predictions A^i for the same attributable, computed by using only the data of February 5.

Both in angles and in proper motion the predictions can be good enough for pointing the telescope and to discriminate this object from others in the same fields, that is by using the triangulation technique the "discovery" could have been a targeted recovery. Note that the uncertainty arises from two mathematically very different sources: the span of the admissible region, and the confidence ellipsoids. In this example, if only the connected component farther from the Earth is considered the span of the predictions from the different VA selected in the admissible region is small, while the size of the ellipsoid is the main error term. On the contrary the other connected component would give predictions for the different VAs spread over a large part of the sky, none of them close to the recovery observations. This means that a targeted recovery after one month was possible only assuming that the object was on an orbit beyond Jupiter, an assumption which was by no means obvious given the available data.

6. Identification and Preliminary Orbits

When using the data from surveys, the correspondence between the very short arcs of observations and the physically distinct objects is not known. This is the problem of *asteroid identification*. Since the arcs are too short,

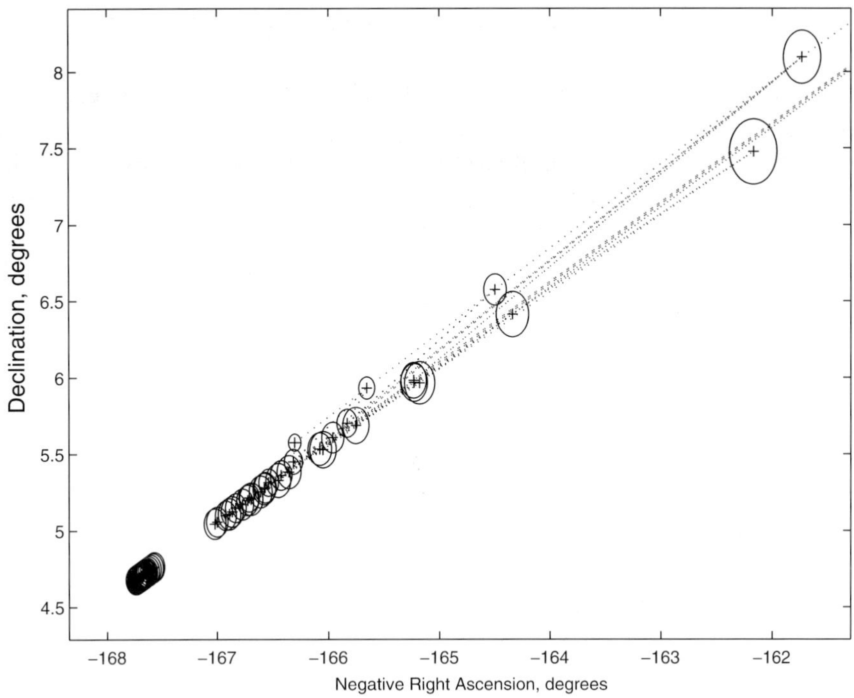

Figure 3. Prediction (propagated triangulation) for the observations of March 3–4, 2000 based only on the prediscovery observations of February 5, projected on the (α, δ) plane. The confidence ellipses are also drawn, but the uncertainty of the prediction is dominated by the unknown location on the $(r, \dot{r})$ plane at the time of the prediscovery attributable.

more than one TSA needs to be used to constrain the orbit, to be able to obtain a least squares solution. However, how do we know whether two TSAs belong to the same object, having an orbit for neither one of the two?

An algorithm has been defined (Milani et al., 2001) to decide whether two arcs of observations are to be identified, provided that an orbit is available (for at least one of the two) with its uncertainty, represented by a covariance matrix: this case of identification is called an *attribution*, hence the term *attributable*. When both observed arcs are TSA, the only available orbits are the ones of the VA defined by the triangulation of the admissible region of the first attributable.[5]

The predicted attributable with uncertainty (A^i, Γ_{A^i}) is computed for each node and it is compared with the attributable computed from the

[5]Of course it is also possible to triangulate the admissible region of the second attributable: the algorithm is not invariant by exchanging the two observed arcs.

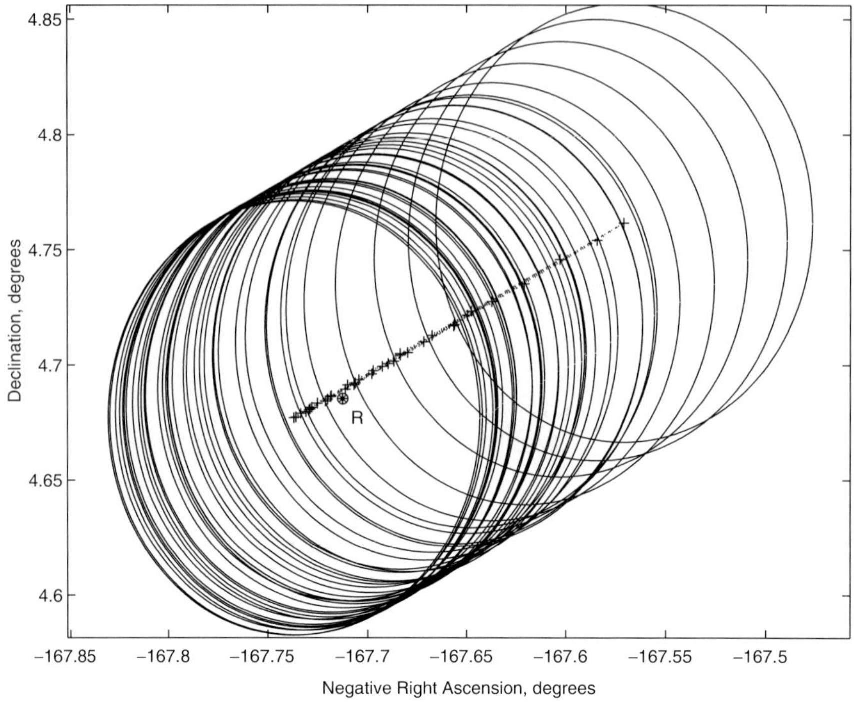

Figure 4. Detail of the previous figure, showing only the predictions from the nodes belonging to the second connected component of the admissible region, the one farther from the Earth. For this subset of VAs, the uncertainty is dominated by the uncertainty of the attributable, as expressed by its covariance matrix and represented here by the confidence ellipses. The recovery attributable as observed is labeled by the letter R.

observations of the second arc (A_1, Γ_{A_1}). By means of the normal matrices

$$C_{A^i} = [\Gamma_{A^i}]^{-1}, \quad C_{A_1} = \left[\Gamma_{A_1}\right]^{-1}$$

the *attribution penalty* K_i is computed as

$$C_1^i = C_{A^i} + C_{A_1},$$
$$C = C_{A^i} - C_{A^i}\left[C_1^i\right]^{-1} C_{A^i} = C_{A_1} - C_{A_1}\left[C_1^i\right]^{-1} C_{A_1},$$
$$K_i \simeq \frac{1}{2}\left[A^i - A_1\right]^T C \left[A^i - A_1\right].$$

If the attribution penalty K_i is low for some VA index i, then the two attributables A_0 and A_1 may belong to the same object. If this is the case, we need a preliminary orbit to be used as first guess to start the differential correction procedure. One possibility is to use X_i as the preliminary orbit, but there are ways to define a better one, taking into account that an optimal "compromise" attributable between the predicted A^i and the observed

A_1 is provided by the algorithms of Milani et al. (2001, 2005b, submitted):

$$A_1^i = \left[C_1^i\right]^{-1} (C_{A_1} A_1 + C_{A^i} A^i).$$

This algorithm has a geometrical interpretation in terms of intersections of the two families of confidence ellipsoids.

At this stage, the identification between the two attributables can only be hypothetical. The values of the attribution penalty to be considered acceptable cannot be very small, because the minimum of the values of K_i for $i = 1, N$ is not the same as the minimum for all possible choices of $(r, \dot{r})$ in $\mathcal{D}(A_0)$: the finite sampling implies an increase of the penalty with respect to the minimum possible. As a result, acceptable values may occur also for couples of attributables which either cannot be fit together, or would give large residuals.

The procedure outlined in this section has to be interpreted just as a step of a multi-stage filtering process to select the proposed identifications, to be confirmed by finding an orbit fitting both sets of observations according to the least square principle.

7. Constrained Differential Corrections

When only two very short arcs are available the classical differential correction procedure, starting from a rough preliminary orbit, may not converge. Even if it converges to a nominal least squares solution, the latter may be poorly determined and may a posteriori turn out to be very far from the true orbit. A good strategy is to seek for a *Line Of Variations* (LOV) orbit, by means of a *constrained differential corrections* algorithm (Milani et al., 2005a).

At a point X in the elements space (at an epoch near the observations) the orbit determination has a *weak direction* $V_1(X)$, corresponding to the long axis of the confidence ellipsoid computed with the normal matrix $C(X)$. The point X is on the LOV if the cost function Q restricted to the hyperplane $H(X)$ normal to the weak direction $V_1(X)$ has a local minimum. The differential corrections can be constrained to corrections lying on the hyperplane $H(X)$. If the iterations converge, then the limit point is on the LOV (Milani et al., 2005a). In other words, the constrained differential corrections are attempting to find a five parameter solution, with the assumed variable along the weak direction. It is often the case, when there are few data, that constrained differential corrections converge, but full differential corrections do not.

This procedure is very effective when combined with the computation of preliminary identification orbits described in the Section 6. For the same example of the previous Figures, namely the attributables formed with the prediscovery data of February 5, 2000 and with the discovery data of March 3–4 for the Centaur 2000 EC_{98}, we have selected the 5 VAs with $\sqrt{K_i} \leqslant 6$, corresponding to the nodes 4, 15 and 25–27 of the triangulation, computed the preliminary orbits and used them to start the constrained differential corrections.

The LOV is represented by a straight line segment on the $(r, \dot{r})$ plane of the first attributable in Figure 5, with the 5 LOV solutions labeled with the node indexes. Note that 3 out of the 5 LOV solutions are outside the admissible region, and indeed they are hyperbolic; nevertheless, one of the elliptic solutions (number 4) is quite close to the "true" solution (as determined a posteriori with additional observations), which is marked with a crossed square.

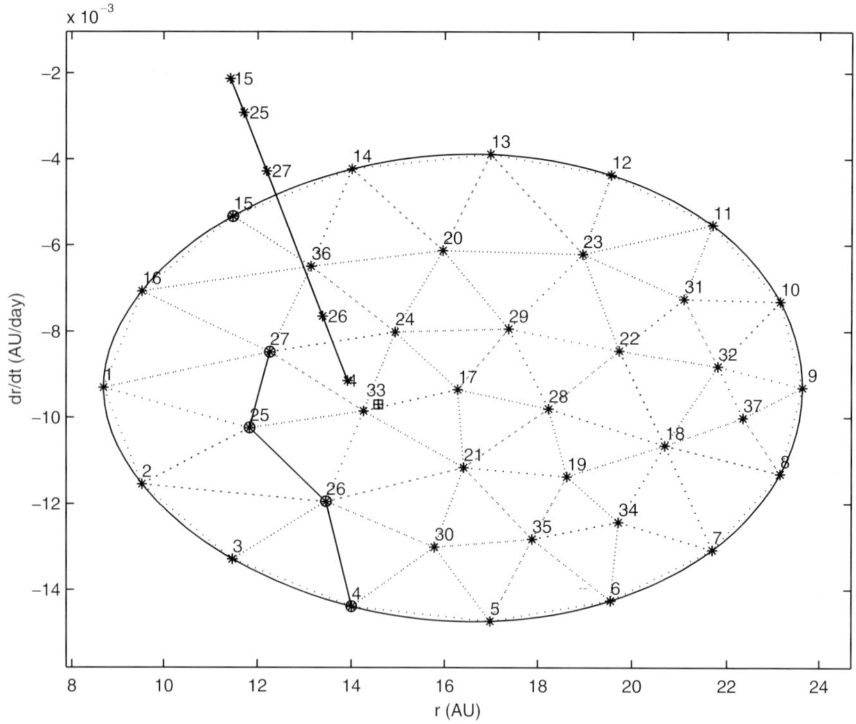

Figure 5. Identification of the discovery observations (March 3–4, 2000) with the preliminary orbits based upon the February 5 prediscovery observations. The sides of the triangulation joining nodes with identification penalty $\leqslant 36$ are marked with a full line, the others are dotted. The true solution is marked with a crossed square.

In this case the attempt to compute a nominal solution, even starting from a comparatively good guess (such as the LOV point number 4), fails. However, the two elliptic solutions (in particular number 4) are good enough to compute useful predictions, allowing identification with additional observations, as described in the next Section.

8. From Multiple Solutions to the Nominal Orbit

When a third very short arc is available, the observations can be compared with the predictions resulting from the LOV solutions. As an example, from LOV solution number 4 of Figure 5 we have computed the predictions for an observation of 2000 EC_{98} on April 4, 2000. The confidence region is not small, although it is narrow.

The attribution of this observation to one of the LOV orbits computed as described in the Section 7 is quite obvious. Differential corrections provide an orbit fitting the three very short arcs with a normalized RMS of 0.35.

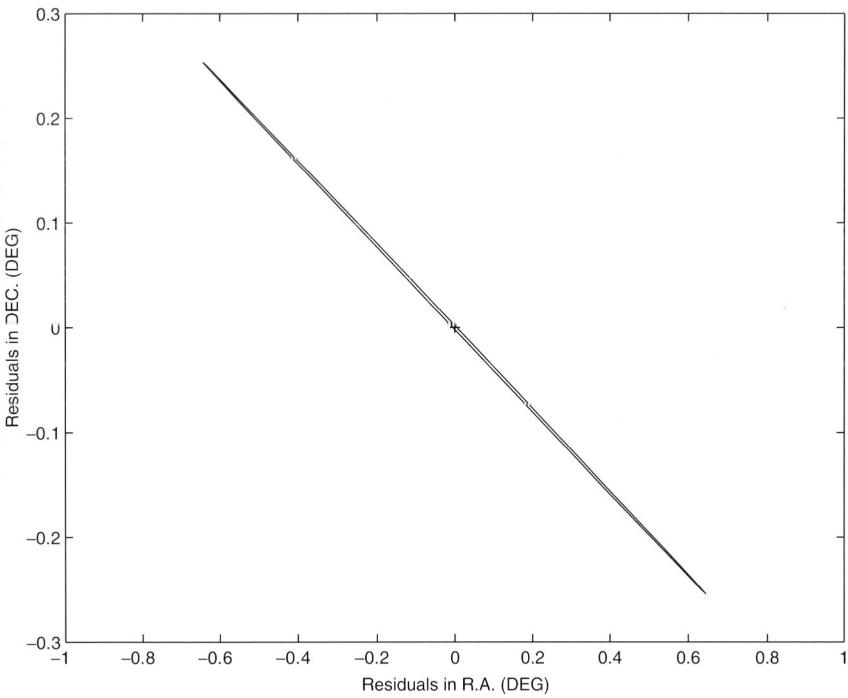

Figure 6. Confidence ellipse on the (α, δ) residuals plane for a recovery observation of 2000 EC_{98} on 4 April 2000, computed by using one of the LOV solutions based upon the observations of February 5 and March 3–4. The cross is the recovery observation.

The orbit is still not tightly constrained by the observations, e.g., the semimajor axis is 11.3 ± 0.8 AU, the eccentricity 0.33 ± 0.15, but the confidence ellipsoid is still a reasonably accurate description of the uncertainty region. As this example shows, the attribution of additional observations becomes easier and easier as more short arcs are available. After three short arcs have been fitted, the orbit determination procedure can follow the classical paradigm, with step by step improvements obtained by full differential corrections.

This could be the new paradigm, replacing the classical one (Gauss' preliminary orbit from three observations followed by full differential corrections). First, the attributables are computed by fitting the observations of the available very short arcs. Let us suppose three attributables belonging to the same object are available: the steps leading to the orbit determination are the following:

1. The admissible region of the first attributable is computed and sampled by a Delaunay triangulation, providing a set of VAs.
2. The predictions for the time of the second arc, computed from the VAs of the first, are compared with the second arc attributable.
3. For the VAs such that the attribution penalty (with the second attributable) is low a preliminary orbit is computed.
4. The above preliminary orbits are used as first guess in constrained differential corrections, providing, when there is convergence, LOV solutions fitting both very short arcs.
5. The LOV orbits are propagated and the predictions are compared with the attributable of the third arc.
6. For the LOV orbits such that the attribution penalty (with the third attributable) is low full differential corrections are used to check the attribution.
7. The orbit resulting from the fit to three very short arcs is used with its covariance, as in the classical procedure, for additional attributions when more observations are available.

In this way the orbit progresses from the stage in which there is an essentially 2-dimensional indetermination (steps 1–3), to a 1-dimensional indetermination (steps 4, 5), to a single nominal least squares solution (steps 6, 7).

9. Conclusions and Future Work

The new procedure, as described in the previous Section, has been rigorously defined, but to test it for reliability and efficiency in processing a large data set of astrometric observations is quite another matter.

Indeed the preceding discussion does not clearly indicate the main difficulty, namely, the false identifications. The problem to be solved is by no means to find an orbit fitting three very short arcs of observations, known to belong to the same object. If this was the case, by selecting one observation from each arc we could use Gauss' preliminary orbit method, then proceed with the classical paradigm. The problem is to compute $\simeq N$ orbits, given $\simeq N$ very short arcs in each one of three separate nights covering the same (large) region in the sky: for the current surveys, the number N is of the order of a few thousands, for the next generation surveys $N > 100,000$. It is computationally impossible to proceed, for each triple formed with one arc per each night, with the classical orbit determination procedure[6]. Thus we need to apply the most computationally intensive tests, such as the differential corrections, to a very small subset of the set of all triples.

The challenge is to define a sequence of filtering stages, with increasing computational load and screening a decreasing number of couples/triples of arcs. Of the steps listed in the previous Section, the ones with computational load linear in N, such as step 1, are not a problem. Step 2 is quadratic in N and requires a very efficient use of the orbit propagator, possibly a simplified one. Step 3 is quadratic, but simple; it should provide a number of couples candidate for identification by far smaller than N^2, because step 4 is the most computationally intensive. Steps 5 and 6 are expensive, but they should be performed only on the confirmed couples. Step 7 needs to be performed only on the confirmed triples. All this would be inefficient if the number of spurious identifications resulting from steps 2–4 was large.

To test the algorithms of the new paradigm, and especially the global data flow and computational load, we need to use first simulated data. Real data are affected by too many errors and we can not know which identifications remain to be found. In a simulation, we have a "ground truth", that is the catalog of objects used as input, thus we can exactly compute the level of completeness of a set of identifications obtained with a given procedure, and also the fraction of false identifications. By training our algorithms with the simulation test cases we can select the most efficient sequence of operations and the best values for the control parameters. After this, we have to test on real data.

So far the individual steps have been tested on both simulated and real data, but the full procedure has been tested only on comparatively small simulations. The next stages of our work will include testing on full scale

[6]For $\simeq 10^{15}$ preliminary orbits and a similar number of differential corrections, the CPU time with a current processor would be between 10^{12} and 10^{13} s.

simulations of the next generation surveys, and on the real data of the current surveys. Only after these tests are successfully concluded we will be able to claim we have established a new paradigm for the conversion of astrometric data into catalogs of orbits, and a new link between astrometry and celestial mechanics.

Acknowledgements

For this research we would like to acknowledge the essential contributions from O. Arratia, S.R. Chesley, M. de' Michieli Vitturi, G. Gronchi, M.E. Sansaturio, G. Tommei. We also thank R. Jedicke, G.B. Valsecchi and A. Boattini for giving important insight on the next generation surveys. We thank the organizers of the Humboldt Colloquium for the opportunity to present, as *Eichhorn Lecture*, our long term research program on asteroid orbit determination. Z.K. acknowledges support from the Ministry of Science and Environmental Protection of Serbia through project No 1238.

References

Carpino, M., Milani, A. and Chesley, S. R.: 2003, 'Error statistics of asteroid optical astrometric observations', *Icarus*, **166**, 248–270.

Danby, J. M. A.: 1989, *Fundamentals of Celestial Mechanics*, Willmann-Bell, Richmond.

Gauss, C. F.: 1809, *Theory of the Motion of the Heavenly Bodies Moving about the Sun in Conic Sections*, reprinted by Dover publications, 1963.

Goldader, J. D. and Alcock, C.: 2003, 'Constraining recovery observations for Trans-Neptunian objects with poorly known orbits', *Publ. Astron. Soc. Pacific*, **115**, 1330–1339.

Milani, A., Sansaturio, M. E. and Chesley, S. R.: 2001, 'The asteroid identification problem IV: Attributions', *Icarus* **151**, 150–159.

Milani, A., Gronchi, G. F., de' Michieli Vitturi, M. and Knežević, Z.: 2004, 'Orbit determination with Very Short Arcs. I Admissible Regions', *CMDA*, **90**, 59–87.

Milani, A., Sansaturio, M. E., Tommei, G., Arratia, O. and Chesley, S. R.: 2005a, 'Multiple solutions for asteroid orbits: computational procedure and applications', *Astron. Astrophys.* **431**, 729–746.

Milani, A., Gronchi, G. F., Knežević, Z., Sansaturio, M. E., and Arratia, O.: 2005b, 'Orbit determination with Very Short Arcs. II Identifications', *Icarus*, submitted.

Tholen, D. and Whiteley, R. J.: 2003, 'Short Arc Orbit Computations', Preprint.

Virtanen, J., Muinonen, K. and Bowell, E.: 2001, 'Statistical ranging of asteroid orbits', *Icarus*, **154**, 412–431.

Celestial Mechanics and Dynamical Astronomy (2005) 92:19–28
DOI 10.1007/s10569-005-2630-2

ON THE STABILITY REGIONS OF THE TROJAN ASTEROIDS

RUDOLF DVORAK and RICHARD SCHWARZ
University of Vienna, Institute for Astronomy, Türkenschanzstr. 17, A-1180 Wien, Austria, e-mails: {dvorak, schwarz}@astro.univie.ac.at

(Received: 13 September 2004; revised: 14 December 2004; accepted 17 January 2005)

Abstract. The orbits of fictitious bodies around Jupiter's stable equilibrium points L_4 and L_5 were integrated for a fine grid of initial conditions up to 100 million years. We checked the validity of three different dynamical models, namely the spatial, restricted three body problem, a model with Sun, Jupiter and Saturn and also the dynamical model with the Outer Solar System (Jupiter to Neptune). We determined the chaoticity of an orbit with the aid of the Lyapunov Characteristic Exponents (=LCE) and used also a method where the maximum eccentricity of an orbit achieved during the dynamical evolution was examined. The goal of this investigation was to determine the size of the regions of motion around the equilibrium points of Jupiter and to find out the dependance on the inclination of the Trojan's orbit. Whereas for small inclinations (up to $i = 20°$) the stable regions are almost equally large, for moderate inclinations the size shrinks quite rapidly and disappears completely for $i > 60^0$. Additionally, we found a difference in the dynamics of orbits around L_4 which – according to the LCE – seem to be more stable than the ones around L_5.

Key words: difference L_4 and L_5, stability regions, trojans

1. Introduction

The quite complicated dynamics of Trojans was discussed by several authors since the discovery of the first Trojan in 1906 (Achilles by Max Wolf in Heidelberg); especially in recent years numerical and analytical work has been accomplished. One of the first estimations of the stability regions around the equilibrium points was made by Rabe (1967) in the framework of the restricted three-body problem. Érdi in many papers (e.g. 1988, 1997) studied the motion of the Trojans also with analytical methods in the model of the spatial elliptic restricted three-body problem and even took into account partly Saturn's perturbations. Using numerical methods Milani (1993, 1994) could show that some of the real Trojans are on chaotic orbits. An extensive study via numerical simulations was undertaken by Levison et al. (1997) who found that the Trojans undergo a slow dispersion in Gigayears time scales. In simple dynamical models also analytical and semianalytical methods can be applied which lead to estimates of the stable regions too small compared to the real ones (e.g. Celletti and Giorgilli, 1991; Beaugé and

Roig, 2001; Skokos and Docoumetzidis, 2001; Efthymiopoulos, 2005). Also possible escapes from the Trojan cloud were discussed by different groups (e.g. Pilat-Lohinger et al., 1999; Dvorak and Tsiganis, 2000; Marzari and Scholl, 2002; Tsiganis et al., 2000) in connection with chaotic orbits. The goal of this investigation is twofold: first we wanted to find out how the largeness of the stable areas around the Lagrangian equilibrium points of Jupiter changes with the inclination of the asteroids (e.g. Schwarz and Gyergyovits, 2003; Brasser et al., 2004). Second we wanted to check why there is a different number of real Trojans around L_4 and L_5[1]. In fact this difference is significant ($N(L_4)/N(L_5) = 5/3$) and it is something which we do not understand up to now. In Figure 1 we show a histogram with respect to the inclinations and also one with respect to the eccentricities, where the difference in the distribution for L_4 and L_5 Trojans is well visible. Although we have some indications, our results cannot confirm that the reason for the actual difference is due to a different dynamical structure.

2. Dynamical Models and Numerical Methods

It is evident that the influence of the terrestrial planets is very small on asteroids around the stable Lagrangian equilibrium points. In numerical simulations the necessary time step – when we include inner planets – would be in the order of 1/10 (including Mars) down to 1/80 (including Mercury) of the time step needed for an integration of the OSS. Therefore any numerical work on the dynamics of Trojans is undertaken by adding the masses of the inner planets to the Sun, which is a way of taking into account their influence on the motion of the other bodies in the Solar system. In our case we tested three models, namely

1. SUN + JUPITER + massless asteroids (= SJA), which is the spatial elliptic restricted three-body problem
2. SUN + JUPITER + SATURN + massless asteroids (= SJS)
3. SUN + outerplanets (JUPITER to NEPTUNE) + massless asteroids (the outer Solar system = OSS).

As integration method for solving the equations of motion we used two different methods:

- On one hand we used the Lie-integrator with recurrence formulae for the Lie-terms which can also be utilized for high eccentric orbits due to the automatic step size (e.g. Hanslmeier and Dvorak, 1984; Lichtenegger,

[1]By September, 10th 2004 the numbers are for Jupiter Trojans L4: 1060, L5: 628.

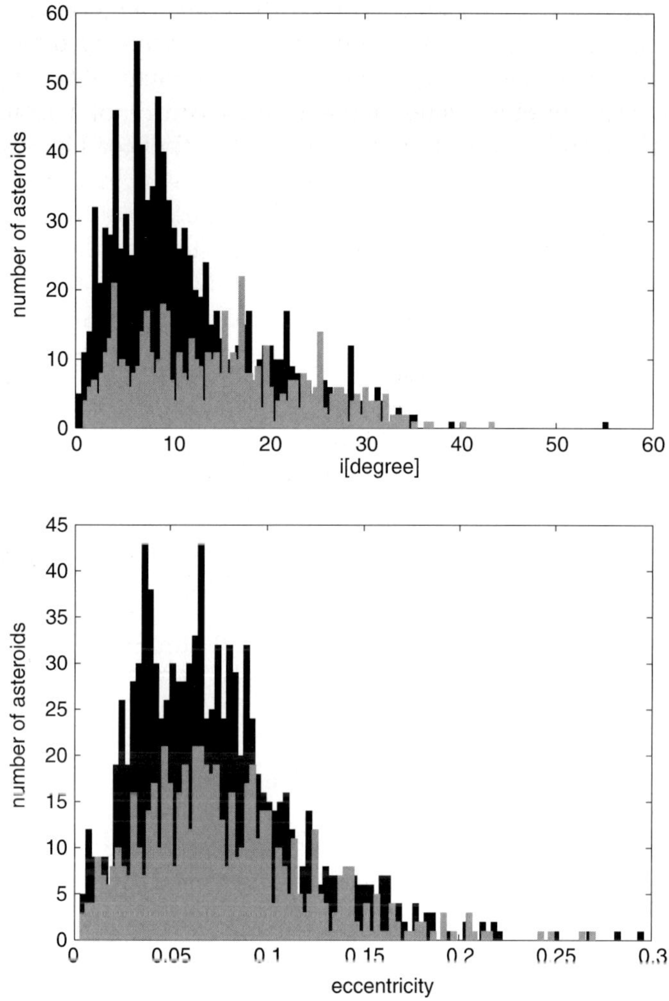

Figure 1. Distribution of all real L_4 and L_5 Trojans with respect to their inclination (top) and their eccentricity (bottom); L_5 Trojans are depicted in light grey.

1984). We already used this method for many numerical simulations and compared it to other methods (e.g. Tsiganis et al., 2000).

– On the other hand we used the program *orbit9*, developed by Milani[2] (1999), a high order Runge Kutta method. This software also computes the Lyapunov Characteristic Exponents (=LCE), respectively, the Lyapunovtime (=LT) which are essential for the determination of the chaoticity of an orbit (e.g. Froeschlé, 1984).

[2]For detailed information see: http://copernico.dm.unipi.it/~ milani/propel5/node3.html

Additionally to the LCEs we computed the maximum eccentricity during the evolution of an orbit to determine its stability and also the region of stable motion around the two equilibrium points. Our stability criterion for a Trojan was, that the eccentricity should not exceed $e = 0.5$; this is a good measure which we tested and compared to other definitions like crossing the line of syzygy.[3]

The integration times were slightly different for the different runs in the three models; we used always at least 10^7 years, but for the LCEs we used always 10^8 years.

We present the results of the numerical experiments as follows:

- the largeness of the stable region around the equilibrium points depending on the initial semimajor axes and the synodic longitudes with respect to the initial inclination in the model SJS (Figure 2);
- the state of chaoticity of the orbits close to L_4 and L_5 depending on initial eccentricity and inclination via the LT in the model OSS (Figures 3 and 4);
- the extension in the synodic longitude of the stable zones for a fixed value of the semimajor axis $a_{\mathrm{Trojan}} = a_{\mathrm{Jupiter}}$ depending on the inclination in the OSS (Figure 5);
- the comparison of L_4 and L_5 Trojans in all three models in an initial condition diagram eccentricity versus LT (Figure 6).

3. Stable Regions Around L_4 and L_5

We determined the extension of the stable regions with the Lie-integrator and the maximum eccentricity of an orbit as stability parameter. In the model of the SJS we integrated the equations of motion from 0° to 360° for a grid of $\Delta\lambda = 1°$ in the synodic longitude and semimajor axis of $4.9AU < a_{\mathrm{Trojan}} < 5.6AU$ with $\Delta a = 0.02$ AU for massless fictitious Trojans. We set the initial inclination to $0° < i_{\mathrm{Trojan}} < 60°$ with $\Delta i = 5°$ and the initial eccentricity to zero; Ω and ω of the Trojans were set to the respective orbital elements of Jupiter. In Fig. 2 we can see that the size of the stable region for small inclinations $0° < i < 15°$ is almost the same and it is also equally large for both Lagrangian points. There is only a small decrease in the size with decreasing inclination of the Trojan's orbit $20° \leqslant i \leqslant 35°$ (not shown here). For high inclinations $40° < i < 55°$ (see also Figure 2) the size shrinks very fast and the stable region disappears completely for $i = 60°$. The size of the stable region (number of stable orbits) was determined with a least square fit: $N(i) = a \cdot i^2 + b \cdot i + c$ with: $a = -0.0046 \pm 0.0007; b = 0.0800 \pm 0.0352;$

[3] Alignment of Sun, Jupiter and the Trojan.

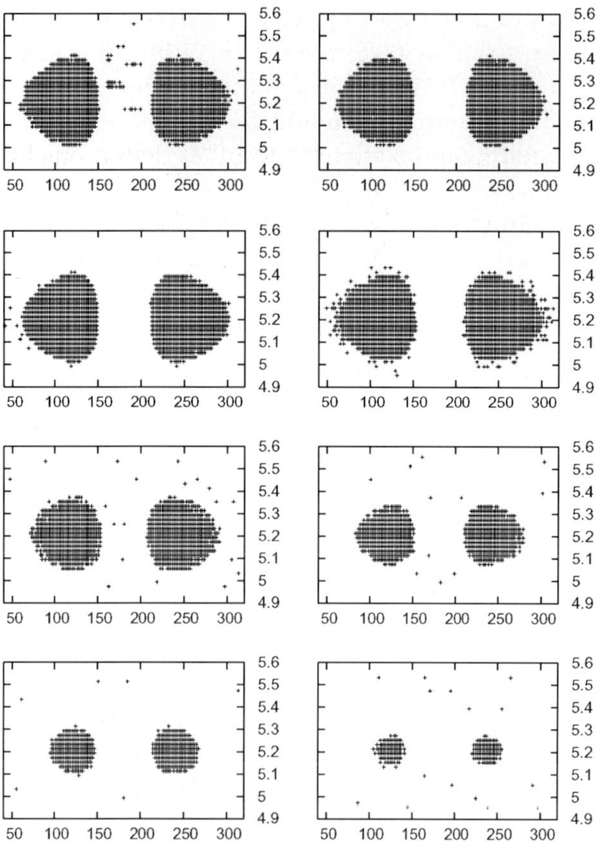

Figure 2. Largeness of the stable regions around L_4 and L_5 in the dynamical model SJS synodic longitude versus initial semimajor axis (in AU). Eight different initial inclinations of the fictitious Trojans are shown: $i = 0°$ and $i = 5°$ (first row), $i = 10°$ and $i = 15°$ (second row), $i = 40°$ and $i = 45°$(third row), $i = 50°$ and $i = 55°$ (fourth row); in the synodic coordinates the position of Jupiter is at $\lambda = 180°$ and a = 5.2 AU. Points indicate stable orbits.

$c = 9.13406 \pm 0.3782$. N corresponds to the percentage of the stable orbits out of the grid in initial conditions specified above.

4. Lyapunov-Times for Fictitious L_4 and L_5 Trojans Depending on i_{ini} and e_{ini}

With the forementioned program *orbit9* we integrated for a grid $\Delta e = 0.01$ between $0 < e < 0.2$ and $\Delta i = 1.°75$ between $0° < i < 34°$ fictitious Trojans around L_4 and L_5 with a semimajor axis $a = a_{\text{Jupiter}}$.[4] In the dynamical model

[4]The angle λ was chosen such that for L_4 Trojans the actual position was 60° ahead of Jupiter's mean longitude and for L_5 Trojans 60° behind.

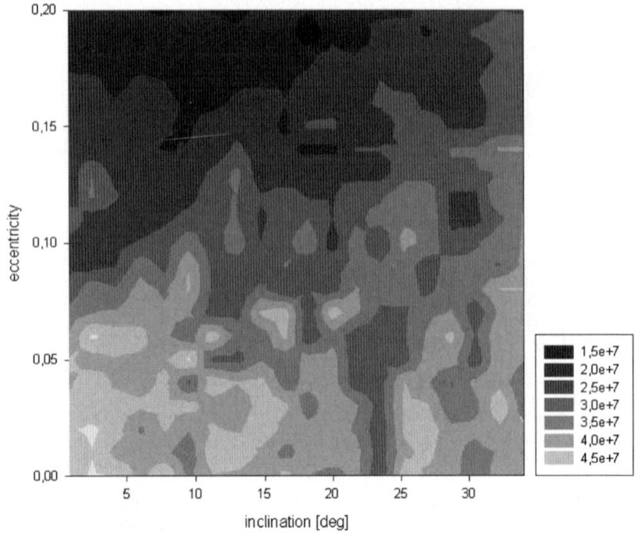

Figure 3. Initial condition diagram i_{ini} versus e_{ini} for fictitious L_4 Trojans showing LTs in the dynamical model OSS; the initial semimajor axes were set to the one of Jupiter.

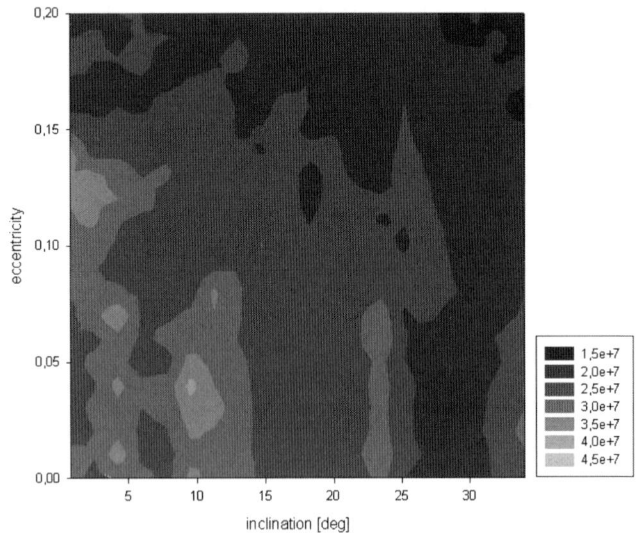

Figure 4. Caption like in Figure 4 but for L_5 Trojans.

OSS we fixed the integration time to 100 Million years, computed the LTs for each orbit and plotted it in the respective inclination versus eccentricity diagram. Globally there is the expected tendency to more chaotic orbits for larger eccentricities AND larger inclinations; this is true for both equilibrium

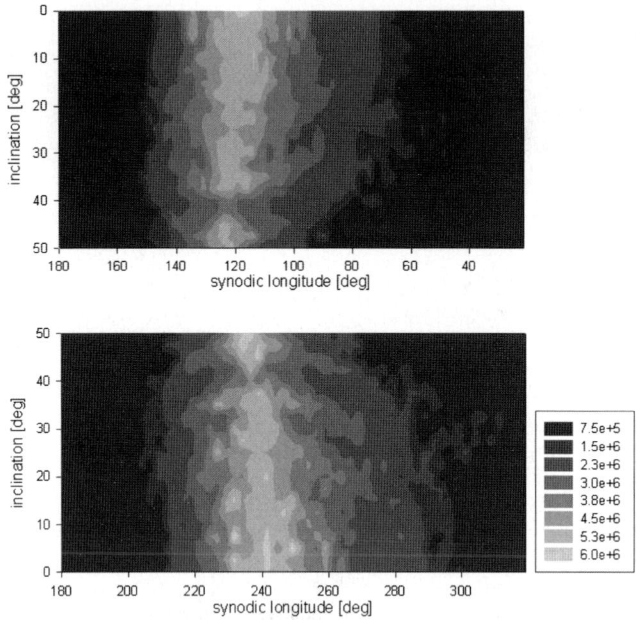

Figure 5. Initial condition diagram i_{ini} versus the synodic longitude λ for fictitious L_4 (lower graph) and L_5 (upper graph) Trojans showing the LTs in the dynamical model OSS. The initial semimajor a_{ini} of the fictitious Trojan with initially a circular orbit was set to the one of Jupiter, which is located at $\lambda = 180°$.

points. Furthermore we can see that in the L_4 diagram (Figure 3) even for large inclinations and small eccentricities the LT is relatively large; this is not the case for the L_5 Trojans (Figure 4). From a comparison of these two figures it seems that there is more chaos around L_5; this is the confirmation of results of a former study (Schwarz et al., 2004). In addition we checked the validity of these results by computations of the orbits of fictitious Trojans in the two simpler models SJA and SJS; this comparison will be discussed in the final chapter.

5. Extension of the Stable Zone Depending on the Synodic Longitude of the Trojan

In a further step of our investigation we checked the extension of the stable region with respect to the initial synodic longitude λ in the dynamical model OSS with *orbit9*. The initial conditions were the same as in the former runs ($a_{\mathrm{ini}} = a_{\mathrm{Jupiter}}$), but we varied λ from $\lambda = 0°$ to $\lambda = 360°$ with a grid of $\Delta\lambda = 1°$ and varied the inclination from $i = 0°$ to $i = 50°$ with $\Delta i = 2.°5$. The position

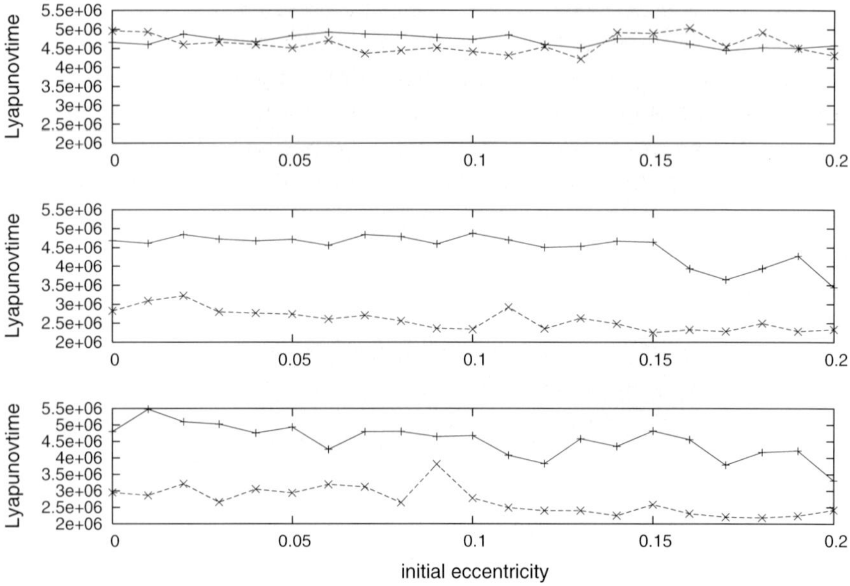

Figure 6. Comparison of L_4 and L_5 Trojans in the three dynamical models SJA (upper graph), SJS (middle graph) and OSS (lower graph) for $i_{\text{ini}} = 30°$. LT is plotted versus the initial eccentricity for the L_4 (solid line) and the L_5 Trojans (dashed line).

of Jupiter (Figure 5) is at $\lambda = 180°$, the two Lagrangian points are at 120° and 240° for L_5 respectively, L_4. Again we can see a slightly different structure with more chaotic orbits in the L_5 region (upper graph).

6. Comparison of the Three Dynamical Models SJA, SJS and OSS

We directly compared the three dynamical models SJA, SJS and OSS[5] for a cut of the former Figures 3 and 4 with three different inclinations ($i = 10°, 20°$ and 30°) and a time interval of 10^7 years, where we varied the eccentricities from e = 0 to e = 0.2 with a grid of $\Delta e = 0.01$. Figure 6 shows the respective results for $i = 30°$; there we have plotted how the LT depends on the initial eccentricity of the fictitious body close to L_4 and to L_5. In the upper graph for model SJA we see that the two lines almost coincide and that the LT is around 5×10^6 for all different eccentricities. In the middle graph for model SJS we can see that the LT for all orbits around L_5 is lower than in the model SJA (dashed line) and differs significantly from L_4. The same effect is visible

[5]Note that taking a more realistic model we introduce additional degrees of freedom and thus new resonances appear (see Robutel et al., 2005).

in the lower graph when we compare the dynamical models OSS and SJS[6]. In the model SJA there is no decreasing LT visible and the lines are almost straight; in the two models SJS and OSS there is a small dependance on the eccentricity visible (slight decrease with e_{ini}). It should be emphasized that the cut for $i = 30°$ is representative and we got similar results for other inclinations. We conclude from this comparison that the effect of different chaoticity of the orbits for the Lagrangian points is an effect which already appears when we include Saturn in the model. But, the difference in the chaotic behaviour of the L_4 and the L_5 orbits – found with the aid of the LCE – could be due to a biased choice of initial conditions for Trojans around the two equilibrium points. We therefore can not yet claim that there is in fact a difference in the dynamics between the two stable Lagrangian points for more realistic models including other planets.

7. Conclusions

In this investigation we established the size of stable regions of fictitious Trojans around the equilibrium points of Jupiter for different dynamical models: the elliptic restricted problem, a model including also Saturn and one where the outer planets with their mutual perturbations and their gravitational force on the fictitious Trojans were fully taken into account. This goal was achieved with long term numerical experiments for fictitious Trojans around L_4 and L_5 for a chosen grid of initial conditions. We have tested different initial eccentricities and inclinations of these Trojans and determined their stability with a straightforward check of their maximum eccentricity (using results of the Lie-series integration). The major point of our investigation is that we can see how the stability regions shrink with larger inclinations and that they finally disappear completely for $i = 60°$. As proper tool for determining the dynamical state we also computed the LCE (respectively, the LT) which gave us an estimation of the chaoticity of an orbit depending on the inclination and the eccentricity of the Trojans. For that reason we used a programm provided by Milani et al. (loc.cit) for time scales of 10^8 years which gives a good estimate of the LCE. We also pointed out a possible difference in the stability between the two Lagrangian points, but this will be a topic of further investigations.

[6]We did not find this difference when we used the maximum eccentricity. The LCE is a more sensitive tool.

Acknowledgements

We thank the referees for the most useful comments which helped to improve the paper. R. Schwarz wants to acknowledge the support by the Austrian FWF (Project P16024-no5).

References

Beaugé, C. and Roig, F.: 2001, 'A semianalytical model for the motion of the trojan asteroids: Proper elements and families', *Icarus* **153**, 391–415.

Celletti, A. and Giorgilli, A.: 1991, 'On the stability of the Lagrangian points in the spatial restricted problem of three bodies', *Celest. Mech. Dynam. Astron.* **50**, 31–58.

Brasser, R., Heggie, D. C. and Mikkola, S.: 2004, 'One to one resonance at high inclination', *Celest. Mech. Dynam. Astron.* **88**, 123–152.

Dvorak, R. and Tsiganis, K.: 2000, 'Why do Trojan ASCs (not) escape?', *Celest. Mech. and Dyn. Astron.* **78**, 125–136.

Efthymiopoulos, C.: 2005, 'Formal integrals and Nekhoroshev stability in a mapping model for the Trojan asteroids, *Celest. Mech. Dynam. Astron.* **92**, 31–54.

Érdi, B.: 1988, 'Long periodic pertubations of Trojan asteroids', *Celest. Mech. Dynam. Astron.* **43**, 303–308.

Érdi, B.: 1997, 'The Trojan Problem', *Celest. Mech. Dynam. Astron.* **65**, 149–164.

Froeschlé, C.: 1984, 'The Lyapunov characteristic exponents – applications to celestial mechanics', *Celest. Mech.* **34**, 95.

Hanslmeier, A. and Dvorak, R.: 1984, 'Numerical Integrations with Lie-series to celestial mechanics', *A&A* **132**, 203.

Levison, H., Shoemaker, E. M. and Shoemaker, C. S.: 1997, 'The dispersal of the Trojan asteroid swarm', *Nature* **385**, 42–44.

Lichtenegger, H.:1984, 'The dynamics of bodies with variable masses', *Celest. Mech.* **34**, 357–368.

Marzari, F. and Scholl, H.: 2002, 'On the Instability of Jupiter's Trojans', *Icarus* **159**, 328–338.

Milani, A.:1993, 'The Trojan Asteroid Belt: Proper Elements, Stability, Chaos and Families', *Celest. Mech. Dynam. Astron.* **57**, 59–94.

Milani, A.:1994, 'The Dynamics of the Trojan asteroids', *IAU Symp. 160:Asteroids, Comets, Meteors* **160**, 159.

Pilat-Lohinger, E., Dvorak, R. and Burger, Ch.: 1999, 'Trojans in stable chaotic motion' *Celest. Mech. Dynam. Astron.* **73**, 117–126.

Rabe, E.: 1967, 'Third-order stability of the long-period Trojan librations', *Astron. Journal* **72**, 10.

Robutel, P., Gabern, F. and Jorba, A.: 2005, 'The observed Trojans and the global dynamics around the Lagrangian points of the Sun-Jupiter system', *Celest. Mech. Dynam. Astron.* **92**, 55–71.

Schwarz, R., and Gyergyovits, M., and Dvorak, R.: 2004, 'On the stability of high inclined L_4 and L_5 Trojans' *Celest. Mech. Dynam. Astron.*, **90**, 139–148.

Schwarz, R., and Gyergyovits, M.: 2003, 'Stability of Trojans with high inclined orbits' In: F. Freistetter, R. Dvorak and B. Érdi (eds.), Proceedings of the 3rd Austrian Hungarian workshop on Trojans and related topics, Eòtvòs University Press, Budapest.

Skokos, C., Docoumetzidis, A.: 2001, 'Effective stability of the Trojan asteroids' *A&A* **367**, 729–736.

Tsiganis, K., Dvorak, R., Pilat-Lohinger, E.: 2000, 'Thersites: a 'jumping' Trojan ?' *A&A* **354**, 1091–1100.

Celestial Mechanics and Dynamical Astronomy (2005) 92:29–52
DOI 10.1007/s10569-004-4495-1

FORMAL INTEGRALS AND NEKHOROSHEV STABILITY IN A MAPPING MODEL FOR THE TROJAN ASTEROIDS

CHRISTOS EFTHYMIOPOULOS
Research Center for Astronomy, Academy of Athens, Soranou Efessiou 4, 115 27 Athens, Greece, e-mail: cefthim@cc.uoa.gr

(Received: 18 July 2004; revised: 4 October 2004; accepted: 6 October 2004)

Abstract. A symplectic mapping model for the co-orbital motion (Sándor et al., 2002, *Cel. Mech. Dyn. Astr.* **84**, 355) in the circular restricted three body problem is used to derive Nekhoroshev stability estimates for the Sun–Jupiter Trojans. Following a brief review of the analytical part of Nekhoroshev theory, a direct method is developed to construct formal integrals of motion in symplectic mappings without use of a normal form. Precise estimates are given for the region of effective stability based on the optimization of the size of the remainder of the formal series. The stability region found for $t = 10^{10}$ yrs corresponds to a libration amplitude $D_p = 10.6°$. About 30% of asteroids with accurately known proper elements (Milani, 1993, *Cel. Mech. Dyn. Astron.* **57**, 59), at low eccentricities and inclinations, are included within this region. This represents an improvement with respect to previous estimates given in the literature. The improvement is due partly to the choice of better variables, but also to the use of a mapping model, which is a simplification of the circular restricted three body problem.

Key words: Nekhoroshev stability, Trojan asteroids

1. Introduction

In recent years, considerable attention has been paid to the so-called *Nekhoroshev stability estimates* for non-linear Hamiltonian dynamical systems of the form

$$H(J,\phi) = H_0(J) + \epsilon H_1(J,\phi), \tag{1}$$

where (J,ϕ) are action angle variables, and the Hamiltonian satisfies appropriate non-degeneracy and analyticity conditions. The Nekhoroshev time T (Nekhoroshev, 1977; Benettin et al., 1985), or time of practical stability, is defined by the relations

$$|J(t) - J(0)| < \epsilon^{\alpha} \quad \text{for all } t \leqslant T, \text{ with } T = \mathrm{O}\left(\exp\left(\left(\frac{\epsilon_0}{\epsilon}\right)^b\right)\right), \tag{2}$$

where α, b and ϵ_0 are parameters depending on the number of degrees of freedom and the particular form of H_0 and H_1. The Nekhoroshev estimates apply to all the orbits in an open domain of initial conditions, independently of whether a particular orbit is regular, i.e., lays on a KAM torus, or chaotic. Regular orbits are stable for infinite times, thus they always remain confined on their respective KAM tori. On the other hand, chaotic orbits diffuse, in general, away from the domain of their initial conditions in the action space in a time which is finite, although it can be extremely long. The Nekhoroshev formula (2) predicts a finite time of stability for both regular and chaotic orbits. But this is only a lower bound, i.e., an underestimate of the real time of stability. This is due to the fact that the Nekhoroshev formula applies globally, i.e., for all the initial conditions within a particular domain of phase space. Nevertheless, the Nekhoroshev estimates are obtained by precise analytical methods, while most methods to determine the details of diffusion for particular orbits are subject to numerical uncertainties.

In many applications, the Nekhoroshev time T is fixed. For example, in celestial mechanics T is equal to the age of the solar system. In this case, the techniques of the analytical part of the Nekhoroshev theorem can be used to provide estimates of the *size* of the region of practical stability, i.e., the size of a domain in the actions for which the Nekhoroshev time is equal to T. This possibility is linked to the *asymptotic properties* of formal series, the construction of which is at the core of the analytical part of Nekhoroshev theorem. Precisely, the analytical part of the theorem is based on the construction of a Birkhoff normal form for the Hamiltonian H by a series of consecutive near-identity canonical transformations $(I^{(i)}, \theta^{(i)}) \to (I^{(i+1)}, \theta^{(i+1)}) \to \cdots$, where the variables $(I^{(i+1)}, \theta^{(i+1)})$ differ by corrections of order $O(\epsilon^{i+1})$ with respect to the variables $(I^{(i)}, \theta^{(i)})$ and $(I^{(0)}, \theta^{(0)}) = (J, \phi)$. After N successive normalizations the Hamiltonian has the form

$$H(I^{(N)}, \theta^{(N)}) = Z_N(I^{(N)}, \theta^{(N)}) + R_N(I^{(N)}, \theta^{(N)}), \tag{3}$$

where the part Z_N (normal form), is an integrable Hamiltonian containing terms up to order ϵ^N. In the non-resonant case Z_N depends on $I^{(N)}$ alone, while in the resonant case Z_N depends on $I^{(N)}$ and on resonant linear combinations of the angles $\theta^{(N)}$. In both cases, one may define approximate integrals $\Phi^{(N)}$ of the Hamiltonian (3). In the non-resonant case, the integrals are as many as the numbers of degrees of freedom and they coincide with the actions $\Phi^{(N)} = I^{(N)}$. In resonant cases, in general, there are fewer integrals than the number of degrees of freedom, and they are defined by relations of the form $\Phi^{(N)}_{\text{res}} = k \cdot I^{(N)}$, where k is an integer vector spanning the plane normal to a vector n satisfying the resonance condition $n \cdot \omega(I^{(N)}) = 0$. In both cases, the integrals Φ are approximate. The speed of their variations is determined by the size of the remainder $R^{(N)}$, since we have

$$\frac{d\Phi^{(N)}}{dt} = \{\Phi^{(N)}, H\} = \{\Phi^{(N)}, R^{(N)}\}. \tag{4}$$

Now, the asymptotic properties of the formal series allow one to prove that the size of the remainder $||R_N||$, as a function of N, initially decreases as N increases, giving the impression that the Hamiltonian is transformed, step by step, to an integrable Hamiltonian in the new variables. However, the decrease of the size of the remainder stops at a particular order N_{opt}, which is called the optimal order of truncation of the formal series. For $N > N_{\text{opt}}$, the size of the remainder increases with N, tending to infinity as $N \to \infty$. The minimum remainder, is obtained for $N = N_{\text{opt}}$. It can be shown (e.g., Giorgilli, 1999) that the size of the optimal remainder is estimated by

$$||R_{\text{opt}}|| = \frac{1}{\text{O}(\exp(1/\epsilon^b))}. \tag{5}$$

It follows that the time derivative (4) is exponentially small with respect to $1/\epsilon$, i.e., we obtain Nekhoroshev's formula.

If the Hamiltonian (1) has an elliptic equilibrium point, then in the neighborhood of this point the effective perturbation is the distance ρ from the equilibrium point. In such a case, we can obtain exponential estimates depending on ρ, namely

$$||R_{\text{opt}}|| = \frac{1}{\text{O}(\exp(1/\rho^b))}, \tag{6}$$

which yield an exponentially long Nekhoroshev time $T = \text{O}(\exp(1/\rho^b))$ (Giorgilli, 1988; Lochak, 1992; Fassò et al., 1998; Niederman, 1998). If the time T is fixed, say the age of the solar system, the size of the region of practical stability ρ can be determined. In practice, the size ρ is determined directly from the norm of the optimal remainder $||R_{\text{opt}}||$ found by calculating the normal form with the help of a program performing computer algebra operations. The results are rigorous on the condition that one can determine rigorously the radius of convergence of the Hamiltonian expansion and the bounds on the round-off errors generated by the computer manipulator (e.g., Celletti et al., 2000). In practice, even without such rigorous bounds, the results are in general quite precise, because the roundoff error does not have a serious influence on the sizes of the dominant terms in the series, which are determined mainly by the mechanism of accumulation of small divisors.

An alternative way to produce formal integrals with exponentially small remainders is the *direct* method due to Whittaker (1916), Cherry (1924) and Contopoulos (1960). Consider an isochronous Hamiltonian of the form

$$H(p, \phi) = \omega_* \cdot p + \epsilon H_1(p, \phi) \tag{7}$$

i.e., for which the point $p = 0$ is an elliptic equilibrium point. We consider formal integrals given as series in the powers of ϵ, namely $\Phi = \Phi_2 + \Phi_3 + \cdots$, where the terms Φ_i are of order ϵ^{i-2}, i.e., of order $i/2$ in the actions p. The integrals can be calculated without any transformation of the Hamiltonian to a normal form, by just solving, at successive orders, the equation $\{\Phi, H\} = 0$. This yields the recursion scheme for the ith formal integral, $i = 1, \ldots, n$ given by

$$\Phi_2 = \omega_{*i} p_i, \quad \{\Phi_r, H_2\} = -\sum_{j=2}^{r-1} \{\Phi_j, H_{r-j}\} \tag{8}$$

If we truncate the integral Φ at order N, we obtain a truncated series $\Phi^{(N)}$ which represents an approximate integral of the Hamiltonian H. The time derivative of the truncated series is given by

$$\frac{\mathrm{d}\Phi^{(N)}}{\mathrm{d}t} = R_N = \sum_{j=N+1}^{\infty} U_j, \tag{9}$$

where

$$U_r = \sum_{k=2}^{N} \{\Phi_k, H_{r-k}\}, \quad r > N \tag{10}$$

It can be shown that the series R_N, called the remainder of the integral, is convergent (Giorgilli, 1988). In most cases the convergence is rather fast, so that the size of R_N does not differ much from the size of the first order term of the reainder U_{N+1}. At any rate, the size of R_N can be estimated analytically. In a recent work (Efthymiopoulos et al., 2004) it is shown that the patterns of accumulation of small divisors in the Birkhoff normal form and in the integrals found by the direct method are similar, so that the size of remainders, as estimated by the two methods, grow, as a function of N, by essentially the same law. It follows that the optimal remainder, as calculated by the direct method, has an exponentially small size, thus it leads to an exponentially small rate of variation of the values of Φ.

The purpose of the present paper is to demonstrate the applicability of the formal integrals calculated by a direct method by considering a problem that represents an example of application of Nekhoroshev stability estimates in a system of physical interest: the problem of the Sun–Jupiter Trojan asteroids.

The best Nekhoroshev result obtained so far as regards the stability of the Trojans is the estimate of the size of the region of effective stability given by Giorgilli and Skokos (1997), and by Skokos and Dokoumentzidis (2001). The size of the region of stability found by these authors, for T fixed to 10^9 revolutions of Jupiter, is about 10 times smaller than the size of the region within which most real asteroids are observed. This is already a realistic result compared to the previous estimates on the same problem, which were giving a size of about 10^4 km (Simo, 1989; Celletti and Giorgilli, 1991). Nevertheless, this result is obtained by considering a simple model for the Trojan motion, namely the Sun–Jupiter circular restricted three body problem (CRTBP). This model cannot capture the reach resonant structure of the phase space in the neighborhood of the triangular points, since most resonant phenomena are produced by secular perturbations induced either by the elliptic motion of Jupiter or by other planets (especially Saturn). The details of this structure, as revealed by elaborate numerical methods (Levison et al., 1997; Dvorak and Tsiganis, 2000; Tsiganis et al., 2000, 2002), are beyond the scope of the present theory of Nekhoroshev estimates.

Some numerical results published in this issue (Robutel et al., 2005), obtained with frequency analysis of orbits in the full problem, indicate the existence of resonant strips responsible for the slow chaotic diffusion of the asteroids, which are generated by secular resonances. At the border of the stability zone these strips overlap. However, most strips do not penetrate into a region, which extends up to ~5.3 AU for eccentricities not larger than 0.05, inclinations $\sim I_{\mathrm{Jupiter}}$. Now, this region is about seven times larger than the region of stability determined analytically by Giorgilli and Skokos (1997) and twice larger than the region found in the present paper. Thus, theory has still much to do to improve the estimates even with the CRTBP (which, numerically, gives overestimated stability regions extending up to $a = 5.4$ AU), before adding more degrees of freedom (which will reduce the estimates). We conjecture that the results on Nekhoroshev stability derived by the CRTBP model are physically relevant in a region devoid of chaotic strips due to secular resonances. Thus, the question of optimizing the analytical Nekhoroshev estimates represents both a mathematically and physically interesting problem. The addition of more degrees of freedom is, of course, an obvious next step in our study.

Section 2 gives the general setting of the direct method of construction of formal integrals for symplectic mappings without use of a normal form. It also explains how to obtain precise Nekhoroshev estimates of the region of effective stability by calculating the norm of the remainder series of the formal integrals. Section 3 presents the application of this method to the problem of the Trojan asteroids. This allows one to obtain an estimate of the size of the region of effective stability expressed in terms of proper elements. The final outcome is that Nekhorshev stability, for 10^{10} yrs, is guaranteed in a region corresponding to a libration amplitude $D_p \simeq 10.6^\circ$. This is the angle corresponding to the amplitude of libration of the critical argument of the asteroid (e.g., Erdi, 1988). About one third of the asteroids with published proper elements by Milani (1993) are included in this region of stability. A correction factor to this result is estimated as 1.28, giving $D_p \simeq 13.5^\circ$. Section 4 contains the main conclusions of the present study and suggestions for future research.

2. Nekhoroshev Stability for Mappings

2.1. FORMAL INTEGRALS FOR MAPPINGS. THE DIRECT METHOD

We shall consider first the construction of formal integrals in non-linear symplectic mappings by a method, which is the discrete analog of the direct method of Whittaker (1916), Cherry (1924) and Contopoulos (1960). The method is easily exemplified in the case of 2D symplectic mappings. Consider the non-linear symplectic mapping given by the equations

$$x' = F(x, y),$$
$$y' = G(x, y) \tag{11}$$

with an elliptic point at the origin. Let $\omega/2\pi$ be the rotation number at the origin. Then, by a standard linear symplectic transformation to new complex conjugate canonical variables z, z_*, the mapping takes the form

$$z' = e^{i\omega}[z + F_2(z, z_*) + F_3(z, z_*) + \cdots], \tag{12}$$

where $F_s(z, z_*)$, $s = 2, 3, \ldots$ are complex polynomial functions of order s in the variables (z, z_*). Furthermore, if the mapping equations (12) are truncated at order N, the symplecticity condition is satisfied also up to terms of order N.

The linearized map $z' = e^{i\omega}z$ possesses the exact integral

$$\Phi_2(z, z_*) = zz_*, \tag{13}$$

which represents circles around the origin. We look for a formal integral of the non-linear mapping (12) starting with the term (13), namely

$$\begin{aligned}\Phi(z, z_*) &= \Phi_2(z, z_*) + \Phi_3(z, z_*) + \cdots \\ &= zz_* + a_{30}z^3 + a_{21}z^2z_* + a_{12}zz_*^2 + a_{03}z_*^3 + \cdots \\ &= \sum_{p,q\geqslant 0, p+q=2}^{\infty} a_{pq}z^p z_*^q.\end{aligned} \tag{14}$$

For $p + q = 2$ we have $a_{pq} = \delta_{pq}$. For $p + q > 2$, the unknown coefficients a_{pq} can be specified by a recursive scheme. The series (14) must satisfy the following homological equation (Bazzani and Marmi, 1991)

$$\Phi(z', z'_*) = \Phi(z, z_*). \tag{15}$$

Equation (15) takes the form

$$\sum_{m,n\geqslant 0, m+n=2}^{\infty} a_{mn}z'^m z'^n_* = \sum_{p,q\geqslant 0, p+q=2}^{\infty} a_{pq}z^p z_*^q. \tag{16}$$

Applying the mapping Equation (12) to each monomial term $z'^m z'^n_*$ yields

$$z'^m z'^n_* = \sum_{p+q=m+n}^{\infty} A_{mnpq} z^p z^q_*, \tag{17}$$

where the sum is over all positive integers p, q.

The coefficients A_{mnpq} are known quantities from the mapping equations. In particular, for the same orders $m + n = p + q$, we have

$$A_{mnpq} = \begin{cases} e^{i(p-q)\omega} & \text{if } m = q, \quad n = p, \\ 0 & \text{otherwise.} \end{cases} \tag{18}$$

Then, the left-hand side of Equation (16) takes the form

$$\sum_{m+n=2}^{\infty} a_{mn} z'^m z'^n_* = \sum_{m+n=2}^{\infty} a_{mn} \sum_{p+q=m+n}^{\infty} A_{mnpq} z^p z^q_*$$

$$= \sum_{p+q=2}^{\infty} \left(\sum_{m+n=2}^{m+n=p+q} a_{mn} A_{mnpq} \right) z^p z^q_*. \tag{19}$$

Equating the last expressions in Equations (16) and (19) yields a relation between the coefficients

$$\sum_{m+n=2}^{m+n=p+q} a_{mn} A_{mnpq} = a_{pq}. \tag{20}$$

This is a linear system of equations for the unknown coefficients a_{pq}. Notice that the coefficients a_{pq} of any given order $s = p + q$ depend on the known coefficients A_{mnpq} and on the coefficients a_{mn} of orders $m + n$ smaller or equal to s. Therefore, the system (20) can be solved step by step for successive orders. In fact, from Equation (18) it follows that the system (20) is diagonal. Namely

$$[1 - e^{i(p-q)\omega}] a_{pq} = \sum_{m+n=2}^{m+n<p+q} a_{mn} A_{mnpq}. \tag{21}$$

A term $z^p z^q_*$ is called 'resonant' if the quantity $1 - e^{i(p-q)\omega}$ in the left-hand side of (21) is equal to zero. If a term is not resonant, then Equation (21) can be solved immediately yielding

$$a_{pq} = \frac{\sum_{m+n=2}^{m+n<p+q} a_{mn} A_{mnpq}}{1 - e^{i}(p-q)\omega}. \tag{22}$$

If however a term is resonant, then the construction cannot in general proceed, unless the right-hand side of (21) (numerator of (22)) is also zero, in which case the coefficient a_{pq} remains unspecified and it can be assigned arbitrary values. In this case the resonant term is called 'arbitrary'.

We distinguish two cases:

(a) Non-resonance. If $\omega/2\pi$ is irrational, then $1 - e^{i(p-q)\omega}$ becomes zero iff $p = q$. This condition corresponds to terms of the form $z^p z_*^p$ which are powers of the second order term Φ_2. In this case, by applying a theorem of Birkhoff (1920) it is possible to show the existence of a solution of (21) that defines a non-resonant formal integral series. Therefore, the numerator of (22) for $p = q$ is always zero, and terms of the form $z^p z_*^p$ appear with arbitrary coefficients (see Bazzani and Marmi, 1991 for details).

(b) Resonance. If $\omega/2\pi = n/m$, with n, m relatively prime integers, then $1 - e^{i(p-q)\omega}$ becomes zero if $p - q = Km$, with K integer. In this case we have the appearance of zero divisors in (22), and the formal construction cannot proceed further.

In the resonant case, however, it is still possible to proceed by an algorithm which is the discrete analog of the 'resonant third integral' of Contopoulos (1963). This algorithm is as follows. Starting the construction as previously with $\Phi_2 = zz_*$, yields a zero divisor in (22) with non-zero numerator for $p = m, q = 0$ or for $p = 0, q = m$, meaning that the coefficients of the terms z^m and z_*^m, of degree m, cannot be specified and the construction cannot proceed further.

On the other hand, due to the resonant rotation angle $\omega/2\pi = 1/m$, there is one more real-valued isolating integral of the linearized mapping $z' = e^{i2\pi/m}z$. This is the *phase* integral

$$S_{0,m} = z^m + z_*^m. \tag{23}$$

Starting the construction with the terms $S_{0,m}$ yields a zero divisor in Equation (22), in the resonant terms $S_{1,m+1} = z^{m+1}z_* + zz_*^{m+1}$, of degree $m + 2$.

Now, starting a construction with the *fourth order* term $\Phi_{2,2} \equiv \Phi_4' = z^2 z_*^2$ it is easy to prove that this term cannot produce a resonant term of degree m, but only resonant terms of degree $m + 2$, namely $S_{1,m+1} = z^{m+1}z_* + zz_*^{m+1}$. Thus, if the terms $\Phi_{2,2}$ and $S_{0,m}$ are combined, i.e., if the construction starts with both terms present, it is possible to eliminate the resonant terms $S_{1,m+1}$, i.e., to render the corresponding numerator in Equation (22) equal to zero. The correct series starts with the terms

$$\Phi_{\text{res}}^{(4)} = b_{2,2}\Phi_{2,2} + \cdots + S_{0,m} + \cdots, \tag{24}$$

where the value of the coefficient $b_{2,2}$ is chosen by the requirement that the numerator of (22) for $p = m + 1, q = 1$ (or $p = 1, q = m + 1$) is zero. Thus, the solution of Equation (21) for the terms $S_{1,m+1}$ is an arbitrary constant. This method eliminates the resonant terms of order $m + 2$.

In the same way it is possible to eliminate all the resonant terms of subsequent orders. A careful analysis of the contributions of the various resonant terms to resonant terms of higher order reveals the appropriate algorithm, which is given by the following scheme:

- Start the construction with $S_{0,m}$
- For $q = 2, \ldots, N$, solve a linear system for the coefficients of the terms $\Phi_{q,q}$ and $S_{q-2j,q+j(m-2)}$ ($j = 1$ to $[(q-1)/2]$) so as to eliminate the terms $S_{q+1-2j,q+1+j(m-2)}$ with ($j = 1$ to $[(q-1)/2] + 1$)

The same algorithm eliminates also all the terms of the form $\Phi_{q,q}$, up to ordet N, as in the non-resonant construction.

This method is easy to implement by a computer program. In fact, the algorithm is faster than the algorithm of the resonant normal form, because it avoids any canonical transformation from old to new variables, and then back to the old variables, in terms of which the final integral is expressed.

2.2. DETERMINATION OF THE REGION OF EFFECTIVE STABILITY

Let $\Phi^{(N)}(z, z_*) = \Phi_2(z, z_*) + \Phi_3(z, z_*) + \cdots + \Phi_N(z, z_*)$ be a formal integral of the mapping (12) truncated at order N. The homological equation (15) is satisfied up to order N, i.e.,

$$\Phi^{(N)}(z', z'_*) = \Phi^{(N)}(z, z_*) + R^{(N+1)}(z, z_*), \tag{25}$$

where the remainder $R^{(N+1)}$ contains terms starting at order $N + 1$, namely

$$R^{(N+1)} = U^{(N+1)}_{N+1} + U^{(N+1)}_{N+2} + \cdots \tag{26}$$

We shall define the norm of a polynomial as the sum of the moduli of the polynomial coefficients

$$\left\| \sum_{k,l=1}^{r} a_{kl} z^k z_*^l \right\| = \sum_{k,l=1}^{r} |a_{kl}| \tag{27}$$

and the polynomial norm at radius ρ as

$$\left\| \sum_{k,l=1}^{r} a_{kl} z^k z_*^l \right\|_\rho = \sum_{k,l=1}^{r} |a_{kl}| \rho^{k+l}. \tag{28}$$

The series

$$||R^{(N+1)}||_\rho = ||U^{(N+1)}_{N+1}||_\rho + ||U^{(N+1)}_{N+2}||_\rho + \cdots \tag{29}$$

is absolutely convergent in a disk $\rho < R$ with $R > 0$ (Bazanni and Marmi, 1991). The size of $||R^{(N+1)}||$ can be bounded from above as follows: the remainder up to order M is given by the finite sum

$$R^{(N+1,M)} = U_{N+1}^{(N+1)} + U_{N+2}^{(N+1)} + \cdots + U_{M}^{(N+1)}. \tag{30}$$

Assuming that the mapping (11) is analytical, the expansion (12) is convergent in a disk of non-zero radius around the origin. This implies that there is a positive constant r_F such that $||F_n||r_F^n \leqslant 1$ for all n with $1 \leqslant n < \infty$. Then, for all $n \geqslant N$, and for all ρ with $0 \leqslant \rho \leqslant r_F$, the following inequality holds (Bazanni and Marmi, 1991):

$$||U_{n+1}^{(N+1)}||_\rho \leqslant \frac{\rho}{r_F} \frac{n}{(n-N+1)} ||U_{n+1}^{(N+1)}||_\rho. \tag{31}$$

The ratio $b_n = n/(n-N+1)$ is a decreasing sequence of n, having its maximum $b_N = N$ at $n = N$. Thus, the following bound holds also:

$$||U_{n+1}^{(N+1)}||_\rho \leqslant \frac{N\rho}{r_F} ||U_{n+1}^{(N+1)}||_\rho \tag{32}$$

implying that, for ρ sufficiently small, the subsequence $||U_n^{(N+1)}||$ for $n = N, N+1, \ldots$ is bounded from above by a converging geometrical sequence with ratio $N\rho/r_F$. Thus, the remainder series is absolutely convergent. However, the bound (32) is far from optimal. A much better bound is found by applying the inequality (31) for $n = M$. Thus we find

$$||R^{(N+1)}||_\rho \leqslant ||R^{(N+1,M-1)}||_\rho + ||U_M^{(N+1)}||_\rho \sum_{k=0}^{\infty} (g_{M,\rho}^{(N+1)})^k, \tag{33}$$

where

$$g_{M,\rho}^{(N+1)} = \frac{M\rho}{r_F(M-N+1)}.$$

Let ρ_* be such that $g_{M,\rho_*}^{(N+1)} < 1$. Then, for all $\rho < \rho_*$ we have

$$||R^{(N+1)}||_\rho \leqslant ||R^{(N+1,M-1)}||_\rho + ||U_M^{(N+1)}||_\rho \frac{1}{1 - g_{M,\rho_*}^{(N+1)}}. \tag{34}$$

Define now the constant B_{ρ_*} as

$$B_{\rho_*} = \frac{||R^{(N+1,M-1)}||_{\rho_*} + ||_M^{(N+1)}||_{\rho_*} 1/(1 - g_{M,\rho_*}^{(N+1)})}{||U_{N+1}||\rho_*^{N+1}}. \tag{35}$$

Then, the following lemma holds:

$$\text{for all } \rho < \rho_*, \quad ||R_{N+1}||_\rho \leqslant B_{\rho_*} ||U_{N+1}^{(N+1)}||\rho^{N+1}. \tag{36}$$

Equation (36) determines a rigorous upper bound for the size of the remainder $||R_{N+1}||_\rho$ in terms of the size of the leading term of the remainder $||U_{N+1}^{(N+1)}||$.

The next step is to define appropriate bounds on the values of the truncated integral Φ_N. Since $\Phi_N(z, z_*) = zz_* + O(|z|^3) = \rho^2 + O(\rho^3)$, with $\rho = |z|$, and Φ_N is real-valued, it follows that there exists a radius ρ_0 such that, for all z with $|z| < \rho_0$, $\Phi_N(z, z_*) > 0$. Furthermore, the following inequality holds obviously

$$\text{for all } \rho < \rho_0, \quad \rho^2 - \sum_{k=3}^{N} ||\Phi_k||\rho^k \leqslant \Phi_N(z, z_*) \leqslant \rho^2 + \sum_{k=3}^{N} ||\Phi_k||\rho^k. \tag{37}$$

Let now $I_N > 0$ denote a *fixed value* of the function $\Phi_N(z, z_*)$. Define the maximum radius $\rho_{\max}(I_N)$ as

$$\rho_{\max}(I_N) = \max\{\rho : \Phi_N(\rho e^{i\phi}, \rho e^{-i\phi}) = I_N, 0 \leqslant \phi < 2\pi\} \tag{38}$$

and the minimum radius

$$\rho_{\min}(I_N) = \min\{\rho : \Phi_N(\rho e^{i\phi}, \rho e^{-i\phi}) = I_N, 0 \leqslant \phi < 2\pi\}. \tag{39}$$

Since the value $I_N = 0$ corresponds to the origin, we have $\rho_{\max}(0) = 0$ while, by continuity arguments, the function $\rho_{\max}(I_N)$ is monotonically increasing in a neighborhood of the origin with $\rho_{\max}(I_N) = I_N^{1/2} + O(I_N)$. Select I_N so that $\rho_{\max}(I_N) < \rho_0$. Define the coefficient

$$A = \frac{1}{\rho_{\max}^2(I_N)} \sum_{k=3}^{N} ||\Phi_k||\rho_{\max}^k(I_N). \tag{40}$$

Then, the following inequality holds

$$\text{for all } \rho_{\min}(I_N) \leqslant \rho \leqslant \rho_{\max}(I_N), \quad \rho^2 - \sum_{k=3}^{N} ||\Phi_k||\rho^k \geqslant (1 - A)\rho^2, \tag{41}$$

which, in view of Equation (37) implies that

$$\text{for all } \rho_{\min}(I_N) \leqslant \rho \leqslant \rho_{\max}(I_N), \quad I_N > (1 - A)\rho^2. \tag{42}$$

A comment is in order to clarify the meaning of the above, rather technical, inequalities. When one works with the normal form approach, the level curves of the truncated integrals I_N after N Birkhoff normalizations represent exact circles, i.e., $\Phi_N(\tilde{z}, \tilde{z}_*) = \tilde{z}\tilde{z}_*$, where $(\tilde{z}, \tilde{z}_*)$ are the new canonical variables after N successive canonical transformations. Thus the relation $I_N = \rho^2$ is exact. This one-to-one correspondence between level values of the integrals and the respective radii ρ allows to convert the variations of the integrals ΔI_N to variations $\Delta\rho$ determining immediately the size of the region of effective stability in the phase space of the new canonical variables. This is no longer true with integrals Φ_N calculated with a direct method, because in that case the level curves $\Phi_N(z, z_*)$ are deformed circles. Thus one has to define annuli $\rho_{\min}(I_N) \leqslant \rho \leqslant \rho_{\max}(I_N)$, which contain the level curve $\Phi_N(z, z_*) = I_N$. The norm of the remainder (Equation (36)) is given in terms of ρ rather than I_N. By using the coefficient A an 'effective radius' $\sqrt{I_N/(1-A)}$ is defined such that the size of the remainder can be expressed in terms of the level value I_N rather than ρ. Precisely, for all I_N such that $\rho_{\max}(I_N) < \rho_*$ the following inequality holds

$$\text{for all } \rho_{\min}(I_N) \leqslant \rho \leqslant \rho_{\max}(I_N) < \rho_*,$$

$$||R_{N+1}||_\rho \leqslant B_{\rho_*} ||U_{N+1}|| \left(\frac{I_N}{1-A} \right)^{(N+1)/2}. \tag{43}$$

Finally, consider a discrete time $t = 1, 2, \ldots$, corresponding to the number of iterations of the mapping (12). Let $\Phi_N(t) \equiv \Phi_N(z_t, z_{*t})$. Initially, an orbit is on the level curve $I_N = \Phi_N(0)$. But due to Equation (25), at every iteration the orbit changes level curve by moving to a new integral curve I'_N, which differs from I_N by R_{N+1}. The value of R_{N+1} is real, but it can be either positive, or negative. Let us now suppose that after t steps, the orbit has reached the level curve $\Phi_N(z, z_*) = I'_N$, with $I'_N > I_N$. The maximum value of the variation $I'_N - I_N$ is found by replacing the remainder R_{N+1} in Equation (25) by its upper limit given by Equation (43). This yields

$$I'_N - I_N \leqslant t B_{\rho_*} \left\| U_{N+1}^{(N+1)} \right\| \left(\frac{I'_N}{1-A} \right)^{(N+1)/2}. \tag{44}$$

Equation (44) implies that the minimum time t needed for an orbit to reach a level curve $\Phi_N(z, z_*) = I'_N$, starting from the level curve $\Phi_N(z, z_*) = I_N$, is given by

$$t_{\min} = \frac{(I'_N - I_N)(1-A)^{(N+1)/2}}{B_{\rho_*} \left\| U_{N+1}^{(N+1)} \right\| (I'_N)^{(N+1)/2}}. \tag{45}$$

Equation (45) always underestimates the Nekhoroshev time of stability, because the expression in the denominator is an upper bound of the real size of the remainder which overestimates the speed of variations ΔI_N. Indeed, Equation (45) implies that the time to reach infinity $I'_N \to \infty$ is equal to zero, which is obviously wrong. This is due to the fact that the size of the remainder during the whole excursion from I_N to $I'_N \to \infty$ is estimated by the value of the denominator of Equation (45), when $I'_N \to \infty$, which is infinite. To circumvent this problem, we follow the approach of Giorgilli and Skokos (1997), namely, we find the maximum of (45) with respect to I'_N, which corresponds to the least possible underestimate of the Nekhoroshev time. The maximum of (45) is at $I'_N = (N+1)I_N/(N-1)$, which yields the optimal Nekhoroshev time

$$t_{nk} = \frac{2(1-A)^{(N+1)/2}}{(N+1)B_{\rho_*}||U_{N+1}^{(N+1)}||(I'_N)^{(N-1)/2}}. \tag{46}$$

Solving Equation (46) with respect to I'_N yields the level curve $\Phi_N(z, z*) = I_N$ such that no initial condition in the interior of this curve may travel a distance larger than $\Delta I_N = I'_N - I_N$ within the time t_{nk}. This is given by

$$I_N = \left(\frac{N-1}{N+1}\right)\left(\frac{2(1-A)^{(N+1)/2}}{(N+1)B_{\rho_*}||U_{N+1}^{(N+1)}||t_{nk}}\right)^{2/(N-1)}. \tag{47}$$

The region of effective stability defined by Equation (47) is a function of the order of truncation N of the formal integral. This reflects the fact that the size of the remainder depends on N. The asymptotic character of this dependence (e.g., Bazzani and Marmi, 1991) is expressed in the estimate $||U_{N+1}^{(N+1)}|| \sim N!$ which, by use of Stirling's formula, implies

$$I_N \sim \left(\frac{1}{t_{nk}}\right)^{2/N}\left(\frac{c}{N}\right)^2. \tag{48}$$

For large N, I_N decreases as $1/N^2$, but for small N, I_N increases with N essentially as $1/\sqrt[n]{t_{nk}^2}$. It follows that there is an optimal order of truncation $N = N_{\rm opt}$ at which I_N has its maximum value. This defines the optimal region of Nekhoroshev stability for the given time of stability. In the actual calculations, we do not make any use the asymptotic formula (48), but only of Equation (47). Namely, the region of stability is determined by evaluating I_N up to the order $N = N_{\rm opt}$, where I_N reaches its maximum value. Thus all the above described inequalities are fulfilled.

3. Application to the Trojan Asteroids

3.1. THE MAPPING MODEL

When seeking to determine a region of Nekhoroshev stability, a crucial factor is the choice of appropriate variables which should be such so as to fit better the shape of the region that needs to be described. A natural choice of variables for the Trojan motion are the canonical variables

$$x = \sqrt{\frac{\alpha}{\alpha'}} - 1, \quad \tau = \lambda - \lambda',$$

$$x_2 = \sqrt{\frac{\alpha}{\alpha'}}\left(\sqrt{1-e^2} - 1\right), \quad \varpi, \tag{49}$$

where α, e are the semi-major axis and eccentricity of the asteroid, α' is the semi-major axis of Jupiter, τ is the critical argument, i.e., the difference between the mean longitude of the asteroid and of Jupiter, and ϖ is the longitude of the pericenter of the asteroid. These variables have the advantage that the pair of action–angle variables x, τ are immediately translated in the motion of the asteroid in configuration space. Namely, x gives the amplitude of librations perpendicularly to the circle of the 1:1 coorbital motion, while $\tau - \tau_0$ measures the synodic libration around the equilibrium values $\tau_0 = \pi/3$ (for L4) or $5\pi/3$ (for L5). If both librations are considered nearly harmonic, then, following Érdi (1988), the amplitude of librations is measured by the parameter D_p given by

$$\Delta\alpha = \alpha - \alpha' \simeq 2\alpha' x \simeq \sqrt{3\mu}\alpha' D_p \sin\phi, \quad \Delta\tau = \tau - \tau_0 \simeq D_p \cos\phi, \tag{50}$$

where ϕ is the phase of a libration, which is defined by the position of the guiding center of the motion of the asteroid along a tadpole-shaped orbit. According to Equation (50) the parameter D_p is the amplitude of libration of the critical argument. For most asteroids the amplitude of libration D_p ranges from a few degrees up to about $D_p \simeq 35°$ (Milani, 1993; Érdi, 1997; Levison et al., 1997), while in general D_p decreases as the eccentricity increases.

Now, the period of oscillations of x_2 is shorter than the period of oscillations in the plane (x, τ) by a factor $\simeq 1/13$. Thus, if $H(x, x_2, \tau, \varpi)$ is a 2 degrees of freedom Hamiltonian describing the motion in the variables (49) in a rotating frame, a good first approximation of the stability region is obtained by averaging H with respect to the short period terms. The resulting Hamiltonian $\langle H\rangle(x, \tau, x_2)$ is of one degree of freedom, with x_2 as a parameter. In this approximation, the width of the stability region coincides with the width of the separatrix passing through the unstable Lagrangian point at $\tau = \pi$. This gives $D_p \simeq 75°$.

However, the averaging process ignores all the resonant phenomena producing chaos near the separatrix. To reintroduce these phenomena, Sándor et al. (2002) used the averaged Hamiltonian $\langle H\rangle$ to produce a mapping for the Trojan motion, based on the method of Hadjidemetriou (1991). The mapping of Sándor et al. reproduces the characteristics of the Poincaré surface of section of the original Hamiltonian. In particular, the mapping has the same fixed points and with the same stability as the Hamiltonian.

In the case of zero eccentricity $e = 0$, the mapping reads

$$\begin{aligned} x_{n+1} &= x_n + 2\pi\mu \sin\tau_n \left(1 - \frac{1}{(2 - 2\cos\tau_n)^{3/2}}\right), \\ \tau_{n+1} &= \tau_n + 2\pi\left(\frac{1}{(1 + x_{n+1})^3} - 1\right). \end{aligned} \tag{51}$$

Figure 1(a) shows the phase portrait of the mapping (51). The invariant KAM curves of this mapping correspond to librations around L_4. Each invariant curve defines approximate proper elements corresponding to the amplitudes of libration $D_p = \tau_{\max} - \tau_{\min}$, and $d_p = \alpha_{\max} - \alpha_{\min}$, where α is defined in terms of x by Equation (49). Figure 1(b) shows D_p as a function of d_p, calculated from Figure 1(a), by considering 10 invariant curves of Figure 1(a) with initial conditions $x = 0$ and $\tau = \pi/3 + n\Delta\tau$ with $n = 1, 2, \ldots, 10$ and $\Delta\tau = \pi/60$. Notice that the points of Figure 1(b) are almost on a straight line with slope $\simeq (1/0.273)$(rad/AU). The theoretical value given by Érdi (1988) is $(1/0.2783)$(rad/AU).

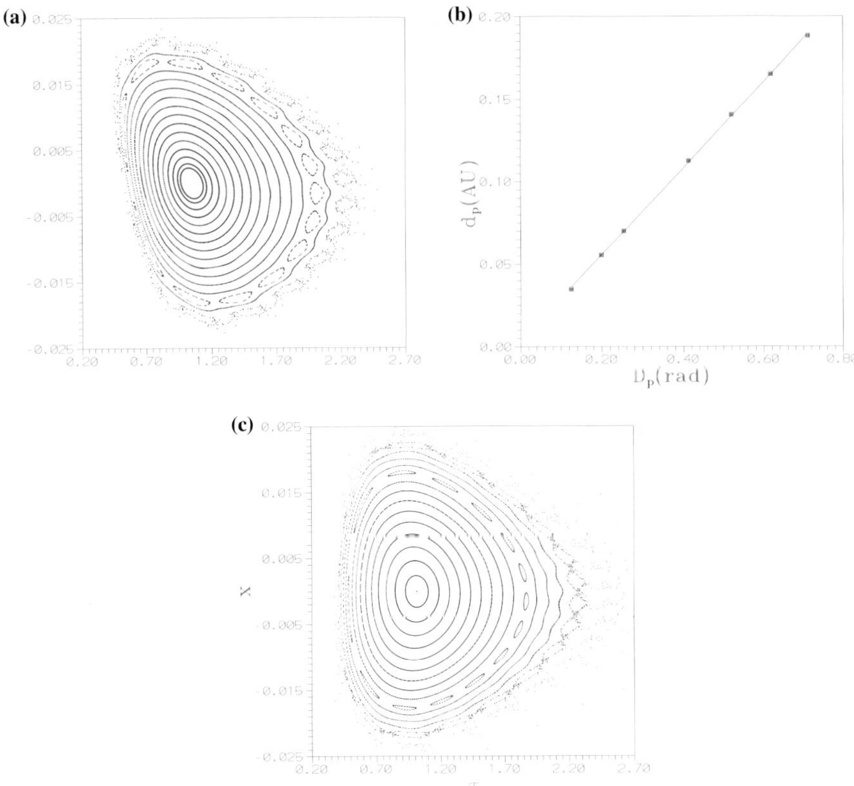

Figure 1. (a) The phase portrait of the mapping (51). (b) The relation d_p versus D_p (see text for definitions) as found by the invariant curves of the mapping (51). (c) The Poincaré surface of section for the equations of motion under the full Hamiltonian of the planar circular restricted three-body problem. The section is given by the variables (τ, x) when the surface $\varpi - \lambda' = 0$ is crossed by an orbit at the positive sense. The Jacobi constant is taken equal to $E = -1.49948$, which is very close to the Jacobi constant at $L4$, $E_{L4} = -1.49952$. This corresponds to eccentricities $e < 0.04$.

It should be stressed that a large fraction of the KAM tori of Figure 1(a) are destroyed, when the eccentricity of Jupiter, or other secular perturbations, are turned on. Thus, it makes no sense to try to define a region of effective stability by identifying its limit as the outermost KAM curve of Figure 1. This is what we *avoid* doing by use of Nekhoroshev theory, i.e., we avoid determining whether an orbit is of KAM type or not. Precisely, the power of Nekhoroshev theory is in the fact that the size of the remainder of formal series at any order is determined by the accumulation of small divisors corresponding to near-resonances of similar order. The sequences of small divisors a_k, as a function of the order k can be determined by the relation

$$a_k = \min\{\exp(i(m-n)\omega) - 1, m, n > 0, m + n = k\}, \tag{52}$$

where ω is the rotation number of the central fixed point.

Figure 1(c) shows the Poincaré surface of section for the exact Hamiltonian model of the CRTBP (Szebehely, 1967), at the Jacobi constant $E = -1.49948$, which is very close to the value at L4 $E_{L4} = -1.49952$. The maximum eccentricity, at the outer invariant curve is $e = 0.04$, thus the surface of section can be compared with the mapping phase portrait, which corresponds to $e = 0$. The mapping portrait is angularily deformed with respect to the Hamiltonian portrait. This phenomenon is an artifact of the method to produce the mapping. However, the extent of the stability region is the same in the two portraits, and the resonant chains of same multiplicity are at approximately the same distances from the center. The rotation number of the central point of Figure 1(c) can be used to compare the sequences of divisors found for the mapping model (51) with the corresponding sequences for the Hamiltonian model. The comparison is shown in Figure 2(a), for even orders, and Figure 2(b), for odd orders. The sequences are similar, in general, but there is a small divisor appearing in the mapping model at order 37, which appears much later (at higher orders) for the Hamiltonian model. However, this divisor does not affect the results on Nekhoroshev stability, because it was checked numerically that it does not produce dominant terms in the formal series up to order ~50 because of the 'retardation effect' (Turchetti, 1989).

We conclude that the mapping (51) is a good model of the planar CRTBP for orbits with low proper eccentricities. On the other hand, the present model is of limited use in the case of orbits with high proper eccentricities. Let us note, however, that the ratio d_p/D_p remains remarkably constant in this case too, as well as in the case where the proper elements are determined either by the elliptic restricted three body problem (Érdi, 1988), or by precise numerical integrations including the effects of the major planets. For example, the ratios the proper elements D_p and d_p for 177 asteroids published

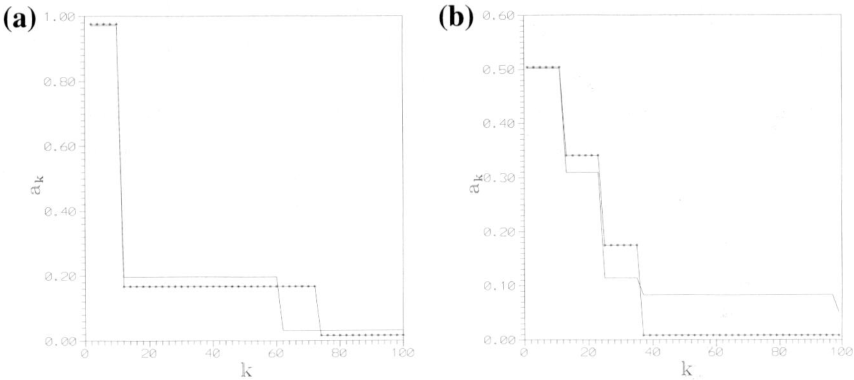

Figure 2. The sequences of divisors a_k defined by Equation (52) for the mapping model (51) (line with dots) and for the mapping defined by the surface of section of the full Hamiltonian of the planar CRTBP (line without dots). (a) Even orders. (b) Odd orders.

by Milani (1993), in a full model with all major planets, respect the costancy of the ratio D_p/d_p, despite the fact that the individual variations exhibited by D_p or d_p for particular asteroids are large.

3.2. RESULTS

To calculate a formal integral for the mapping (51), the mapping has to be written first in the form (12). The mass parameter is set $\mu = 00095387536$. The mapping (51) is Taylor expanded around the elliptic point $x_0 = 0$, $\tau_0 = \pi/3$. Setting $x = x_0 + u$, $\tau = \tau_0 + v$ the polynomial expansion of the mapping is given in the variables u, v by

$$u' = u + 0.013485095v + \text{h.o.t},$$
$$v' = -18.849555921u + 0.745811943v + \text{h.o.t.} \tag{53}$$

The expansion (53) is truncated at order 60, meaning that the mapping (53) satisfies the symplectic condition also up to terms of order 60. The size of the first neglected term increases with the distance from the elliptic equilibrium. At the border of the stability the error is $\sim 10^{-19}$, which is below the computer precision.

By a standard procedure, the linearized mapping is diagonalized so that its invariant ellipses are transformed to invariant circles. The diagonalizing transformation is given in terms of the complex conjugate variables z, z_* by

$$\frac{u}{\sqrt{2}} = 0.08043775836(z + z_*) + 0.02096888502i(z - z_*),$$
$$\frac{v}{\sqrt{2}} = 0.00054271899(z + z_*) - 3.10785164655i(z - z_*). \tag{54}$$

After the above transformation, the mapping takes the standard form

$$z' = e^{i2\pi\omega} z + \text{h.o.t}, \tag{55}$$

where the rotation number is equal to $\omega = -0.0811163650236938$. This value is quite close to the value of the rotation number $\omega_2/\omega_1 = -0.0807256266212$ of the frequencies ω_1, ω_2 found by the diagonalization of the Hamiltonian of Giorgilli and Skokos (1997, Equation (4)). In particular, we notice that the rotation number is negative, namely the two first frequencies of the system have opposite sign. This is characteristic of the Lagrangian points L4 and L5.

A formal integral for the mapping (55) is calculated by a computer program as described in Section 2. Then, the estimate of the effective region of stability can be obtained as explained in Subsection (3.1), after the constants A and $B_{\rho*}$ have been specified. To specify these constants we proceed in the following way. Setting the Nekhoroshev time equal to $t_{nk} = 10^9$ periods of Jupiter, we obtain first an approximation to the radius of the region of stability by finding the maximum value of the quantity

$$\rho_N = \left(\frac{1}{N||U_N||t_{nk}}\right)^{1/(N-1)} \tag{56}$$

with respect to N. This maximum value is set equal to ρ_*, while the order $N = N_{\text{opt}}$, where the maximum occur is set as the optimal order of truncation. Then, the value of ρ_* is used to calculate B_{ρ_*} by Equation (35), and A by Equation (40), where $\rho_{\max}$ is set equal to ρ_*. In fact, in all the calculations we have multiplied the value of the coefficient B_{ρ_*} by a factor 2, called a 'safety factor'. By this method the values of both A and B_{ρ_*} are overestimated, because ρ_* is always larger than the effective radius corresponding to the level value I_N found by Equation (47). This makes the results precise although not necessarily optimal.

In the particular case of the mapping (55) we find the parameters

$$r_F = 0.11, \quad N_{\text{opt}} = 38, \quad A = 0.1896, \quad B_{\rho*} = 7.83$$

and by the criterion (47), the region of effective stability for $t_{nk} = 10^9$ is identified as the interior of the closed curve $\Phi_{38}(z, z_*) = I_{38}$ with

$$I_{38} = 4.31587 \times 10^{-4}.$$

This result can now be expressed in terms of the proper elements D_p and d_p. It is known that the librations around the fixed point are not exactly symmetric

(Namouni and Murray, 2000). Thus, the amplitude of the libration D_p is defined as

$$D_p = \frac{v_{\max} - v_{\min}}{2}, \tag{57}$$

where $v_{\min}$ and $v_{\max}$ are the minimum and maximum value of v, respectively on the curve $\Phi_{38}(u, v) = I_{38}$, where the integral Φ_{38} is expressed in terms of the old variables u, v. Similarly, by defining $u_{\min}$ and $u_{\max}$ we find the amplitude of oscillations

$$x_p = \frac{u_{\max} - u_{\min}}{2}, \tag{58}$$

which gives the amplitude of oscillations of the semi-major axis

$$d_p = [(x_p + 1)^2 - 1]\alpha' \tag{59}$$

with $a' = 5.2037$AU, i.e., the semi-major axis of Jupiter. The final results are

$$D_p \leqslant 10.6^\circ, \quad d_p \leqslant 0.0512 \text{ AU} \tag{60}$$

This result represents an improvement over previously obtained Nekhoroshev estimates of the region of effective stability (Giorgilli and Skokos, 1997; Skokos and Dokoumentzidis, 2001). The region of stability where real asteroids are observed extends to $D_p \simeq 33^\circ$, meaning that the region given in Equation (60), by analytical methods, has a size equal to about one third the real size of the observed region of stability. Previous estimates were giving a size smaller by a factor 10 for most asteroids, and up to a factor 30 in the worst case (Giorgilli and Skokos, 1997). It should be stressed, however, that the mapping model used here is also a simplification of the Hamiltonian problem, which reproduces approximately the dynamics only at low proper eccentricities.

The improvement made by the present estimates can be checked also by finding the proportion of real asteroids which are in the region of stability defined by Equation (60). Milani (1993) published proper elements for 174 asteroids in the 1:1 resonance with Jupiter calculated by numerical integration of the orbits in a model including Jupiter and the other major planets. These elements were reconfirmed by Beaugé and Roig (2001) up to a marginal difference in the values of D_p. There are 54 out of 174 asteroids for which $D_p < 10.6^\circ$. This corresponds to a proportion of about one third of the asteroids for which Nekhoroshev stability is guaranteed. However,

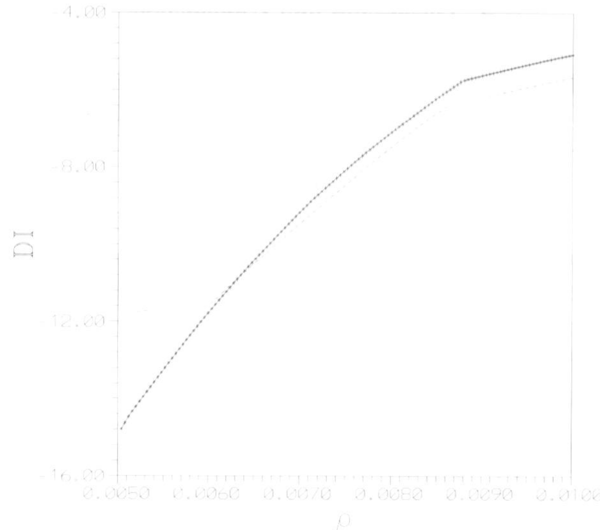

Figure 4. The average variation ΔI of the value of the formal integral, truncated at order 38, as a function of the distance ρ along a line of initial conditions ($v = 0, u = \rho$). The dashed line is a numerical evaluation along the orbits, while the solid line is the analytical estimate by the norm of the remainder.

every step may have opposite signs, thus adding to a total variation much smaller than their absolute sum. An elementary example is the random walk model. After N random steps, the maximum possible variation is $N \times DI$, but the mean variation is only $N^{1/2} \times DI$. For $N = 10^{10}$ (age of the Universe) the two estimates differ by five orders of magnitude. On the other hand, the maximum variation provides a rigorous upper bound of the variation, while the mean variation, although more realistic, is not a rigorous estimate.

This kind of problem cannot be faced by the present method to obtain Nekhoroshev stability estimates, but requires a detailed theory of the diffusion at the border of the stability region. However, even the present estimates are realistic and demonstrate the usefulness of Nekhoroshev theory.

In conclusion the present paper:

(1) Provides an algorithm for the construction of formal integrals in symplectic mappings by a direct method, i.e., without use of the normal form. The algorithm is given for both the non-resonance and resonance cases.
(2) Determines a method to obtain precise analytical Nekhoroshev estimates on a region of effective stability for a given time of stability.
(3) Implements the above methods in the case of a symplectic mapping for coorbital motion in the circular restricted three body problem, derived by Sándor et al. (2002). By this method, the effective stability for the Trojan

asteroids, over 10^9 periods of Jupiter, is demonstrated theoretically in a region corresponding to a maximum amplitude of libration $D_p = 10.6°$. About 30% of asteroids with known proper elements, at low proper eccentricities and inclinations, are included in this region.

(4) A further improvement requires (a) a more complete model (elliptic + other major planets), and (b) the use of a detailed theory of diffusion at the border of the stability region. These are topics of current research.

Acknowledgements

I am grateful to Professors G. Contopoulos and N. Voglis, as well as to Drs. Z. Sándor, Ch. Skokos, and K. Tsiganis for stimulating discussions on the problem of stability of the Trojan asteroids. This research was supported in part by the Research Committee of the Academy of Athens, and by an Archimedes grant of the Ministry of National Education.

References

Bazzani, A. and Marmi, S.: 1991, *Nuovo Cimento* **106B**, 673.
Beaugé, C. and Roig, F.: 2001, *Icarus* **153**, 391.
Benettin, G., Galgani, L. and Giorgilli, A.: 1985, *Cel. Mech.* **37**, 1.
Birkhoff, G. D.: 1920, *Acta Math.* **43**, 1.
Celletti, A. and Giorgilli, A.: 1991, *Cel. Mech. Dyn. Astr.* **50**, 31.
Celletti, A., Giorgilli, A. and Locatelli, U.: 2000, *Non-linearity* **13**, 397.
Cherry, T. M.: 1924, *Proc. Cambridge Phil. Soc.* **22**, 325, 510.
Contopoulos, G.: 1963, *Astron. J.* **68**, 763.
Contopoulos, G.: 1960, *Z. Astroph.* **49**, 273.
Dvorak, R. and Tsiganis, K.: (2000), *Cel. Mech. Dyn. Astr.* **78**, 125.
Efthymiopoulos, C., Giorgilli, A. and Contopoulos, G.: 2004, *J. Phys. A* **37**, 10831 10858.
Érdi, B.: 1988, *Cel. Mech.* **43**, 303.
Érdi, B.: 1997, *Cel. Mech. Dyn. Astr.* **65**, 149.
Fassò, F., Guzzo, M. and Benettin, G.: 1998, *Commun. Math. Phys.* **197**, 347.
Giorgilli, A.: 1988, *Ann. Inst. H. Poincaré* **48**, 423.
Giorgilli, A.: 1999, In: C. Simo (ed.), *Hamiltonian Systems with Three or More Degrees of Freedom*, Kluwer, Dordrecht.
Giorgilli, A. and Skokos, Ch.: 1997, *Astron. Astrophys.* **317**, 254.
Hadjidemetriou, J. D.: 1991, In: A. E. Roy (ed.), *Predictability, Stability and Chaos in N-Body Dynamical Systems*, Plenum Press, New York, pp. 157–175.
Levison, H., Shoemaker, E. M. and Shoemaker, C. S.: 1997, *Nature* **385**, 42.
Lochak, P.: 1992, *Russ. Math. Surv.* **47**, 57.
Milani, A.: 1993, *Cel. Mech. Dyn. Astron.* **57**, 59.
Namouni, F. and Murray, C. D.: 2000, *Cel. Mech. Dyn. Astr.* **76**, 131.
Nekhoroshev, N. N.: 1977, *Russ. Math. Surv.* **32**(6), 1.
Niederman, L.: 1998, *Non-linearity* **11**, 1465.

Robutel, P., Gabern, F. and Jorba, A.: 2005, 'The observed Trojans and the global dynamics around the Langvangian points of the Sun-Jupiter system', *Celest. Mech. Dynam. Astron.* **92**, 55–71.
Sándor, Z., Érdi, B. and Murray, C.: 2002, *Cel. Mech. Dyn. Astr.* **84**, 355.
Sándor, Z. and Érdi, B.: 2003, *Cel. Mech. Dyn. Astr.* **86**, 301.
Szebehely, V. G.: 1767, *Theory of Orbits*, Academic Press, New York.
Simó, C.: 1989, 'Memorias de la Real Academia de Ciencias y Artes de Barcelona', **48**, 303.
Skokos, Ch. and Dokoumetzidis, A.: 2001, *Astron. Astrophys.* **367**, 729.
Turchetti, G.: 1989, In: G. Turchetti (ed.), 'Conference on Non-linear Dynamics', World Scientific, Singapore.
Tsiganis, K., Dvorak, R. and Pilat-Lohinger, E.: 2000, *Astron. Astrophys.* **354**, 1091.
Tsiganis, K., Varvoglis, H. and Hadjidemetriou J. D.: 2002, *Icarus* **155**, 454.
Whittaker, E. T.: 1916, *Proc. Roy. Soc. Edinburgh* **37**, 95.

Celestial Mechanics and Dynamical Astronomy (2005) 92:53–69
DOI 10.1007/s10569-004-5976-y

THE OBSERVED TROJANS AND THE GLOBAL DYNAMICS AROUND THE LAGRANGIAN POINTS OF THE SUN–JUPITER SYSTEM

P. ROBUTEL[1], F. GABERN[2] and A. JORBA[2]

[1]*Astronomie et Systèmes Dynamiques, IMCCE, CNRS UMR 8028, 77 Av. Denfert-Rochereau, 75014 Paris, France, e-mail: robutel@imcce.fr*

[2]*Departament de Matemàtica Aplicada i Anàlisi, Universitat de Barcelona, Gran Via 585, 08007 Barcelona, Spain, e-mails: gabern@mat.ub.es, angel@maia.ub.es*

(Received: 02 July 2004; revised: 20 October 2004; accepted: 09 November 2004)

Abstract. In this paper, we make a systematic study of the global dynamical structure of the Sun–Jupiter L_4 tadpole region. The results are based on long-time simulations of the Trojans in the Sun, Jupiter, Saturn system and on the frequency analysis of these orbits. We give some initial results in the description of the resonant structure that guides the long-term dynamics of this region. Moreover, we are able to connect this global view of the phase space with the observed Trojans and identify resonances in which some of the real bodies are located.

Key words: Arnold web, Frequency Map analysis, Trojan asteroids, resonances

1. Introduction

The long-term stability of the Jovian Trojan asteroids is a classical problem of dynamical astronomy. In the literature, this question is usually approached using analytical (see Giorgilli et al., 1989; Celletti and Giorgilli, 1991; Giorgilli and Skokos, 1997; Skokos and Dokoumetzidis, 2000; Gabern and Jorba, 2001) or numerical methods. Among the numerical studies, two points of view are prominent:

(1) *Local approach.* Definition and computation of proper elements and proper frequencies. A synthetic theory for the proper elements was first established by Milani (1993, 1994). Later on, Beaugé and Roig developed a semi-analytical theory for Trojan proper elements (Beaugé and Roig, 2001). The question of Trojan proper frequencies was also tackled by (Gabern et al., 2004).
(2) *Global approach.* Since the work by (Levison et al., 1997), where the spatial distribution of the escape times was studied, different attempts to describe the global dynamics of the co-orbital region have been done: (Michtchenko et al., 2001; Nesvorny and Dones, 2002 and Tsiganis et al., 2005).

In this paper, we try to combine these two different approaches by placing several hundreds of observed Trojans in the corresponding global dynamical background.

First, we show some initial results of a global and systematic study of the tadpole region near the Jupiter co-orbital L_4 point (Robutel et al., in preparation). We describe some of the families of resonances that are fundamental to understand the complicated structure underlying the 1:1 mean motion resonance between Jupiter and a Trojan. The computations are based on the Frequency Map analysis (Laskar, 1999) and the results rely on an estimate of the chaoticity of some relevant slices of the phase space. These global pictures of the dynamical structure around the L_4 co-orbital region and the knowledge of the specific values of the fundamental frequencies allow us to identify the resonances that arrange the main structures of the phase space.

Once these global dynamical pictures are obtained, a natural question is what information they provide concerning the real Trojan asteroids. In this regard, we explain a way of superimposing the observed Trojans in these dynamical maps and we are able to identify actual asteroids inside some of the main resonances of the global pictures.

The simulations are based on a direct numerical integration of the Restricted 4-Body Problem defined by Sun, Jupiter, Saturn and the asteroid (SJS model). Gabern et al. (2004) already showed that for studying the Trojan problem, restricted 3-body models are not enough. Moreover, in (Gabern, 2003), some relatively simple semi-analytic four and five body models (3-body models with 2-dimensional quasi-periodic forcing (Gabern and Jorba, 2004) were used to study this problem and also proved not accurate enough for the complete description of the fundamental frequencies of the Trojan asteroids. Thus, more planetary frequencies have to be taken into account.

In (Robutel et al., in preparation), it is shown that the SJS model already captures the main global dynamical structures of the co-orbital regions. Actually, the addition of Uranus and Neptune to the problem, does not affect the main features of the phase space, but just shifts slightly the location of the resonances and makes everything a little bit more unstable (Robutel et al., in preparation).

2. Frequency Map and Global Structure of the Phase Space

In Figure 1, we show a dynamical map of the tadpole region of the leading Lagrangian point L_4 (similar pictures are obtained in the L_5 case). This picture is generated by an integration of 32,000 fictitious Trojans. Their initial semi-major axis and eccentricities are chosen on a grid of 400×80 points belonging to the domain

$$(a, e) \in \mathcal{A} = [5.2035, 5.4030] \times [0.05, 0.30], \qquad (1)$$

where the points in the mesh are equally separated at a distance of $\Delta a = 0.0005AU$, for the a-axis, and $\Delta e = 0.003125$, for the e-axis. The remaining initial elliptic elements are fixed and equal the following values: $\sigma = \lambda - \lambda_5 - \pi/3$, $\sigma_g = \varpi - \varpi_5 - \pi/3$, $\Omega = \Omega_5$ and $I = I_5 + 2°$, where the subscript 5 denotes the elements of Jupiter. The choice of these values is natural if one realizes that, for the Sun–Jupiter Elliptic Restricted Three Body Problem (ERTBP), the elliptic elements of the L_4 point are: $a = a_5$, $e = e_5$, $\sigma = \sigma_g = 0$, $\Omega = \Omega_5$ and $I = I_5$.

The trajectories of these bodies are numerically integrated using a symplectic integrator of the family $SABA_n$ (Laskar and Robutel, 2001) on two consecutive time spans of 5 Myears each. Then, using the Frequency Map Analysis method (Laskar, 1990, 1999), two determinations of their fundamental frequencies are associated to every Trojan (one for each time span).

If we assume that the motion of Jupiter and Saturn is quasi-periodic (which is a very natural assumption on the 10 Myears considered here; Laskar, 1990; Robutel and Laskar, 2000) the orbit of this planetary system lies on a five-dimensional invariant torus, with fundamental frequencies $(n_5, n_6, g_5, g_6, s_6)$. The two first frequencies are the proper mean motions (frequencies associated to the orbital motion) of Jupiter and Saturn, respectively; while the other three are the secular frequencies of the Sun Jupiter–Saturn system (see Table I).

In these conditions, the motion of the asteroid can be seen as a 3-degrees of freedom Hamiltonian system with quasi-periodic forcing. This implies that a quasi-periodic trajectory is parametrized by eight fundamental frequencies. Five correspond to the quasi-periodic forcing and the remaining three

TABLE I
Fundamental frequencies.

Planet freq.	(″/year)	Asteroid freq.	Min. (″/year)	Max. (″/year)
n_5	109254.63165	ν	7000	9500
n_6	43995.34975	g	250	450
g_5	4.02760	s	−50	10
g_6	28.00657			
s_6	−26.03912			

The first two columns show the fixed frequencies used for the planets, subscript 5 stands for Jupiter and 6 for Saturn. In the last three columns, we show the minimum and maximum values (for the initial conditions considered) of the frequencies of the test-particles.

characterize the dynamics of the Trojan (Jorba and Villanueva, 1997). These three fundamental frequencies (ν, g, s) are respectively the proper libration frequency (connected to the libration in the 1:1 Mean Motion Resonance (MMR) with Jupiter), the proper precession perihelion frequency of the asteroid and the one corresponding to its node. They are the image of the Frequency Map, which can be defined as (Laskar, 1999)

$$F_{\theta_0}\colon (a, e, I) \longrightarrow (\nu, g, s), \qquad (2)$$

where $\theta_0 = (\lambda_0, \varpi_0, \Omega_0)$ is the fixed phase vector given above.

In general, the Frequency Map is a correspondence from an action space to a frequency space (Laskar, 1999). Even though this is not exactly the case here, our choice of initial phases θ_0 makes the elements (a, e, I) very close to action variables (Robutel et al., in preparation). Thus, we can assume that, at least inside regular regions of the phase space, the map F_{θ_0} defines a one-to-one correspondence between the domain $\mathcal{D} = \mathcal{A} \times \mathcal{S}$ (where A is given by (1) and $\mathcal{S} = I_5 + [0°, 38°]$ is the interval of inclinations considered) and its image $\mathcal{F} = F_{\theta_0}(\mathcal{D})$. The last three columns of Table I give the extremes of this frequency domain $\mathcal{F}$. For theoretical reasons discussed in (Robutel et al., in preparation), the frequency set $\mathcal{F}$ is a representative domain in the sense that the fundamental frequencies of a given Trojan belong to $\mathcal{F}$, no matter which their initial phases are. This assumption is satisfied for the large majority of the observed Trojans.

Therefore, the determination of the fundamental frequencies ν, g and s is the key point of our study. Besides giving an estimation of the diffusion rate used to detect instabilities (Laskar, 1990; Robutel and Laskar, 2001), it allows us to study the dynamical structures of the frequency space. Indeed, it is in this space (see Figure 2) that phenomena associated to resonances become clear and are quite easy to identify (Robutel et al., in preparation). In addition, the fundamental frequencies are considered in Section 3 as proper elements (Milani, 1993), and are used to locate observed Trojans on our dynamical maps (i.e., Figure 1).

The bodies remaining inside the co-orbital region during the whole integration (10 Myears) are colored, in Figure 1, according to the relative variation of the proper libration frequency $\delta\nu = (\nu_1 - \nu_2)/\nu_1$. Being ν_1 (ν_2) the determination of the proper libration frequency, ν, on the 1st (2nd) time-interval. The color code goes from dark gray, corresponding to motion close to quasi-periodic ($\delta\nu < 10^{-7}$), to light gray, for strongly irregular motion ($\delta\nu > 10^{-2}$). The black points (i.e., the top-right corner of the plot) correspond to initial conditions that escape from the co-orbital region before the 10 Myears integration ends.

Given this color code, it is clear from Figure 1 that the instability increases with the distance from the L_4 equilibrium point (placed in the left-bottom

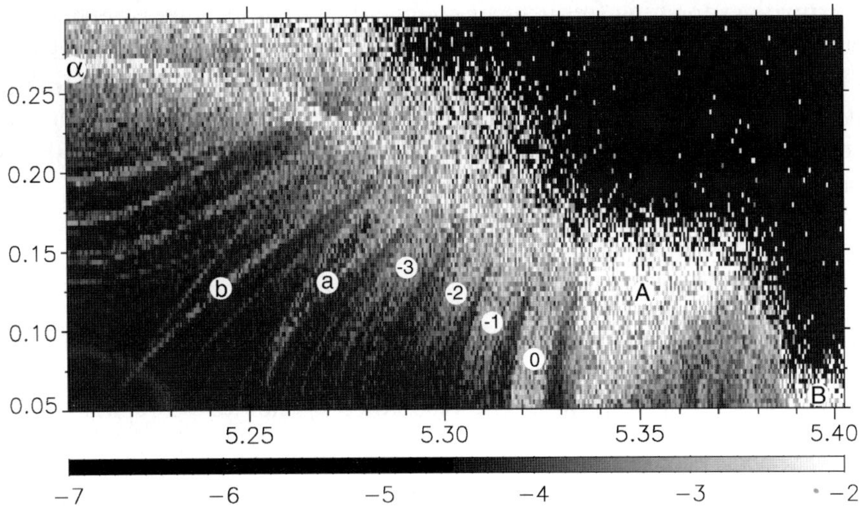

Figure 1. Dynamical map of the tadpole region surrounding L_4. It corresponds to a section of the phase space in the plane (a_0, e_0) where the other initial elliptic elements are fixed (for instance, the inclination is $I = I_5 + 2°$). The gray code corresponds to the temporal variation of v in logarithmic scale: $\log \delta v$ (units: Myears^{-1}). The symbols inside the plot indicate main resonant structures. See text for more details.

corner in this coordinates), ending in the black region (near the top-right corner). This leads to split this part of the phase space into three different domains: the escape domain (black), the high diffusion domain, defined arbitrarily by $\delta v > 10^{-3}$ (light gray to white), and the long-term stability domain, with $\delta v < 10^{-4,5}$ (dark gray).

By comparing this figure (and also Figure 3) with the ones obtained by Tsiganis et al. (2005), we note that a strong correlation exists between the value of δv and the escape time from the co-orbital regions. More precisely, our high diffusion domain ($\delta v > 10^{-3}$) corresponds to an escape time of about $10^7 - 10^8$ years, while the escape time inside the long-term stability region is larger than 10^9 years.

These two domains are strongly interpenetrated. Indeed, Figure 1 shows a kind of Arnold web made of unstable tongues (light gray) generated by resonances. These structures are practically isolated at a small distance of L_4 and they have a trend to overlap when the distance increases, until complete overlapping is reached close to the border of the black area. Each one of these tongues is generated by resonances that can be split in four different families.

Family 1: The first two families belong to a larger class of resonances, which is the one that gathers the secondary resonances between the proper libration frequency v and a linear combination of planetary mean motions. The first family involves the high frequency n_5, while the second family is related to the short period combination $n_5 - 2n_6$. Although this distinction in two different sets is quite arbitrary, it becomes natural if one realizes that the resonances related to n_5 already arise in the ERTBP (or even in the circular one), while, of course, the second family of resonances only can appear when the perturbation of two planets is considered.

Indeed, in the case of the RTBP (circular and planar cases) the secondary resonances associated to n_5 take the form:

$$p\nu - n_5 + g = 0, \text{ with } p \in \{12, 13, 14\}. \tag{3}$$

Since $g \ll \nu \ll n_5$ (see Table I), the resonances (3) are isolated and do not generate significant chaotic behaviors. On the contrary, as soon as a non zero eccentricity is given to Jupiter, the resonance (3) has to be replaced by the multiplet

$$p\nu - n_5 + qg = 0, \quad \text{where } q \text{ is an integer} \tag{4}$$

that generates, by overlapping, a large chaotic zone. It is worth to mention that this kind of secondary resonances can be seen as the overlapping of the 1:1 MMR with the high order resonances ($p : p\pm1$). By overlapping, these narrow resonances that accumulate far before reaching the stable an unstable L_3 manifolds (surfaces that separate tadpole orbits from horseshoe orbits (Nesvorny et al., 2002)), generate strong global instability.

The number of harmonics associated to the first family of resonances still increases when Saturn's perturbation is taken into account. Indeed, because of the additional secular frequencies of the planetary system g_5, g_6 and eventually s_6, the resonant relation defining the first family becomes:

$$p\nu - n_5 + qg + q_5g_5 + q_6g_6 = 0, \text{ with } q + q_5 + q_6 = 1. \tag{5}$$

Two of the most important contributions of this family are visible in Figure 1. The large gap above the white region indicated by "B" is generated by (5) with $p = 14$, while the V-shaped (large light gray to white) region labeled by "A" is associated to $p = 13$.

Family 2: The second family in the class of secondary resonances, that we will call *secondary three body resonance*, plays an important role in the Trojan swarms. This type of resonance appears when the libration frequency ν and the high frequency $n_5 - 2n_6$ are close to commensurability. Among all the possible combinations, the ones that generate large chaotic regions are given by:

$$5\nu - 2(n_5 - 2n_6) + pg + p_5g_5 + p_6g_6 = 0, \text{ with } p + p_5 + p_6 = -2. \tag{6}$$

Nesvorny and Dones (2002) first mentioned the possibility of instabilities generated by this family.

In the same way as in *Family 1*, because the secular frequency g is more than fifteen times larger than g_5 and g_6, a given p defines a multiplet of resonant harmonics. The four widest regions associated to this family are indicated in Figure 1 by the numbers 0 to -3 corresponding to the value of the integer p. It is important to mention that these resonances do not come from the direct action of Saturn, but from the short period perturbations of Jupiter's orbit due to Saturn (the same type of effect appears in (Ferraz-Mello, 1997) for a different problem).

Indeed, the frequencies corresponding to $n_5 - 2n_6$ and to the Great Inequality $2n_5 - 5n_6$ are associated with terms of large amplitude in the quasi-periodic approximation of Jupiter's eccentricity. For small Trojan eccentricity, resonant regions corresponding to different members of this family are isolated between them. But for larger Trojan eccentricities, these regions overlap in the neighborhood of the long white half-arch labeled by "α".

Family 3: This white arch (that cuts the vertical axis at about $e = 0.265$) corresponds to the location of the secular resonance $s = s_6$. This resonance, which influence on the Trojans was already suggested by (Yoder, 1979) and studied by Bien and Schubart (1984), Milani (1993), Dvorak and Tsiganis (2000) and Marzari and Scholl (2002) (among others) is known to induce very strong instabilities in the neighborhood of the long-term stability domain. Indeed, the majority of the Trojans that enter this secular resonance escape the co-orbital region in a few tenths of million years. Apart from this first order resonance, a lot of other secular resonances are present in the Trojan phase space, especially of the form:

$$qs + q_6 s_6 + p_5 g_5 + p_6 g_6 = 0 \tag{7}$$

with $q + q_6 + p_5 + p_6 = 0$ and $(q + q_6)$ even.

The significance of the secular resonances increase with the initial inclination of the Trojan, but even for high inclinations, these resonances are thin and isolated (Robutel et al., in preparation). For instance, the little quarter of circle in the left bottom of Figure 1, where the diffusion rate is of about 10^{-5}, corresponds to the location of the sixth order secular resonance $2s - 3g_5 + g_6 = 0$.

Family 4: The two most representative members of the last family are associated to the unstable structures denoted by "a" and "b" in Figure 1. They penetrate deeply inside the long-term stability region. These structures, and the other thin curves in the left side of "b", are generated by the Great Inequality. From the "frequencies point of view", this phenomenon leads, inside the resonance, to the relations:

$$4g + (2n_5 - 5n_6) + q_5 g_5 + q_6 g_6 + r_6 s_6 = 0 \text{ with } q_5 + q_6 + r_6 = -1. \tag{8}$$

Contrarily to what happens in *Family 1* and *Family 2*, the relation (8) does not contain the libration frequency v and, thus, it is not a secondary resonance of the usual type. Although these structures seem to be very narrow and isolated, they play an important role in the slow diffusion process that drives Trojans from the long time stability inner regions to the short time stability boundary. An example of this transport along resonances is shown in Figure 2 and discussed below.

The interest of this family (and to a Less extent of *Family 2*) is enhanced by the fact that some observed Trojans seem to evolve inside these resonances (see Section 3) and, consequently, may be subject to long-term transport phenomena.

In order to illustrate that it is in the frequency space where the dynamical phenomena associated to resonances can be easily interpreted, we show in Figure 2 the projection on the (g, s) plane of the image of $\mathcal{A} \times (I_5 + 2°)$ by

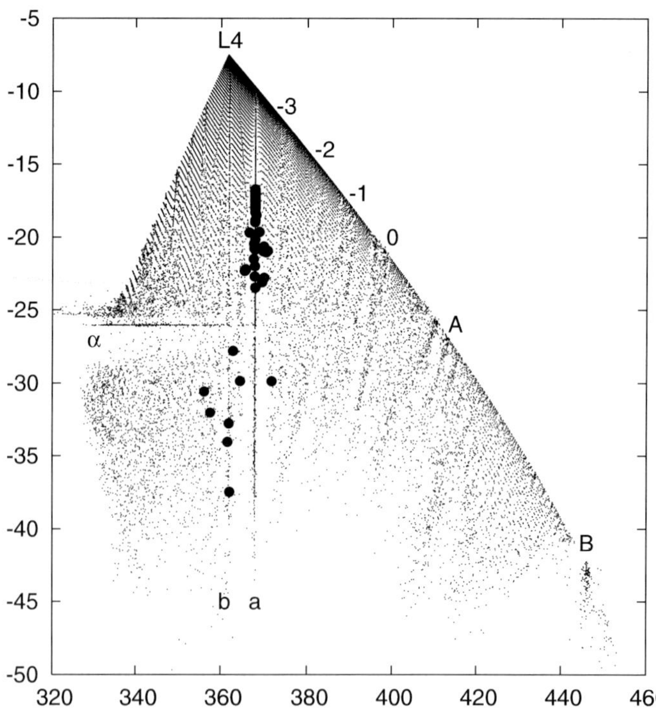

Figure 2. Projection on the (g, s) plane (units: ″/year) of the image of $(a, e, I) \in \mathcal{A} \times (I_5 + 2°)$ by the Frequency Map (2). It is in this space where resonances of the third and fourth family are easy to identify. The labels indicate the same resonant structures shown in Figure 1. The thick black points correspond to a path of a fictitious body that, starting in the resonance "a" (near $s \approx -17$), is subject to long-term transport along this resonance. This body is ejected from the co-orbital region after 355 Myears of integration time. See text for more details.

the Frequency Map F_{θ_0}. This plot is the counterpart of Figure 1 in terms of frequencies.

Figure 2 is made up of a union of curves (more or less smooth) that are the image of the lines e_0 = constant by the Frequency Map. The triangular shape of the picture is explained in the following way: the upper vertex of the triangle corresponds to the L_4 point, and the right and left edges correspond, respectively, to the line $e = 0.05$ (eccentricities lower limit) and to $a = 5.2035$ AU (semi-major axis lower bound). These curves are smooth in regular regions, but singularities arise in chaotic zones. Singularities of the Frequency Map are directly correlated with instabilities of the corresponding trajectories (see Laskar, 1999 for more details).

The resonances of *Family 3* and *Family 4* are very easy to identify in this plot. The vertical lines, where frequencies accumulate near g = constant, correspond to *Family 4*. In particular, the straight lines marked as "a" and "b" (these resonances are also shown in Figure 1 with the same labels) belong to *Family 4* and clearly are of the form g = constant. Horizontal lines correspond to *Family 3* and are of the type s = constant. In particular, the important resonance $s = s_6$ (seen in Figure 1 as an arch denoted by α) is easily identified in the central part of the picture. Note that above this horizontal line the dynamics is more regular than below of it, where the chaotic and escaping dynamics prevails.

In Figure 2, straight lines with finite (and different from zero) slope also appear. They correspond to resonances of the type $pg + qs + 2n_5 - 5n_6 + p_5g_5 + p_6g_6 + q_6s_6 = 0$ with $p + q + p_5 + p_6 + q_6 = 3$ and $q + q_6$ even. These resonances are not discussed in this paper (see Robutel et al., in preparation for more information).

Of course, resonances belonging to *Family 1* and *Family 2* do not appear in Figure 2 as clearly as the secular ones. Indeed, to identify these resonances, one needs to look at the projection on the (ν, g) plane. Nevertheless, these secondary resonances appear in Figure 2 as singularities of the Frequency Map and we have marked them with the same labels as in Figure 1: "A", "B", "0", "−1", "−2" and "−3".

In order to illustrate the influence of *Family 4* on the long-term transport from the inner libration region to the outer unstable zone, we have performed an integration of a fictitious body with an initial condition satisfying the $4g + (2n_5 - 5n_6) - g_5 = 0$ resonance relation (label "a" in Figures 1 and 2). This particle follows a path along this resonance (downward in Figure 2), crosses the secular resonance $s = s_6$ and, finally, escapes the co-orbital region. This path in the frequency space is generated by computing the fundamental frequencies of the body for every consecutive interval of 5 Myears and superimposing them (using thick black points) on Figure 2. This allows to follow its long-term diffusion in the frequency space. During the first

215 Myears, the fictitious Trojan stays inside the resonance "a" with the secular frequency s evolving in the range $[-19, -16.72]$ ($''$/year) without a well-defined trend. Then, the trajectory leaves the resonance "a" and wanders around it for the next 100 Myears (crossing the resonance several times). Meanwhile, the frequency s decreases from $-19''$/year to $-23.5''$/year. Finally the path crosses downward the $s = s_6$ horizontal line and the fictitious Trojan enters the large chaotic zone associated to this resonance, where it remains from the 315 to the 355 Myears of integration time, before being ejected. Other examples of diffusion along resonances and connections between them are presented in Robutel et al. (in preparation) showing the implication of the resonant structure in the long-term dynamics of the Trojan asteroids.

The dynamical role of the four families of resonances presented above depends on the value of the initial inclination. This is particularly true for *Family 3* and *Family 4*. As we see in Figure 3, when the inclination is small, *Family 3* resonances seem to be negligible, except for the important case $s - s_6 = 0$. Then, as the inclination becomes larger, new secular resonances appear. At $I \sim 12°$, a small unstable region around L_4 marks the birth of the resonance $3s - s_6 - 2g_5 = 0$. Similarly, other secular resonances appear for $I \sim 16°$, while the latter ones move farther from the Lagrangian point. On the contrary, the influence of *Family 4* decreases when the inclination, I, is increased. The dynamical implications of this phenomena are discussed in Robutel et al. (in preparation).

3. Analysis of the Observed Trojans

Once the global structure is known, it is tempting to locate the observed Trojans in our dynamical maps (in plots similar to Figure 1). In this regard, we downloaded from Bowell (2001) their osculating elliptic elements at the Julian date 2452200.5 (October 10, 2001) to be used as initial conditions for the simulations.

Unfortunately, a direct projection of these initial conditions into the dynamical maps would be meaningless. This is true even if we project in Figure 1 (where recall that $I = I_5 + 2°$) only the Trojans with small initial inclination (Tsiganis et al., 2005).

Indeed, since the initial phases of a given asteroid are, in general, different from θ_0 (the ones that define the Frequency Map (2)), a direct projection would locate the Trojan at the wrong place (because the global dynamical background would be different from the actual one). Although this procedure works rather well when the goal is to locate a body on a low resolution map (Tsiganis et al., 2005), it fails when we ask for an accurate position of the

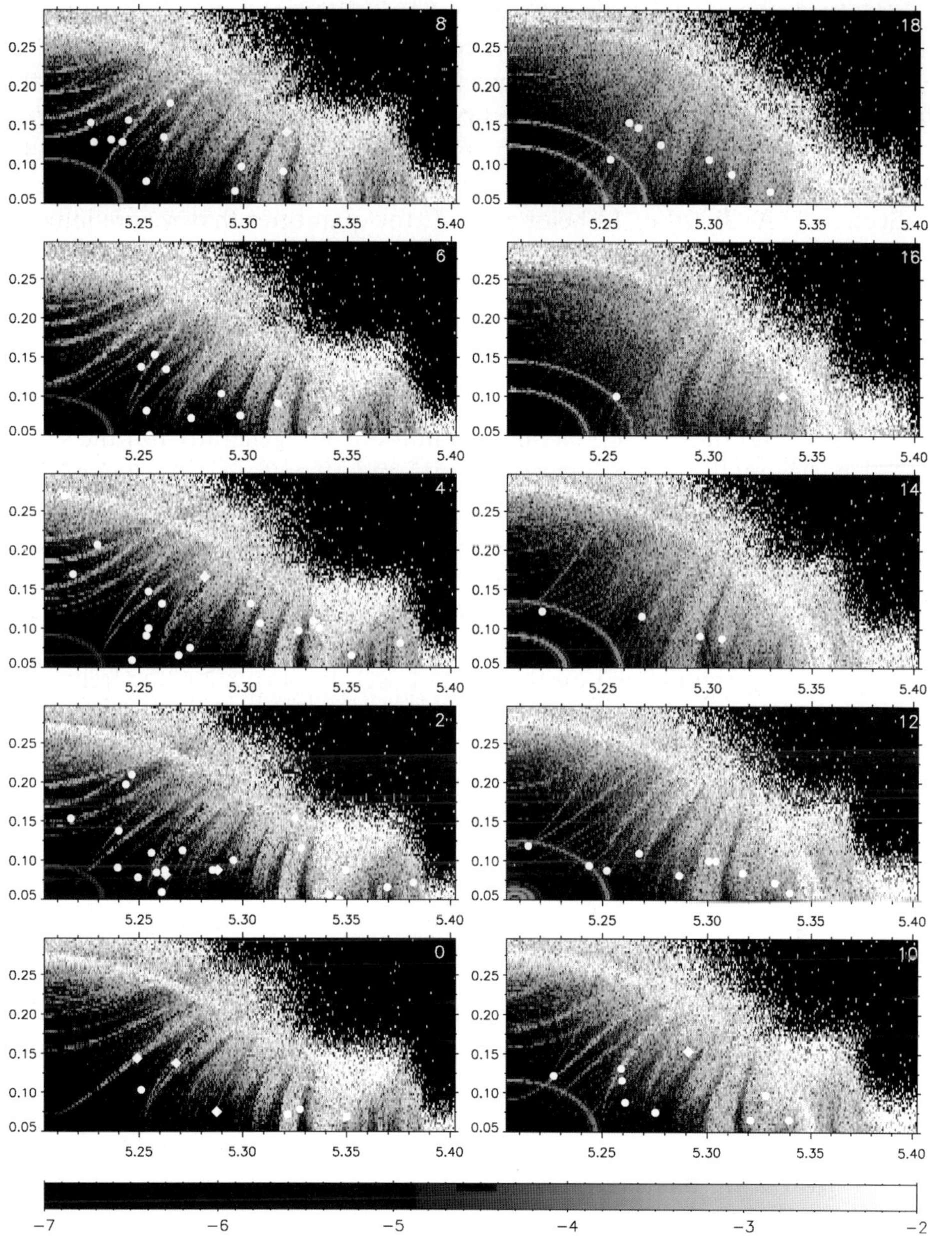

Figure 3. Dynamical maps of the tadpole region surrounding L_4 for inclinations going from I_5 (left-bottom corner) to $I_5 + 18°$ (top-right corner). The axis correspond to the semi-major axis a_0 and eccentricity e_0 of the particle. The white dots represent the "proper elements" (a^*, e^*) of the observed Trojans for which the distance $d_{j,0}$ defined by (10) is smaller than 1″/year. The white diamonds are associated to the Trojans satisfying the previous condition that moreover are closer than 0.1″/year of one of the resonances given in Tables III–V. See text for details.

asteroid, which is precisely what we need to connect the real Trojans to the narrow dynamical structures like the resonant families defined in Section 2.

To get rid of this "phase inconvenient", what we do is to look at the frequency space $\mathcal{F}$. For every observed Trojan j with initial conditions (a_j, e_j, I_j) and $\theta_j = (\lambda_j, \varpi_j, \Omega_j)$, we compute its basic frequencies $(\nu_j, g_j, s_j) = F_{\theta_j}(a_j, e_j, I_j)$, in the same way as it was done for fictitious particles in Section 2. As $F_{\theta_j}(a_j, e_j, I_j)$ belongs to $\mathcal{F}$, the quantities that we are looking for are given by:

$$(\tilde{a}_j, \tilde{e}_j, \tilde{I}_j) = F_{\theta_0}^{-1} \circ F_{\theta_j}(a_j, e_j, I_j), \tag{9}$$

This expression is well defined when the Hamiltonian system governing the motion of the Trojans is integrable. Two trajectories having respectively $(\tilde{a}_j, \tilde{e}_j, \tilde{I}_j, \theta_0)$ and $(a_j, e_j, I_j, \theta_j)$ as initial conditions are generally not the same, but they lie on the same invariant torus and, therefore, are dynamically equivalent. Thus, our choice (9) is natural in the sense that a frequency vector is not equivalent to a trajectory but to an invariant torus.

In practice, two main difficulties have to be overcome. The first one comes from the fact that our Hamiltonian is not integrable on an open subset of the phase space, even though it is integrable on the Cantor set of invariant tori (this makes F_{θ_0} well defined on this Cantor set). In particular, the frequency map F_{θ_0} has singularities (asymptotically in the neighborhood of the resonances), and, thus, F_{θ_0} is probably not invertible. The second is that only a finite number of points of $\mathcal{F}$ is known: The domain of initial conditions $\mathcal{D}$ is sliced in 20 planes (a, e) for fixed inclinations $I–I_5 \in \tilde{\mathcal{S}} = \{0°, 2°, 4°, \ldots, 38°\}$. For each of these planes, we consider 32,000 initial conditions (see Section 2). Thus, $\mathcal{D}$ is replaced by the discrete set $\tilde{\mathcal{D}}$ containing 640,000 points. The discrete frequency space $\tilde{\mathcal{F}} = F_{\theta_0}(\tilde{\mathcal{D}})$ contains less than 420,000 points, because an important fraction of initial conditions of $\tilde{\mathcal{D}}$ correspond to escaping or very chaotic trajectories, for which the quasi-periodic approximation is meaningless. Then, for a given actual Trojan, we can approximate the elements $(\tilde{a}_j, \tilde{e}_j, \tilde{I}_j)$, defined in (9), by the quantities (a_j^*, e_j^*, I^*) such that the expression

$$d_{j,0} = \left\| F_{\theta_j}(a_j, e_j, I_j) - F_{\theta_0}(a^*, e^*, I^*) \right\|_2 \tag{10}$$

is minimal in the grid $\tilde{\mathcal{D}}$ ($\| \cdot \|_2$ denotes the euclidean norm in $\mathbb{R}^3$).

In Figure 3, we plot these "reference elements" (a_j^*, e_j^*), corresponding to the observed L_4 and L_5 Trojans for which the Euclidean norm described above is smaller than $1''$/year, on the global dynamical maps around the L_4 point (this is justified by the fact that no significant differences are found between the dynamical structures of the L_4 and L_5 tadpole regions). Only the first 10 of 20 dynamical maps considered are shown.

These pictures show that the majority of the Trojans are inside the long-time stability region (this was already noticed by many authors: Michtchenko et al., 2001; Nesvorny and Dones, 2002 and Tsiganis et al., 2005, for instance) and also suggest that some actual Trojans may stay inside (or very close to) some of the resonances described in Section 2. This is rather easy to check when one has the basic frequencies of the observed asteroid.

In Table II, we show the actual Trojans that, for the SJS system, lie very close (at a distance smaller than 0.1″/year) to some resonance corresponding to *Family 1* (5), up to order 40. The first and second columns display the catalog number and the name of the particular asteroid. In the last column of the table, we give the distance to the exact resonance in ″/year. The remaining columns are devoted to the multiplet that defines the particular resonance inside the family.

In Table III, we show the Trojans that, up to order 16, are at a distance smaller than 0.1″/year of some resonance corresponding to *Family 2* (6). The last column shows the inclination in which the corresponding asteroid is found in Figure 1 (blank means that the asteroid is at a distance larger that 1 ″/year of any of our dynamical maps and, thus, it is not drawn).

TABLE II

Family 1.

Cat. num.	Name	p	q	q_5	q_6	dist. (″/year)
1749	Telamon	13	−12	3	10	2.42080e−02
5259	Epeigeus	13	−12	6	7	2.26021e−02
20739	1999XM193	13	−13	4	10	4.05032e−03

Actual Trojans at a distance smaller than 0.1″/year of the resonance $pv - n_5 + qg + q_5 g_5 + q_6 g_6 = 0$.

TABLE III

Family 2.

Cat. num.	Name	p	p_5	p_6	dist. (″/year)	$I - I_5$
4035	1986WD	0	6	−8	3.30536e−03	
5023	Agapenor	−1	2	−3	1.02691e−02	
9430	Erichthonios	−2	6	−6	1.54293e−02	2°
9817	Thersander	0	5	−7	2.60215e−02	
11554	Asios	−1	−3	2	9.33537e−02	
13862	1999XT160	−1	7	−8	9.60684e−02	
15536	2000AG191	−2	4	−4	3.51934e−02	
24426	2000CR12	−1	−3	2	8.88632e−03	
24508	2001BL26	−2	6	−6	2.07783e−02	0°

Actual Trojans at a distance smaller than 0.1″/year of the resonance $5v - 2(n_5 - 2n_6) + pg + p_5 g_5 + p_6 g_6 = 0$.

In Table IV, we show some of the real Trojans that are at a distance smaller than 0.1″/year of some representative secular resonance of *Family 3* (7). The computations are done up to order 14, and only 10 of the 46 actual cases found are shown. Note that some Trojans (i.e., 7119 Hiera) may even be very close to a double resonance. Double resonances were already suggested as a possible explanation for the "stable chaos" of some asteroids found in (Milani, 1993) and mentioned by Dvorak and Tsiganis (2000), for high-order secular resonances.

In Table V, we show some of the actual Trojans that are at a distance smaller than 0.1″/year of some secular resonance related to the Great Inequality and belonging to *Family 4* (8). The computations are done up to order 24 and, even though some Trojans are found inside high order resonances, only the cases close to the largest ones (i.e., $(-1,0,0)$, $(0,-1,0)$ and $(-2,1,0)$) are shown. Note that some actual asteroids (for instance, 4057 Demophon, 5233 1988RL10, 5907 1989TU5, 17423 1988SK2 and 18228 Hyperenor) are really very close (distance $< 2 \times 10^{-3}$″/year) to one of these resonances. Actually, in some cases, it can be numerically shown that these particles are captured by these resonances and the period of libration of the critical angle inside the resonance is about several million years (Robutel et al., in preparation).

Finally, let us note that we have found some remarkable examples that are present at the same time in one of the tables of resonances and in the global dynamical maps of Figure 3 (they are marked with a non-blank last column in Tables III–V). More concretely, in the 0° inclination picture we can identify 24508 2001BL26 (Table III), and 5907 1989TU5 and 17423 1988SK2

TABLE IV

Family 3.

Cat. num.	Name	q	q_6	p_5	p_6	dist. (″/year)	$I - I_5$
1173	Anchises	−5	3	4	−2	9.21333e−02	8°
3391	Sinon	2	−2	1	−1	6.58920e−02	
3451	Mentor	−6	4	2	0	9.94339e−02	
4138	Kalchas	−5	1	6	−2	8.14236e−02	2°
5023	Agapenor	3	−3	2	−2	3.22054e−02	
5126	Achaemenides	−2	−2	7	−3	6.49629e−02	
5130	Ilioneus	−6	4	2	0	6.09600e−02	
7119	Hiera	2	−2	1	−1	6.04987e−02	
7119	Hiera	−7	5	1	1	2.40070e−02	
9818	Eurymachos	1	−3	5	−3	3.38956e−02	

Some examples of actual Trojans at a distance smaller than 0.1″/year of the resonance $qs + q_6 s_6 + p_5 g_5 + p_6 g_6 = 0$, up to order 14. In this case, we only show 10 of the 46 Trojans found inside this type of resonances.

TABLE V
Family 4.

Cat. num.	Name	q_5	p_5	p_6	dist. ($''$/year)	$I - I_5$
4057	Demophon	−1	0	0	1.91160e−03	
4543	Phoinix	−2	1	0	6.37993e−02	16°
5233	1988RL10	−1	0	0	2.73850e−04	
5638	Deikoon	−2	1	0	2.98601e−02	10°
5907	1989TU5	−1	0	0	3.46164e−04	0°
13184	Augeias	0	−1	0	1.31297e−02	
13790	1998UF31	−2	1	0	2.45299e−02	
14518	1996RZ30	−2	1	0	2.58076e−02	
17423	1988SK2	−2	1	0	7.00885e−05	0°
18228	Hyperenor	−1	0	0	1.99257e−03	4°

Actual Trojans at a distance smaller than $0.1''$/year of the resonance $4g{+}(2n_5 - 5n_6) + q_5 g_5 + q_6 g_6 + r_6 s_6 = 0$, for the cases $(-1, 0, 0)$, $(0, -1, 0)$ and $(-2, 1, 0)$.

(Table V); at 2°, 9430 Erichthonios (Table III) and 4138 Kalchas (Table IV); at 4°, 18228 Hyperenor (Table V); at 8°, 1173 Anchises (Table IV); at 10°, 5638 Deikoon (Table V) and at 16°, 4543 Phoinix (Table V).

Some of these bodies are inside unstable structures associated to one of the families described above. These are examples of "stable chaos" (Milani and Nobili, 1992). For instance, 4543 Phoinix is in stable chaos according to (Milani, 1993) and we can locate it in Figure 3 (16° case) inside one of the secular resonances.

Moreover, the three asteroids inside the resonance $4g + (2n_5 - 5n_6) - 2g_5 + g_6 = 0$ that appear in Figure 3 (17423 1988SK2 (0°), 5638 Deikoon (10°) and 4543 Phoinix (16°)) show the displacement of this resonance in function of the initial inclination. This resonance exits the region of stability when the inclination increases. Also, the resonance $4g + (2n_5 - 5n_6) - g_5 = 0$ follows the same evolution. Two examples of asteroids inside this latter case are: 5907 1989TU5 (0°) and 18228 Hyperenor (4°).

A last interesting example that we want to point out is 1173 Anchises (Table IV and Figure 3 (8°)). This asteroid lies inside a region of overlapping and it was already in the list of "stable chaos" in Milani et al. (1997). It is very close to the $s = s_6$ resonance, to a resonance of *Family 2* with $p = -2$ (6) and to a secular resonance.

4. Conclusions

In this paper, we have performed a systematic study of the global dynamical structure of the Sun–Jupiter triangular regions in the SJS model.

Moreover, we have identified and classified in four families the main resonances that form the dynamical skeleton and dictate the long-term dynamics of the Trojan asteroids. In addition, we have shown how to place the actual Trojans in the global dynamical maps in a consistent way and we have been able to associate some of them with particular resonances of the four main families.

The method outlined in this paper seems to be very promising in order to study a particular real asteroid in its dynamical environment. Moreover, once the dynamical maps are computed, it is easy to add new observed Trojans and, if necessary, easy to increase the number of points of the domain $\tilde{\mathcal{F}}$ (see Section 3) to get a better accuracy of the asteroids locations.

On the other hand, inside regular regions, it is also possible to improve the determination of (a^*, e^*, I^*) (Section 3) by using interpolation of the n $(n > 1)$ closest points satisfying (10).

Finally, we cannot end the paper without mentioning that to have really accurate results for the observed Trojans, one should take into account the effect of the four major planets (OSS model). This is planned as future work.

Acknowledgements

We thank the comments of K. Tsiganis on a preliminary version of this paper. A financial support from the PNP-CNRS is acknowledged. F.G. and A.J. have been supported by the MCyT/FEDER Grant BFM2003-07521-C02-01, the CIRIT grant 2001SGR-70 and DURSI. The IBM-SP4 at CINES and Hidra at DSG-UB computing clusters have been widely used.

References

Beaugé, C. and Roig, F.: 2001, 'A semianalytical model for the motion of the trojan asteroids: proper elements and families', *Icarus* **153**, 391–415.

Bien, R. and Schubart, J.: 1984, 'Trojan orbits in secular resonances'. *Celest. Mech. Dynam. Astron.* **34**, 425–434.

Bowell, E.: 2001, 'The asteroid orbital elements database'. For more information, visit the URL http://www.naic.edu/˜nolan/astorb.html.

Celletti, A. and Giorgilli, A.: 1991, 'On the stability of the Lagrangian points in the spatial restricted three body problem', *Celest. Mech. Dynam. Astron.* **50**(1), 31–58.

Dvorak, R. and Tsiganis, K.: 2000, 'Why do Trojan ASCs (not) escape?', *Celest. Mech. Dynam. Astron.* **78**, 125–136.

Ferraz-Mello, S.: 1997, 'A symplectic mapping approach to the study of the stochasticity in asteroidal resonances', *Celest. Mech. Dynam. Astron.* **65**, 421–437.

Gabern, F.: 2003, 'On the dynamics of the Trojan asteroids'. Ph.D. thesis, University of Barcelona. http://www.maia.ub.es/~gabern/.

Gabern, F. and Jorba, A.: 2001, 'A restricted four-body model for the dynamics near the Lagrangian points of the Sun–Jupiter system'. *Discrete Contin. Dyn. Syst. Series B* **1**(2), 143–182.

Gabern, F. and Jorba, A.: 2004, 'Generalizing the restricted three-body problem. the bianular and tricircular coherent problems', *Astron. Astrophys.* **420**, 751–762.

Gabern, F., Jorba, A. and Robutel, P.: 2004, 'On the accuracy of restricted three-body models for the trojan motion', *Discrete Contin. Dyn. Syst.* **11**(4), 843–854.

Giorgilli, A., Delshams, A., Fontich, E., Galgani, L. and Simó, C.: 1989, 'Effective stability for a Hamiltonian system near an elliptic equilibrium point, with an application to the restricted three body problem', *J. Differential Equations* **77**, 167–198.

Giorgilli, A. and Skokos, C.: 1997, 'On the stability of the Trojan asteroids'. *Astron. Astrophys.* **317**, 254–261.

Jorba, À. and Villanueva, J.: 1997, 'On the persistence of lower dimensional invariant tori under quasi-periodic perturbations', *J. Nonlinear Sci.* **7**, 427–473.

Laskar, J.: 1990, 'The chaotic motion of the solar system. A numerical estimate of the size of the chaotic zone', *Icarus* **88**, 266–291.

Laskar, J.: 1999, 'Introduction to frequency map analysis', In: C. Simó (ed.), *Hamiltonian Systems with Three or More Degrees of Freedom*, NATO ASI. Kluwer Academic Publishers, Dordrecht, pp. 134–150.

Laskar, J. and Robutel, P.: 2001, 'High order symplectic integrators for perturbed Hamiltonian systems', *Celest. Mech. Dynam. Astron.* **80**, 39–62.

Levison, H., Shoemaker, E. and Shoemaker, C.: 1997, 'The long-term dynamical stability of Jupiter's Trojan asteroids', *Nature* **385**, 42–44.

Marzari, F. and Scholl, H.: 2002, 'On the instability of Jupiter's Trojans'. *Icarus* **159**, 328–338.

Michtchenko, T., Beaugé, C. and Roig, F.: 2001, 'Planetary migration and the effects of mean motion resonances on Jupiter's Trojan asteroids', *Astron. J.* **122**, 3485–3491.

Milani, A.: 1993, 'The Trojan asteroid belt: proper elements, stability, chaos and families'. *Celest. Mech. Dynam. Astron.* **57**, 59 94.

Milani, A.: 1994, 'The dynamics of the Trojan asteroids'. In: *IAU Symp. 160, Asteroids, Comets, Meteors 1993*, Vol. 160, pp. 159–174.

Milani, A. and Nobili, A. M.: 1992, 'An example of stable chaos in the Solar System', *Nature* **357**, 569–571.

Milani, A., Nobili, A. M. and Knezevic, Z.: 1997, 'Stable chaos in the asteroid belt', *Icarus* **125**, 13–31.

Nesvorny, D. and Dones, L.: 2002, 'How long-live are the hypothetical Trojan populations of Saturn, Uranus, and Neptune?', *Icarus* **160**, 271–288.

Nesvorny, D., Thomas, F. Ferraz-Mello, S. and Morbidelli, A.: 2002, 'A perturbative treatment of the co-orbital motion', *Celest. Mech. Dynam. Astron.* **82**, 323–361.

Robutel, P. and Laskar, J.: 2000, 'Global dynamics in the solar system', In: H. Pretka-Ziomek, E. Wnuk, P. K. Seidelmann, and D. Richardson (eds.), *Dynamics of Natural and Artificial Celestial Bodies*, Kluwer Academic Publishers, Dordrecht, pp. 253–258.

Robutel, P. and Laskar, J.: 2001, 'Frequency map and global dynamics in the solar system I', *Icarus* **152**, 4–28.

Skokos, C. and Dokoumetzidis, A.: 2000, 'Effective stability of the Trojan asteroids', *Astron. Astrophys.* **367**, 729–736.

Tsiganis, K., Varvoglis, H. and Dvorak, R.: 2005, 'Chaotic diffusion and effective stability of Jupiter Trojans', *Celest. Mech. Dynam. Astron.* **92**, 73, 89.

Yoder, C.: 1979, 'Notes on the origin of the Trojan asteroids', *Icarus* **40**, 341–344.

Celestial Mechanics and Dynamical Astronomy (2005) 92:71–87
DOI 10.1007/s10569-004-3975-7

CHAOTIC DIFFUSION AND EFFECTIVE STABILITY OF JUPITER TROJANS

KLEOMENIS TSIGANIS[1], HARRY VARVOGLIS[2] and RUDOLF DVORAK[3]
[1]*Observatoire de la Côte d' Azur, CNRS, B.P. 4229, 06304 Nice Cedex 4, France, e-mail: tsiganis@obs-nice.fr*
[2]*Department of Physics, Aristotle University of Thessaloniki, 54 124 Thessaloniki, Greece*
[3]*Institut für Astronomie, University of Vienna, A-1180 Vienna, Austria*

(Received: 27 April 2004; revised: 13 September 2004; accepted: 29 September 2004)

Abstract. It has recently been shown that Jupiter Trojans may exhibit chaotic behavior, a fact that has put in question their presumed long term stability. Previous numerical results suggest a slow dispersion of the Trojan swarms, but the extent of the 'effective' stability region in orbital elements space is still an open problem. In this paper, we tackle this problem by means of extensive numerical integrations. First, a set of 3,200 fictitious objects and 667 numbered Trojans is integrated for 4 Myrs and their Lyapunov time, T_{L}, is estimated. The ones following chaotic orbits are then integrated for 1 Gyr, or until they escape from the Trojan region. The results of these experiments are presented in the form of maps of T_{L} and the escape time, T_{E}, in the space of proper elements. An effective stability region for 1 Gyr is defined on these maps, in which chaotic orbits also exist. The distribution of the numbered Trojans follows closely the $T_{\mathrm{E}} = 1$ Gyr level curve, with 86% of the bodies lying inside and 14% outside the stability region. This result is confirmed by a 4.5 Gyr integration of the 246 chaotic numbered Trojans, which showed that 17% of the numbered Trojans are unstable over the age of the solar system. We show that the size distributions of the stable and unstable populations are nearly identical. Thus, the existence of unstable bodies should not be the result of a size-dependent transport mechanism but, rather, the result of chaotic diffusion. Finally, in the large chaotic region that surrounds the stability zone, a statistical correlation between T_{L} and T_{E} is found.

Key words: Jupiter Trojans, chaos, 1:1 resonance, effective stability

1. Introduction

Today, almost 100 years since the discovery of (588) *Achilles*, more than 1,200 numbered and multi-oppositioned Trojans are known (see the AstDys database, *hamilton.dm.unipi.it/cgi-bin/astdys/astibo*), orbiting around Jupiter's Lagrangian points. These bodies are in a 1:1 resonance with Jupiter, performing tadpole librations about the stable equilibria of the restricted problem, which are located at a relative mean longitude equal to $\pi/3$ (L_4) or $-\pi/3$ (L_5), with respect to Jupiter. Recently, the stability of these orbits, previously taken for granted, has been put to question.

The numerical results of Milani (1993, 1994) were the first to show that a number of Jupiter Trojans follow chaotic orbits. Although the integration time was small by today's standards (5 Myrs), these experiments have revealed some possible routes of escape at the borders of the Trojan region, related to secular resonances. Subsequently, Levison et al. (1997) were the first to present numerical results, indicating a slow dispersion of the Trojan swarms, on a time scale shorter than the age of the solar system. In their runs it was shown that the region of effective stability (for 4.5 Gyrs) is smaller than the one found analytically by Rabe (1967). This is in fact the most extensive long-term numerical integration of Trojans to date. The possibility of real Trojans escaping from the swarms in the future, due to the destabilizing effect of the ν_{16} and other high-order secular resonances, was shown by Tsiganis et al. (2000a) and Dvorak and Tsiganis (2000). The role of secular resonances was extensively studied by Marzari and Scholl (2002).

Analytical works on the long periodic motion of the Trojans are numerous and a complete list cannot be presented here. We should point out though that several important papers have been published recently, starting from the work of Érdi (1988, 1997). More refined models on the long periodic libration of the Trojans were published by Namouni and Murray (2000) and Nesvorný et al. (2002). The secular effect of additional perturbing bodies or of an oblate planet were studied by Morais (1999, 2001).

Milani (1993) defined and computed (synthetic) proper elements for the Trojans, which allowed him to search for families. Later, Beaugé and Roig (2001) presented a semi-analytic theory for Trojan proper elements, based on an asymmetric expansion of the disturbing function and on the principle of adiabatic invariance. This work, along with an increased sample of real bodies, allowed them to confirm and improve the results of Milani (1993), concerning the existence of at least two robust families (those of *Menelaus* and *Epeios*, both around L_4).

The first attempt to compute the extent of the stability region for Trojan-type motion was made by Rabe (1967), who studied the linearized equations of motion. Nowadays, there exist more refined analytical techniques, based on the construction of Nekhoroshev-type normal forms. The effective stability region is defined as an open domain of initial conditions around a torus of given frequencies (e.g., in our case the resonant torus corresponding to the Lagrangian points $L_{4,5}$), for which the time, τ, needed to change the actions, J, by a given small amount, say $|J(\tau)-J(0)| \leqslant \varepsilon$, is larger than the age of the solar system. This mathematically well defined and quite elegant approach suffers from the limitations of all strongly constrained theories. From the technical point of view, the 1:1 resonance seems to be even more difficult to tackle, compared to other resonances. Also, the models, in which these techniques are applied, are too simplified, to represent a realistic long-term evolution for Jupiter Trojans. As a result, the width of the stability region found by these methods is typically very small, compared to the one which can be determined numerically. These points are discussed in detail in Celletti and Giorgilli (1991), Giorgilli and Skokos (1997) and Skokos and Dokoumetzidis (2001).

Since the work of Levison et al. (1997), most numerical work has been oriented towards short-term numerical integrations of large sets of initial conditions, aiming to unveil the resonant structure of the Trojan swarms, and define a stability region in terms of regular/chaotic motion. Important results on this topic have been published by Marzari et al. (2003) and Nesvorný and Dones (2002). However, the most complete work on this subject was presented in this meeting by Robutel et al. (2005) in the same volume. We should also point out the numerical work of Michtchenko et al. (2001), where the important role of the great inequality on Trojan motion was shown, in an indirect way. In that paper it was shown that, if Jupiter and Saturn ever crossed the 5/2 resonance during their early migration, the Trojans would not have survived. Gomes (1998) also studied the effect of planetary migration on Trojans, showing that a near 2/1-resonant configuration for Jupiter and Saturn would also lead to a fast depletion of the swarms.

One has to take care though, since an asteroid undergoing chaotic motion will not necessarily escape from the Trojan region, within the age of the solar system. We remind the reader that ~30% of the main-belt asteroids follow chaotic orbits with Lyapunov times $T_L \leqslant 10^5$ yrs, but many of them have very stable proper elements over Gyr-long time spans. In this paper, we report the results of extensive numerical experiments, performed with the purpose of defining an 'effective stability' region for Trojan-type orbits and comparing with the distribution of real Jupiter Trojans, in orbital elements space. The term 'effective stability' refers not only to regular (quasi-periodic) orbits, but also to chaotic orbits which, however, can wander at the border of the stability region, without escaping within the lifetime of the solar system. The stability region is defined in terms of two quantities, which provide complementary information: (i) the Lyapunov time, T_L, which measures the (inverse of the) local rate of exponential divergence for chaotic orbits, and (ii) the escape time, T_E, which denotes the time needed for a Trojan to encounter Jupiter within a small distance and escape from the tadpole zone.

The structure of this paper is as follows. The way of appropriately selecting initial conditions for the numerical experiment is described in Section 2. The core of our results is given in Section 3, in the form of grey-scale maps, which allow us to define the stability region in the space of proper elements and compare it with the distribution of real Trojans. It is shown that ~14% of the real Trojans are outside the stability region. This result is confirmed by a 4.5 Gyr integration of the 247 real Trojans found on chaotic orbits with $T_{\mathrm{L}} < 4 \times 10^5$ yrs. By estimating the diameters of the observed bodies, we show that the existence of unstable Trojans cannot be the result of a size-dependent process and that it is most likely the outcome of slow chaotic diffusion. For the chaotic region surrounding the effective stability zone, a power-law statistical correlation between T_{L} and T_{E} is found (Section 4). The conclusions of our study and a discussion on open problems are given in Section 5.

2. Numerical Set-up

The physical model we consider consists of the Sun and the four giant planets (Jupiter, Saturn, Uranus and Neptune), fully interacting through Newtonian pointmass gravitational forces. The Trojan (test-particle) is subjected to the forces of the massive bodies. The equations of motion are numerically integrated, using the 2nd order mixed variable symplectic algorithm (MVS) of Wisdom and Holman (1991), as it is implemented in the SWIFT package (Levison and Duncan, 1994). The time step used in our runs was $\delta_t = 0.1$ yrs, i.e., smaller than $0.01 T_{\mathrm{J}}$, where T_{J} the orbital period of Jupiter. This integration scheme is not appropriate when close encounters between bodies occur. However, in the experiments presented here, we are only interested in calculating the escape time of a Trojan and not in following its subsequent evolution. Thus, as described above, we define the escape time of a Trojan as the time at which it approaches Jupiter within 2 Hill's radii, at which point we stop integrating its orbit.

The initial conditions for the Trojans were selected in such a way so that, for each value of the inclination, *i*, a 'representative plane' of initial conditions was studied. By this we mean that the initial conditions were chosen as to provide a first approximation to *proper elements*, so that a comparison with the real Trojan population could be made. To do this, we used the model of Érdi (1988) for the long-period motion of a Trojan, in the frame of the elliptic restricted three-body problem, as the basis of our selection.

According to Érdi (1988), the first term in the expansion of the long periodic variations of a Trojan's semi-major axis, a, and relative mean longitude, $\lambda - \lambda'$, is given by

$$a - a' = d_f \sin\theta + O(d_f^2)$$
$$\lambda - \lambda' = \pm\pi/3 + D_f \cos\theta + O(D_f^2) \qquad (1)$$

where primed quantities refer to Jupiter. The amplitude of libration in semi-major axis, d_f (in AU), is related to the amplitude of libration, D_f (in radians), of the critical argument, $\sigma = \lambda - \lambda'$, through

$$d_f = \sqrt{3\mu a'} D_f \approx 0.2783 D_f \tag{2}$$

where μ is the ratio of Jupiter's mass to the total mass of the system. In the (σ, $a-a'$) plane, each Trojan orbit has the shape of a topological cycle, centered approximately at the Lagrangian point (at $\pm\ \pi/3$), and the angle θ is the angle between the $a = a'$ axis and the 'position vector' of the Trojan, measured from $\pm\ \pi/3$. During the motion, the angle σ librates approximately between $\pm\ \pi/3 - D_f$ and $\pm\ \pi/3 + D_f$, while θ circulates. In our experiments, we chose initial conditions around L_4, with $\theta = \pi/2$, i.e., $\sigma = \pi/3$ and $a = a' + d_f$.

When $a \to a'$ the values of the forced eccentricity and inclination become equal to the (osculating) values of Jupiter's elements. Érdi's model shows that the libration adds only 2nd-order corrections to these values. Then, by selecting the longitude of pericenter of the Trojan, $\bar{\omega}$, and its eccentricity, e, through

$$\begin{aligned} \bar{\omega} &= \bar{\omega}' + \pi/3 \\ e &= e' + e_f \end{aligned} \tag{3}$$

and the longitude of the node, Ω, and inclination, i, through

$$\begin{aligned} \Omega &= \Omega' \\ i &= i' + i_f \end{aligned} \tag{4}$$

the eccentricity offset, e_f, and the inclination offset, i_f, become approximate proper elements, in the framework of the elliptic restricted three-body problem. Hereafter we will refer to the elements e_f, i_f and D_f as eccentricity, inclination and libration width, dropping the subscripts. This choice of initial conditions implies that all test particles have initially a mean anomaly equal to that of Jupiter, $M = M'$.

Following the above described scheme, we constructed four sets of initial conditions, for $0° \leq i \leq 30°$ with a step of 10°. Each of these sets consisted of 806 orbits, set on a 31 × 26 grid on the (D, e) plane, with $0 \leq e \leq 0.25$ and $0° \leq D \leq 45°$ (the step size being $\delta D = 1°.5$ and $\delta e = 0.01$).

Our first numerical experiment consisted in integrating these 3224 orbits for $t_{int} = 4$ Myrs, along with the variational equations. This time is enough to obtain an estimate of T_L, provided that the latter is actually not larger than ~10% t_{int} (see next section). Subsequently we selected only orbits having $T_L < 400{,}000$ yrs to perform our second experiment, which consisted in integrating these new sets of orbits for $t_{int} = 1$ Gyr.

3. Strong Chaos versus Effective Stability

The solution of the variational equations provides us with a time series for the norm of the deviation vector, $v(t)$, for each orbit. For chaotic orbits this

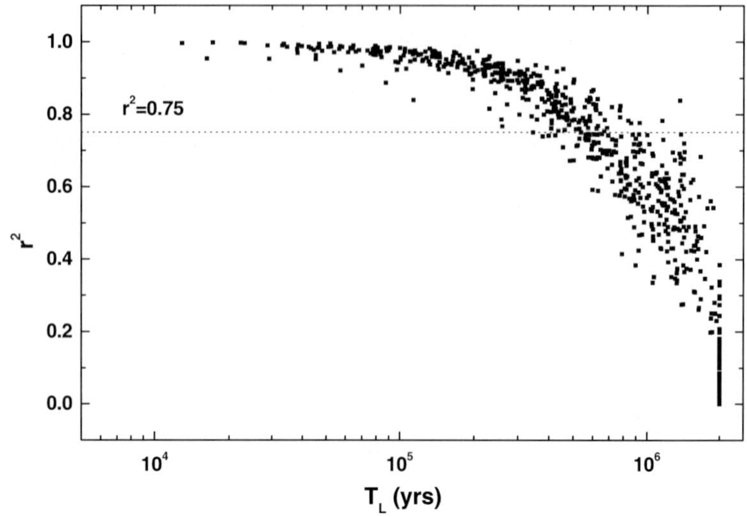

Figure 1. The correlation coefficient of the fit, r^2, as a function of the value of the inverse of the slope, i.e., T_L, for a sample of 667 numbered Trojans. For $T_L > 0.1\ t_{int} \approx$ 400,000 yrs, r^2 drops below 0.75. Had we selected a different value for t_{int}, the functional form would not have changed, but the $r^2 = 0.75$ limit would have shifted towards 0.1 t_{int}.

quantity grows exponentially in time, so we can compute the mean growth rate (i.e. a short time approximation of the Lyapunov Characteristic Number), γ, by performing a linear least-squares fit on the $t - \ln[v(t)/v(0)]$ curve. Then, $T_L = 1/\gamma$. This procedure is followed automatically in our code.

It is known, however, that, in order to have an acceptable estimate using this procedure, the integration time should be long enough for γ to have achieved a constant value (see Milani, 1993). A quantitative check of convergence can be made, and its results are shown in Figure 1, for the sample of 667 numbered Trojans found in the catalogue of proper elements, distributed by the AstDys database. The correlation coefficient, r^2, of the fit, used to obtain the slope, is plotted against the inverse of the slope, i.e., T_L. One can immediately realize that the quality of the fit drops as T_L increases. Since we are forced to set a limit as to what we consider as a 'good approximation,' we arbitrarily chose to accept any value of T_L which was obtained by fit, provided that $r^2 \geq 0.75$. Thus, as can be inferred from the plot, for a measurement yielding $T_L > 400{,}000$ yrs (i.e. longer than about 1/10 of our integration time span), we cannot decide whether the orbit is mildly chaotic or regular. For this reason, we decided to exclude orbits with $T_L > 400{,}000$ yrs from our second experiment of long-term evolution, and concentrate on orbits for which we can claim that they are chaotic, on an acceptable significance level. Concerning our test particle runs, the above described selection method lead us to discard about 25% of them from the

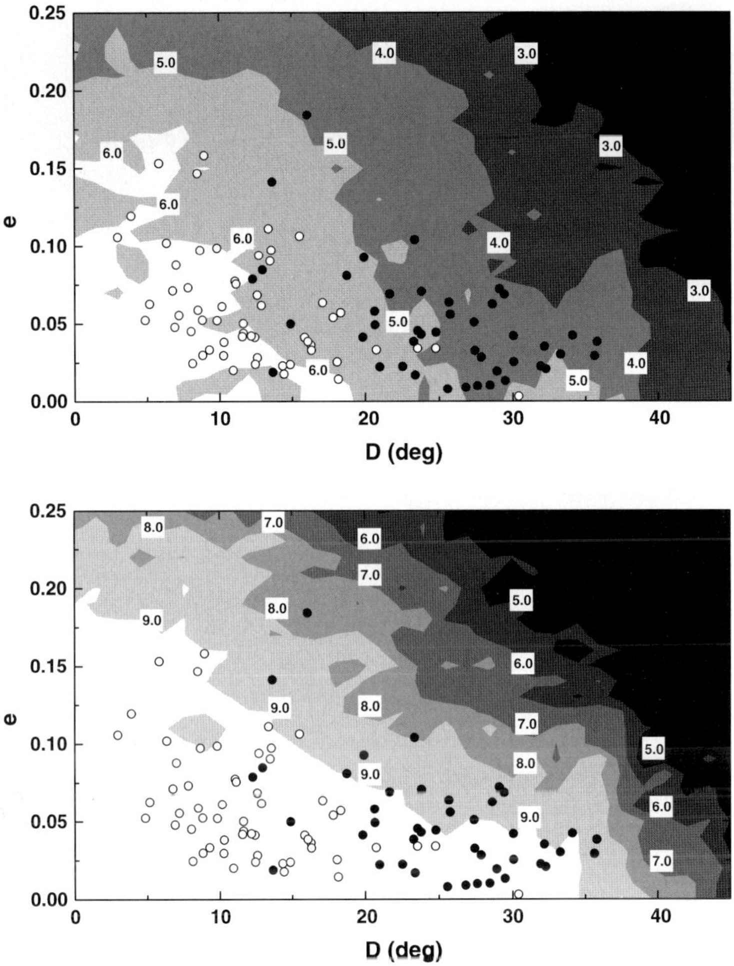

Figure 2. Maps of log T_L (top) and log T_E (bottom) for $i = 0°$. Each grey level corresponds to a change by one order of magnitude. In each panel the real Jupiter Trojans are superimposed, as discussed in the text. The open dots denote the stable Trojans (i.e., $T_L > 400{,}000$ yrs), while the full dots denote the chaotic ones. Note that, in all cases (see also next figures), the distribution of real Trojans follows the shape of the stability zone, but ~14% of the real objects lie outside the $T_E = 1$ Gyr limiting curve.

second experiment. As for the numbered Trojans, 246 bodies (out of 667, i.e. 37%) follow chaotic orbits with $T_L < 400{,}000$ yrs, while the rest could be considered as following stable orbits.

The results of the two runs can be visualized in Figures 2–5. The top panels are grey-scale maps of $\log T_L$, while the bottom panels are grey-scale maps of $\log T_E$, both T_L and T_E measured in years. Each figure corresponds to a different set of initial conditions, i.e., a different initial value of *i*.

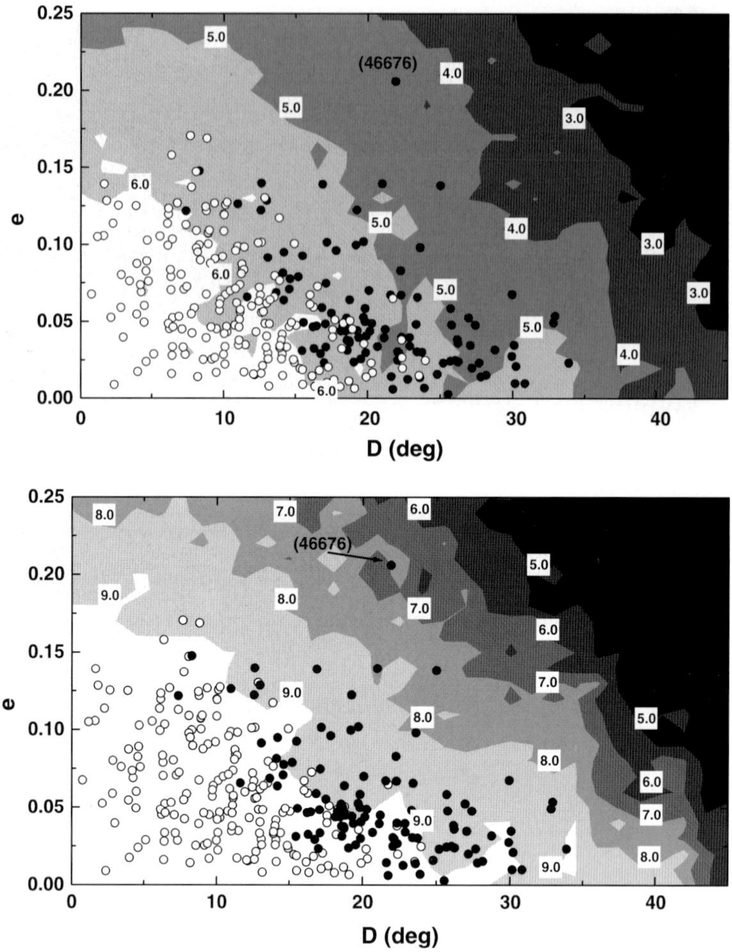

Figure 3. The same as Figure 2, but for $i = 10°$. Also note the existence of real Trojans in highly unstable regions (e.g., object 46676), with $T_L < 10^5$ yrs and $T_E < 10^8$ yrs.

Figure 2 contains the plots for $i_0 = 0°$, Figure 3 for $i_0 = 10°$, Figure 4 for $i_0 = 20°$ and Figure 5 for $i_0 = 30°$. On each panel all numbered Trojans with proper inclination in the range $i_0 - 5° \leq i \leq i_0 + 5°$ are superimposed (obviously, for $i_0 = 0°$, only bodies with $i < 5°$ are shown). The open dots correspond to bodies on regular orbits (i.e., $T_L \geqslant 400{,}000$ yrs), while the full dots correspond to bodies following chaotic orbits. As expected, the chaotic objects are located at greater distances from the nominal location of the Lagrangian point. This projection technique allows us to compare roughly the 1 Gyr effective stability region, defined by the numerical integration of test particles, with the distribution of the real Trojans.

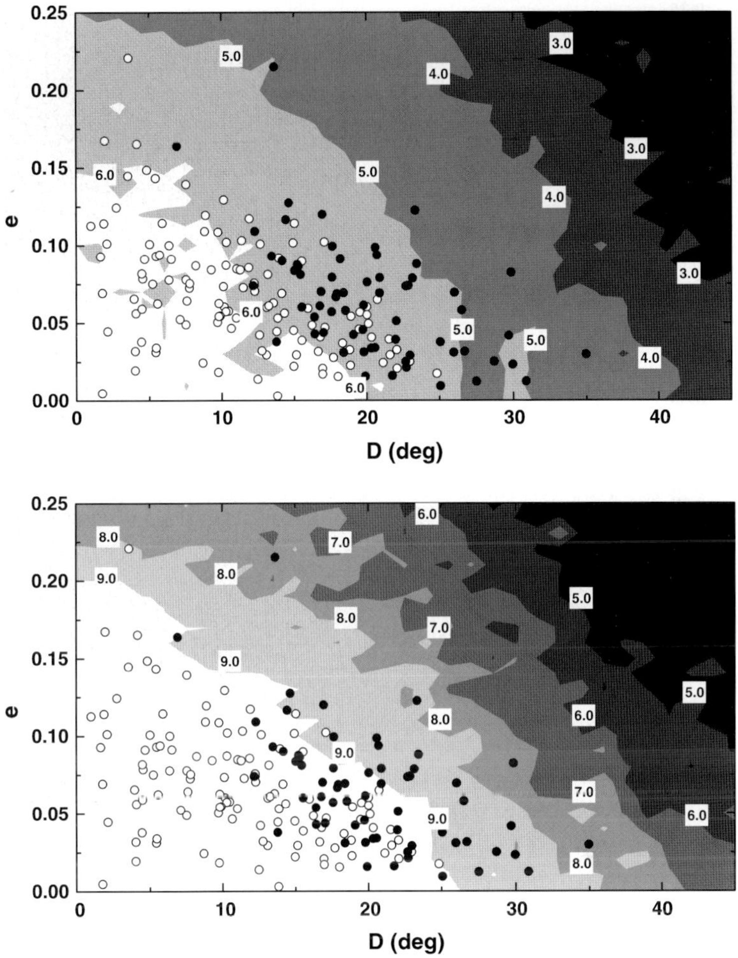

Figure 4. The same as Figure 2, but for $i = 20°$. Note that the extent of the stability region in D shrinks, with respect to the low-inclination cases.

As can be seen in Figures 2–5, ~86% of the real Trojans are projected inside the $T_E \geqslant 10^9$ yrs region, following the shape of the limiting 1 Gyr contour. The remaining 14% (between 12 and 16%, depending on the inclination) is projected outside the stability zone. Note also that ~3% of the real Trojans is projected on highly unstable regions, defined by $T_L \leqslant 10^5$ yrs and $T_E \leqslant 10^8$ yrs. Such an extreme example is asteroid (46676) (see Figure 3). Integrating its nominal orbit we found that it actually escaped from the L_4 region after $T_E = 7.2$ Myrs, a time which agrees well with the location of this object in Figure 3 (the border of the 10^7 yrs contour). The time evolution of its orbital elements is shown in Figure 6. We should note that, a

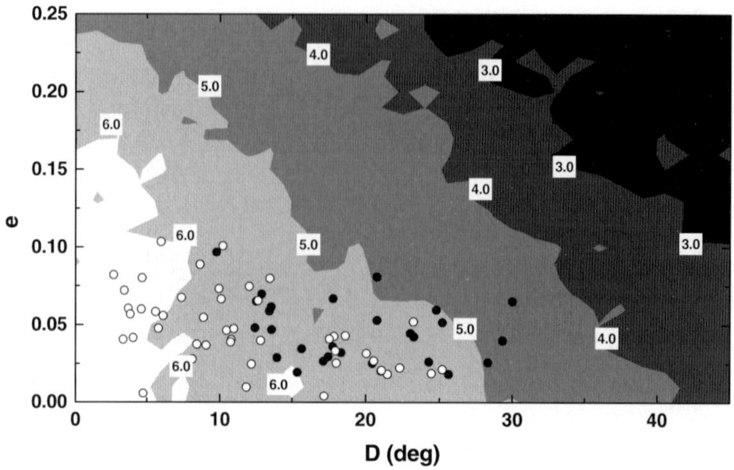

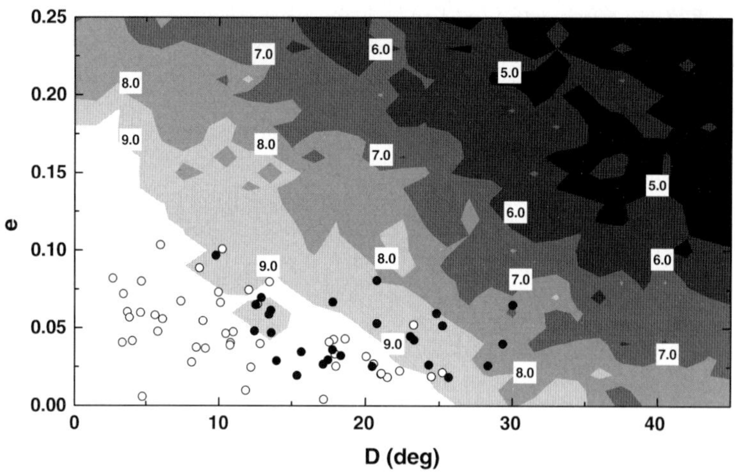

Figure 5. The same as Figure 2, but for $i = 30°$ In this case, $D_{max} = 25°$, for nearly circular orbits.

significant fraction of chaotic orbits with $T_L < 400{,}000$ yrs, does not escape within 1 Gyr. Thus, our region of effective stability also contains chaotic orbits. The extent of the effective stability region, as given from our results, is comparable to the one calculated by Levison et al. (1997), which is smaller than the one initially calculated by Rabe (1967). However, by integrating a much bigger sample of initial conditions, selected as described in Section 2, we are able to demonstrate for the first time the dependence of the size and shape of the effective stability region on the inclination of the Trojans. As can be seen in the figures, for small values of D, the stability region extends up to $e_{max} = 0.2$, for any value of i. On the other hand, for nearly circular orbits,

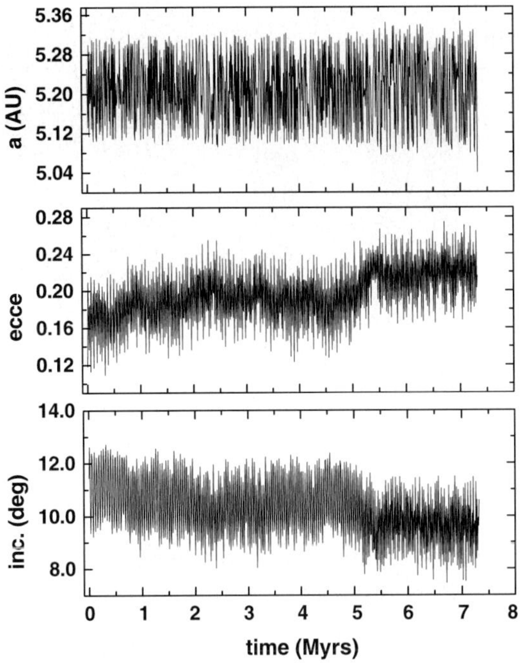

Figure 6. Time evolution of *a, e* and *i* for object (46676). The chaotic evolution of both *e* and *i* is evident. As time progresses, the width of libration in *a* slowly increases, until the asteroid encounters Jupiter.

the stability region shrinks from $D_{max} = 35°$ to $D_{max} = 25°$, as i increases from 0° to 30° It is evident from the plots that the distribution of the real Trojans in the (D, e) plane follows the shape of the effective stability region, as the latter is defined for the corresponding value of i.

Based on these results, we could conclude that ~14% of the real Trojans follow orbits which are not stable over the age of the solar system. This of course has to be shown by direct numerical integration of the numbered Trojans. We performed this experiment, setting our integration time to 1 Gyr. Indeed, out of the 246 numbered Trojans with $T_L < 400{,}000$ yrs, 53 escaped within 1 Gyr, i.e.~8% of the total population (667 objects). This number differs from our initial estimate, mainly due to three limiting factors of the above representation: (i) the uncertainty on the approximate proper elements, defined for the test particles, (ii) the projection of the real Trojans on only four planes, with respect to their proper inclination, and (iii) the dependence of the width of the 'actual' effective stability region (for 4.5 Gyrs), on the integration time.

We decided to extend our integration for the chaotic numbered Trojans, going to 4.5 Gyrs. This resulted in 112 chaotic objects, which amounts to

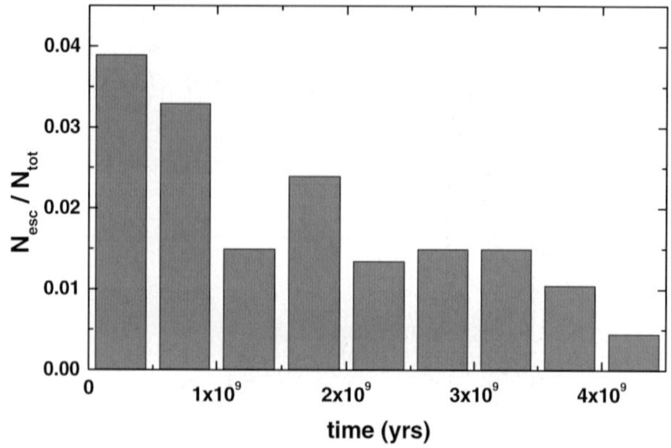

Figure 7. The fraction of escaping Trojans $N_{\rm esc}/N_{\rm tot}$ as a function of time. The size of each bin is 5×10^8 yrs.

17% of the total Trojan population, escaping within the age of the Solar System. The histogram of escape times is shown in Figure 7. One can see that the number of escaping bodies per unit time decays slowly with time. Note also that more than half of the integrated objects follow orbits which, although being chaotic, are stable over the age of the Solar System.

Despite the fact that the existence of unstable Jupiter Trojans has been known for some years, the mechanism by which they may be generated still remains an open issue. There are three main possible mechanisms, one could think of, generating a large fraction of unstable objects: (i) slow chaotic diffusion from the effective stability region, due to secondary resonances (ii) collisions, and (iii) drift due to the Yarkovsky thermal effect (see Farinella and Vokrouhlický, 1999). However, while chaotic diffusion has the same effect no matter what the size of the body is, collisions and thermal forces give size-dependent 'kicks'. Thus, it would be easier for a small body to be transported away from the Lagrangian point (and outside the stability region) by mechanisms (ii) and (iii) than it would be for a large body. Hence, the size distribution of bodies, implanted in the chaotic zone due to mechanisms (ii) or (iii), should be different from the size distribution of regular bodies. In particular, if mechanisms (ii) and (iii) were primarily responsible for replenishing the chaotic region, there should be a lack of large bodies among the currently observed chaotic population.

Using the AstDys catalogue of proper elements, we can derive the size distribution of the Trojans. Fernández et al. (2003) derived the albedo distribution of Jupiter Trojans, showing that it is characterized by a very tight concentration of values around ~0.05. Using this estimate and the values of absolute magnitude, H, reported in the catalogue, an effective diameter, Δ, can be calculated for each body. Then we can derive the distribution of

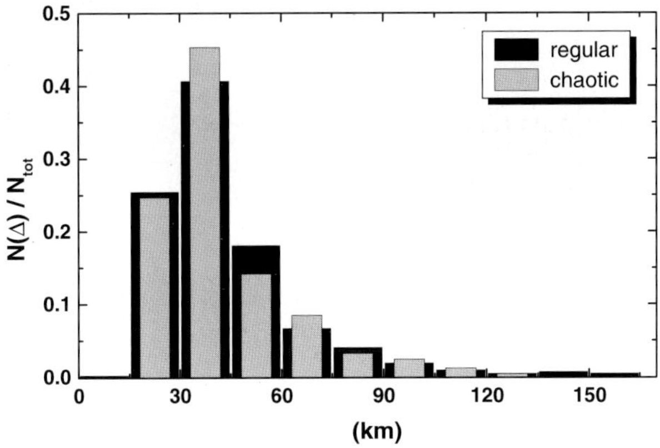

Figure 8. The diameters (Δ) distribution of the regular (black) and chaotic (grey) components of the population of numbered Trojans. The two distributions are normalized with respect to the total number of bodies in each group.

diameters for both the regular and the chaotic population of numbered Trojans. The results are shown in Figure 8. It is evident that the two distributions are nearly identical. This result clearly suggests that the primary mechanism responsible for generating unstable objects is chaotic diffusion through secondary resonances.

It should be noted that the observed Trojans population is only complete up to $H \sim 11$–12, and thus there are only 4 bodies with $\Delta < 10$ km in the catalogue of numbered Trojans. Future observations may reveal a difference in the small–Δ tails of these two distributions, since $\Delta > 10$ km bodies are very little affected by the Yarkovsky effect over 4 Gyrs, but this is not true for $\Delta \sim 1$ km bodies. However, the important element here is that the probability of finding a body as large as 150 km in diameter is the same, in both the chaotic and the stable region.

4. A Statistical Correlation for Escaping Orbits

Figures 2–5 suggest an approximately smooth decay of both T_L and T_E, as we move away from the origin of each plot, i.e. the nominal location of L_4 at $(D, e) = (0, 0)$. This intriguing result lead us to consider the possible existence of a statistical correlation between these two quantities.

Figure 9 is a log–log plot of T_L versus T_E for each test particle. As before, only orbits with good measurements of T_L ($r^2 > 0.75$) were considered. Both quantities are normalized to the mean revolution period of Jupiter, $T_J \approx 11.86$ yrs. The distribution of points on this plot suggests a power-law

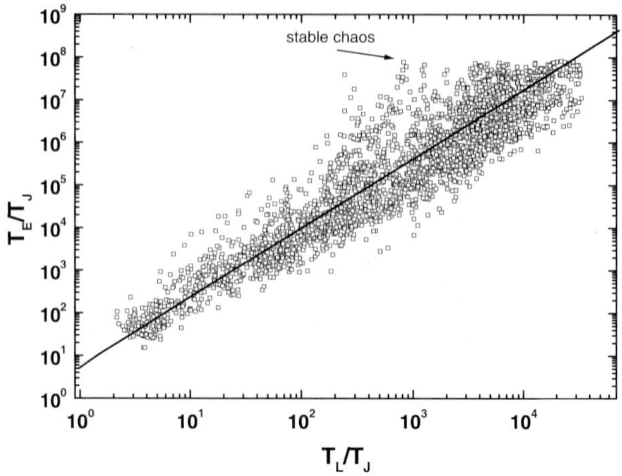

Figure 9. A statistical correlation between T_L and T_E. The least squares line gives the power-law trend. The manifestation of stable chaos, for orbits with $T_L \approx 10^4$ yrs, is indicated by the arrow.

trend, $T_E \approx a T_1{}^b$, between these two quantities. Performing a least-squares fit on the data we found $a = 0.75 \pm 0.08$ and $b = 1.62 \pm 0.03$. The correlation coefficient of the fit is $r^2 = 0.88$.

Although the least-squares line seems to fit the data well, it is evident from the plot that the dispersion of the points is quite large, especially when moving towards large values of T_L. Thus, the simple power-law relation described is not useful in determining T_E by computing T_L from a much shorter integration. On the other hand, as shown in Figure 9, there is a well defined lower envelope of the scatter plot. This enables us to make a meaningful extrapolation, which suggests that orbits with $T_L > 650{,}000$ yrs (top-right corner of the plot), although chaotic, would be most likely stable over the age of the solar system.

The scatter of T_E values, around a given value of T_L, increases with T_L, i.e., when approaching the stability region. In Hamiltonian systems, it is known that, as the border of a chaotic region is approached, the geometry of the phase space becomes highly complex, an effect which leads to temporary confinement of chaotic orbits, starting close to the border. Consequently the escape time of such orbits, corresponding to a given narrow range of T_L values, can vary by several orders of magnitude, depending on the duration of the trapping phase. In terms of asteroid dynamics, this is one manifestation of the *stable chaos* phenomenon, first discovered by Milani and Nobili (1992; see also Tsiganis et al., 2000b, 2002a). We remind the reader that stable chaos refers to orbits characterized by a small value of T_L, but of extremely stable orbital elements and a value of T_E typically larger than

$10^4 \cdot T_L$. Such orbits, which are close to the stability boundary, can be seen on the upper part of Figure 7, for $(T_L/T_J) \sim 10^3$.

5. Conclusions and Discussion

In this paper we presented the results of an extensive numerical experiment on Trojan-type motion. A carefully chosen set of 3224 orbits was integrated for 4 Myrs. From this set, all chaotic orbits with $T_L < 400{,}000$ yrs were selected and integrated for 1 Gyr. This experiment took a few months of CPU time on a custom PC. The purpose was to calculate the two relevant quantities, defining the stability region of the 1:1 resonance with Jupiter – the Lyapunov time, T_L, and the escape time, T_E–in a model containing all four giant planets.

The basic results of this experiment are given in Section 3 (Figures 2–8). An effective stability region for 1 Gyr is defined (see Figures 2–5), in the space of proper elements (D, e, i). For nearly circular orbits, the stability region shrinks in D as i increases while, for small values of D, the maximum extent in e is almost constant. As shown in these figures, the distribution of the numbered Jupiter Trojans follows closely the stability curve, for all values of i. However, about ~14% of the real Trojans was found to lie outside the stability region.

A 4.5 Gyr integration of the orbits of 246 chaotic numbered Trojans confirmed the above estimate. Our result is that 17% of the numbered Trojans follows orbits which are unstable over the age of the Solar System. We note that there is another 20% of Trojans undergoing chaotic motion, but whose orbits are stable over 4.5 Gyrs.

The small escape times of many observed Trojans suggest a constant leakage of bodies from the stability region towards the large chaotic sea. However, the mechanism that generates this unstable population was not known up to now. Analyzing the size distribution of the regular and chaotic components of the Trojan population (Figure 8), we can conclude that the main mechanism, by which bodies are delivered from the outshirts of the stability zone to the chaotic region, is independent of the size of the bodies. Thus, chaotic diffusion, rather than collisions or thermal effects, is at the origin of the unstable population.

Chaotic diffusion is the result of higher-order resonant multiplets (either of mean motion or secular type) inside the 1:1 tadpoles region. Robutel et al. (2005) have identified several types of resonances, which cross the stability region. The reader is referred to Figures 1 and 3 of the paper by Robutel et al. (2005) in this volume. The relevance of each of these types of resonance, concerning the transport of bodies, remains to be assessed.

In Section 4 we have found an approximate power-law statistical correlation between the values of T_L and T_E, for chaotic orbits that escape from the 1:1 resonance. A similar result, with b ≈ 1.7, was reported by Lecar et al. (1992), for asteroids of the outer main belt. Murray and Holman (1997) have shown analytically that such a relationship exists, in the region where all mean motion resonances of order $q = 1$ overlap. On the other hand, Shevchenko (1998) has shown that a relationship of the form, $T_E \sim T_L^2$ can also be found in the immediate vicinity of the border between regular and chaotic motion, around a perturbed principal resonance. It is not easy to give an answer as to which of the two mechanisms we are observing in the case of Trojans. This is obviously related to the study of secondary resonances in the vicinity of the 1:1 resonance, which will be the topic of future work.

Acknowledgements

We would like to thank P. Robutel for the interesting discussions we had, during the 6th Alexander von Humboldt Colloquium. Moreover we would like to thank the referee D. Nesvorný for his constructive criticism and comments which greatly improved the presentation of our work. Part of this work was done during the stay of K.T. in Vienna, and was funded by the Austrian FWF (Project P14375-TPH). K.T. would like to thank the people of the Institute for Astronomy in Vienna for their hospitality. The work of K.T. in O.C.A. is supported by an EC Marie Curie Individual Fellowship, (contract No HPMF-CT-2002-01972).

References

Beaugé, C. and Roig, F.: 2001, 'A semianalytical model for the motion of the Trojan Asteroids: proper elements and families', *Icarus* **153**, 391–415.

Celletti, A. and Giorgilli, A.: 1991, 'On the stability of the Lagrangian points in the spatial restricted problem of three bodies', *Celest. Mech. Dyn. Astr.* **50**, 31–58.

Dvorak, R. and Tsiganis, K.: 2000, 'Why do Trojan ASCs (not) escape?', *Celest. Mech. Dyn. Astr.* **78**, 125–136.

Érdi, B.: 1988, 'Long periodic perturbations of Trojan asteroids', *Celest. Mech.* **43**, 303–308.

Érdi, B.: 1997, 'The Trojan problem', *Celest. Mech. Dyn. Astr.* **65**, 149–164.

Farinella, P. and Vokrouhlický, D.: 1999, 'Semi-major axis mobility of asteroid fragments', *Science* **283**, 1507–1510.

Fernández, Y. R., Sheppard S. S., and Jewitt D.: 2003, 'The Albedo distribution of Jovian Trojan asteroids', *Astron. J.* **126**, 1563–1574.

Giorgilli, A. and Skokos, C.: 1997, 'On the stability of the Trojan asteroids', *Astron. Astrophys.* **317**, 254–261.

Gomes, R. S.: 1998, 'Dynamical effects of planetary migration on primordial Trojan-type asteroids', *Astron. J.* **116**, 2590–2597.

Lecar, M., Franklin, F. and Murison, M.: 1992, 'On predicting long-term orbital instability – A relation between the Lyapunov time and sudden orbital transitions', *Astron. J.* **104**, 1230–1236.
Levison, H. F. and Duncan, M. J.: 1994, 'The long-term dynamical behavior of short-period comets', *Icarus* **108**, 18–36.
Levison, H., Shoemaker, E. M. and Shoemaker, C. S.: 1997, 'The dispersal of the Trojan asteroid swarm', *Nature* **385**, 42–44.
Marzari, F. and Scholl, H.: 2002, 'On the Instability of Jupiter's Trojans', *Icarus* **159**, 328–338.
Marzari, F., Tricarico, P. and Scholl, H.: 2003, 'Stability of Jupiter Trojans investigated using frequency map analysis: the MATROS project', *Mon. Not. R. Astron. Soc.* **345**, 1091–1100.
Milani, A.: 1993, 'The Trojan asteroid belt: proper elements, stability, chaos and families', *Celest. Mech. Dyn. Astron.* **57**, 59–94.
Milani, A.: 1994, The Dynamics of the Trojan Asteroids. IAU Symp. 160: Asteroids, Comets, Meteors, 1993, pp. 159–174.
Milani, A. and Nobili, A. M.: 1992, 'An example of stable chaos in the Solar System', *Nature* **357**, 569–571.
Michtchenko, T. A., Beaugé, C. and Roig, F.: 2001, 'Planetary migration and the effects of mean motion resonances on Jupiter's Trojan asteroids', *Astron. J.* **122**, 3485–3491.
Morais, M. H. M.: 1999, 'A secular theory for Trojan-type motion', *Astron. Astrophys.* **350**, 318–326.
Morais, M. H. M.: 2001, 'Hamiltonian formulation of the secular theory for Trojan-type motion', *Astron. Astrophys.* **369**, 677–689.
Murray, N. and Holman, M.: 1997, 'Diffusive chaos in the outer asteroid belt', *Astron. J.* **114**, 1246–1259.
Namouni, F. and Murray, C. D.: 2000, The effect of eccentricity and inclination on the motion near the Lagrangian points L_4 and L_5; *Celest. Mech. Dyn. Astron.* **76**, 131–138.
Nesvorný, D. and Dones, L.: 2002, How long-lived are the hypothetical Trojan populations of Saturn, Uranus, and Neptune?', *Icarus* **160**, 271–288.
Nesvorný, D., Thomas, F., Ferraz-Mello, S. and Morbidelli A.: 2002, 'A perturbative treatment of the co-orbital motion', *Celest. Mech. Dyn. Astron.* **82**, 323–361.
Rabe, E.: 1967, 'Third-order stability of the long-period Trojan librations', *Astron. J.* **72**, 10–19.
Robutel, P., Gabern, F. and Jorba, A.: 2005, 'The observed Trojans and the global dynamics around the Lagrangian points of the Sun–Jupiter system', *Celest. Mech. Dynam. Astron.* **92**, 55–71.
Shevchenko, I. I.: 1998, 'On the recurrence and Lyapunov time scales of the motion near the chaos border', *Phys. Lett. A* **241**, 53–60.
Skokos, C. and Dokoumetzidis, A.: 2001, 'Effective stability of the Trojan asteroids', *Astron. Astrophys.* **367**, 729–736.
Tsiganis, K., Dvorak, R. and Pilat-Lohinger, E.: 2000a, 'Thersites: a 'jumping' Trojan?', *Astron. Astrophys.* **354**, 1091–1100.
Tsiganis, K., Varvoglis, H. and Hadjidemetriou, J. D.: 2000b, 'Stable chaos in the 12:7 mean motion resonance and its relation to the stickiness effect', *Icarus* **146**, 240–252.
Tsiganis, K., Varvoglis, H. and Hadjidemetriou, J. D.: 2002a. 'Stable chaos in high-order Jovian resonances', *Icarus* **155**, 454–474.
Tsiganis, K., Varvoglis, H. and Hadjidemetriou, J. D.: 2002b, 'Stable chaos versus Kirkwood gaps in the asteroid belt: a comparative study of mean motion resonances', *Icarus* **159**, 284–299.
Wisdom, J. and Holman M.: 1991, 'Symplectic maps for the *n*-body problem', *Astron. J.* **102**, 1528–1538.

Celestial Mechanics and Dynamical Astronomy (2005) 92:89–111
DOI 10.1007/s10569-005-2858-x

ANALYTICAL PROPER ELEMENTS FOR THE HILDA ASTEROIDS I: CONSTRUCTION OF A FORMAL SOLUTION

O. MILONI[1], S. FERRAZ-MELLO[1] and C. BEAUGÉ[2]
[1]*Instituto de Astronomia, Geofísica e Ciências Atmosféricas, Universidade de São Paulo, Rua do Matão 1226, 05508-900 São Paulo, Brasil,*
e-mail: octavio@astro.iag.usp.br
[2]*Observatório Astronómico, Universidad Nacional de Córdoba, Laprida 854, (5000) Córdoba, Argentina*

(Received: 27 September 2004; revised: 7 February 2005; accepted: 28 February 2005)

Abstract. In this paper, we present the mathematical basis for the calculation of proper elements for asteroids in 3:2 mean-motion resonance with Jupiter from their osculating Keplerian elements. The method is based on a new resonant Lie-series perturbation theory (Ferraz-Mello, 1997, 2002), which allows the construction of formal solutions in cases where resonant and long-period angles appear simultaneously. For the disturbing function, we used the Beaugé's expansion (Beaugé, 1996), adapted to include short period terms. In this paper, the theory is restricted to the planar case and only the perturbations due to Jupiter are considered.

Key words: Hilda asteroids, perturbation theory, proper elements

1. Introduction

In a time where most of the work in Solar System dynamics is related to chaotic motion and evolution, proper elements (hereafter PE) are the exact opposite. For a given dynamical system (in our case, a main belt asteroid perturbed by Jupiter), the idea is to construct an integrable approximation of its solution. The proper elements are functions of the "first integrals". Even if the system is chaotic, these "integrals" are still informative when: (i) the chaos is local, or (ii) it has a very long timescale. In either case, PE can be thought as quasi-integrals of motion that vary little over a long time span. In other words, they allow us to search for dynamical structures in the system, which may have remained practically unchanged in hundreds of millions of years.

The origin of this idea dates from the early twentieth century. Hirayama (1918, 1923) (see also Brouwer 1951) used the Laplace-Lagrange secular linear theory to determine PE for main-belt asteroids. His results allowed him to discover regions of accumulation of bodies in the space of proper elements, which were later named *Hirayama families*. These are now believed to be the remnants (fragments) of catastrophic collisions in the asteroid belt. For many years, most of the effort in this area was devoted to the construction of better analytical and numerical determinations of proper elements for asteroids outside mean-motion resonances.

As is common in the three-body problem, the main practical limitation in the construction of analytical theories stems from the limits of convergence of the disturbing function. The first attempt to avoid these problems was undertaken by Williams (1969), who constructed a semi-analytical theory based on the numerical determination of the disturbing potential in a grid of initial conditions. His results not only included new PE, but also a very important determination of secular resonances in the asteroid region. More recently, and as the discovery of new bodies increased the total population to large numbers, Milani and Knežević (1990, 1992, 1994) introduced the Lie series formalism to estimate proper elements up to second (and third in some cases) order of the masses. Lemaître and Morbidelli (1994) developed a semi-analytical theory, similar to that of Williams, but based on a Hamiltonian formalism and introducing terms of second order not present in Williams theory. All theories, whether analytical or semi-numerical, obtain the PE by successive averaging operations: the first one to obtain the secular Hamiltonian, and the second one to derive the proper elements. Finally, in the latest few years, Knežević and Milani (2000) have adopted a new approach, based on purely numerical tools and the construction of synthetic solutions.

For the resonant case, Schubart (1968) analyzed the long-term behavior of Hilda asteroids, via a numerical averaging over short-period terms (synodic angle). He derived three characteristic parameters (eccentricity, inclination and amplitude of librations) for all the real Hilda asteroids known at the time (Schubart 1982a, b).

Milani (1993) calculated synthetic proper element for Trojan asteroids by mean of a purely numerical treatment. More recently Beaugé and Roig (2001) developed a semi-analytical theory to obtain proper elements for the Trojans. Their theory consists, basically, in a succession of averaging operations and a hierarchical distinction of faster and slower angles.

In the theory presented in this paper, we construct analytically the planar invariants of motion for Hilda asteroids by application of the Lie series theory, starting from the osculating elements and ending with the dynamical proper elements. The work is divided in the following parts: (i) passage from osculating to semi-mean[1] orbital elements by averaging over the short periods terms. (ii) Analytical integration of the Hori Kernel

[1]Semi-mean is a nomenclature proposed by Milani and now widely adopted.

(pendulum model) (iii) Application of the resonant Lie series theory to average over the librating resonant angle and obtain the mean elements. (iv) Calculation of the proper actions from the solution of the resulting equations. (In practice, through the elimination of the remaining long-period angular variable).

2. Variables and Equations

2.1. THE PLANAR ASTEROIDAL PROBLEM

Consider an asteroid moving around the Sun in the same plane as Jupiter and in a 3:2 mean-motion resonance with this planet. The planet is considered on an uniformly precessing orbit. Let a, e, ϖ, a_J, e_J, ϖ_J be the osculating semi-major axis, eccentricity and longitude of perihelion of the asteroid and Jupiter, respectively, and let n_J be the mean motion of Jupiter. We will denote by g_J the precession rate of Jupiter, due to the gravitational effects of the other planets. To take into account the time dependence, we work in the extended phase space. In this space, the canonical elements of the asteroid are:

$$\begin{aligned} &\lambda \ , \quad L = \sqrt{\mu a} \\ &\varpi \ , \quad G - L = L(\sqrt{1-e^2} - 1) \\ &t \ , \quad \mathcal{T} \end{aligned} \tag{1}$$

where L, G are the usual planar Delaunay variables and $\mathcal{T}$ is the canonical conjugate of t.

In the 3:2 resonance, we use the standard resonant variables (see, for instance, Ferraz-Mello, 1987):

$$\begin{aligned} \theta_1 &= 3\lambda_J - 2\lambda - \varpi \\ \theta_2 &= 3\lambda_J - 2\lambda - \varpi_J \\ \theta_3 &= \lambda - \lambda_J \end{aligned} \tag{2}$$

$$\begin{aligned} J_1 &= L - G \\ J_2 &= \frac{n_J G + \mathcal{T}}{n_J - g_J} \\ J_3 &= 3L + 2\frac{(g_J G + \mathcal{T})}{n_J - g_J}. \end{aligned} \tag{3}$$

The equations of motion are

$$\begin{aligned} \dot{\theta}_i &= \frac{\partial H}{\partial J_i} \\ \dot{J}_i &= -\frac{\partial H}{\partial \theta_i} \end{aligned} \tag{4}$$

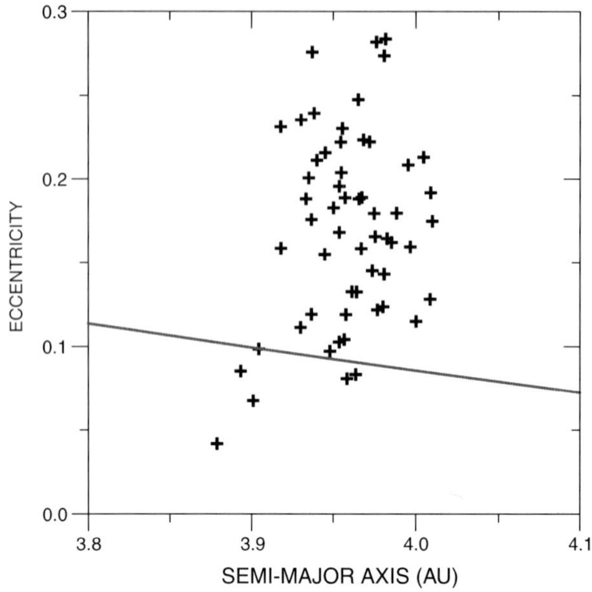

Figure 1. Distribution of actual Hildas in osculating eccentricity and semi-major axis. The solid line represents the limit of convergence of the Laplace expansion of the disturbing function (see Ferraz-Mello, 1994).

where the Hamiltonian is

$$H = -\frac{\mu^2}{2L^2} + \mathcal{T} - \varepsilon \mathcal{R}, \tag{5}$$

R is the disturbing function and $\varepsilon = Gm_J/m_\odot$. For the sake of future calculations, we note that if $\mathcal{T}$ is eliminated from Equation (3) (in the definition of J_3), we obtain $L = J_3 - 2(J_1 + J_2)$.

Since the Hildas have, generally, eccentricities larger than the radius of convergence of the classical expansion in Laplace coefficients (see Figure 1), we use, in this work, the Beaugé's representation, which is not given by infinite series and is thus valid for all values of eccentricity (see the Appendix I for details). We also mention that the asymmetric expansion of Ferraz-Mello (1987) cannot be used because of the large amplitude of libration of some of the actual asteroids. Beaugé's expansion was adapted to include short period terms. It is

$$R = \frac{1}{a_J} \sum_{i=0}^{4} \sum_{j=0}^{15} \sum_{k=0}^{15} \sum_{l=-15}^{15} \sum_{m=-15}^{15} \sum_{n=-15}^{15} R_{ijklmn} \times$$
$$(\alpha - \alpha_{\text{res}})^i e^j e_J^k \cos(l\theta_1 + m\theta_2 + n\theta_3) \tag{6}$$

with

$$
\begin{aligned}
e &= \sqrt{1-\left[1-\frac{J_1}{J_3-2(J_1+J_2)}\right]^2}, \\
a &= \left[J_3-2(J_1+J_2)\right]^2/\mu,
\end{aligned}
\tag{7}
$$

where R_{ijklmn} are constant coefficients, $\alpha = \frac{a}{a_J}$ and $\alpha_{\text{res}} = \frac{2}{3}^{\frac{2}{3}}$. We note that, at variance with the formulation presented in the Appendix, we expand the disturbing function in powers of $\alpha - \alpha_{\text{res}}$ instead of α. Since $\alpha - \alpha_{\text{res}}$ is small near the resonance, the expansion can be limited to order 4 in this parameter.

3. First Averaging: The Semi-Mean Elements

The first averaging is done over the fast angle θ_3 (the mean synodic longitude of the asteroid). It is done using the non-resonant Lie series theory (Hori, 1966; Ferraz-Mello, 1990) up to first order in the small parameter.

We start with the Hamiltonian given by Equation (5) and we search a canonical transformation, via Lie series, leading to the transformed Hamiltonian

$$
H^* = E_{W^{(1)}} H(\theta_1^*, \theta_2^*, \theta_3^*; J_1^*, J_2^*, J_3^*) = \sum_{l=0}^{\infty} \frac{\varepsilon^l}{l!} D^l_{W^{(1)}}(H). \tag{8}
$$

If we assume

$$
H^* = H_0^* + \varepsilon H_1^* + \varepsilon^2 H_2^* + \varepsilon^3 H_3^* + \cdots, \tag{9}
$$

$$
W^{(1)} = W_1^{(1)} + \varepsilon W_2^{(1)} + \cdots, \tag{10}
$$

and introduce these expansions in Equation (8), we obtain the perturbation equations

$$
\begin{aligned}
H_0^* &= \left(-\frac{\mu^2}{2L^2} + T\right)_{(J_1^*, J_2^*, J_3^*)}, \\
H_1^* &= \left(R + \left\{H_0^*, W_1^{(1)}\right\}\right)_{(\theta_1^*, \theta_2^*, \theta_3^*; J_1^*, J_2^*, J_3^*)}, \\
&\ \vdots
\end{aligned}
\tag{11}
$$

We adopt the averaging

$$
H_1^* = \langle R \rangle_{\theta_3^*} \tag{12}
$$

and obtain the first-order Lie generator

$$W_1^{(1)} = \frac{1}{a_J} \sum_{i=0}^{4} \sum_{j=0}^{15} \sum_{k=0}^{15} \sum_{l=-15}^{15} \sum_{m=-15}^{15} \sum_{n=-15}^{15} \frac{R_{ijklmn}}{l\Gamma_1 + m\Gamma_2 + n\Gamma_3} \times$$
$$(\alpha^* - \alpha_{\mathrm{res}})^i e^{*j} e_J^k \sin(l\theta_1^* + m\theta_2^* + n\theta_3^*) \tag{13}$$

where

$$\Gamma_j = \frac{\partial H_0^*}{\partial J_j^*}, \quad j = 1, 2, 3. \tag{14}$$

The functions $e^*(J^*)$ and $a^*(J^*)$ are defined by the same functional relations as $e(J)$ and $a(J)$ (Equation 7).

The semi-mean elements and the osculating elements are related by the canonical transformation generated by $W^{(1)}$, which, at first order is

$$\begin{aligned} \theta_i &= \theta_i^* + \{\theta_i^*, W_1^{(1)}\} \\ J_i &= J_i^* + \{J_i^*, W_1^{(1)}\}. \end{aligned} \tag{15}$$

The *semi-mean Hamiltonian* is

$$H^* = H_0(J^*) - \varepsilon \frac{1}{a_J} \sum_{i=0}^{4} \sum_{j=0}^{15} \sum_{k=0}^{15} \sum_{l=-15}^{15} \sum_{m=-15}^{15} R_{ijklm0} \times$$
$$(\alpha^* - \alpha_{\mathrm{res}})^i e^{*j} e_J^k \cos(l\theta_1^* + m\theta_2^*). \tag{16}$$

4. Second Averaging: The Mean Elements

The *Resonant Lie Series Theory* used to obtain the mean elements consists basically of the followings steps (see, for details, Ferraz-Mello 2003):

- Expand the Hamiltonian H^* about a reference value;
- split the Hamiltonian into two parts:

$$H^*(\theta^*, J^*) = H^*_{\mathrm{pendulum}} + \Delta H^*; \tag{17}$$

- integrate the pendulum in the variables set $\{\theta_1^*, (J_1^* + J_2^*)\}$ and obtain the corresponding angle-action variables (w_1^*, Λ_1^*);
- extend the transformation to include the other degree of freedom via the *Henrard-Lemaître Transformation*;
- obtain the equations of perturbation;
- average over the fast angle w_1^*.

4.1. EXPANSION OF THE HAMILTONIAN ABOUT A REFERENCE VALUE

Once made the first averaging, the resulting Hamiltonian does not depend on the angle θ_3^*. Then, J_3^* is a constant. We fix the value of this constant at the reference value

$$J_3^* = L_{\rm res} = \sqrt{\mu a_{\rm res}} = \left[\frac{2\mu^2}{3n_J}\right]^{1/3}$$

and expand L^* about it. Since $L^* = J_3^* - 2(J_1^* + J_2^*)$, this is equivalent to expand the Hamiltonian in Taylor series around $\xi = J_1^* + J_2^* = 0$.

The unperturbed part can be written as:

$$H_0^* = -\frac{\mu^2}{2}\sum_{i=0}^{\infty}\frac{(i+1)}{(J_3^*)^{(2+i)}}2^i(J_1^* + J_2^*)^i + 3n_J(J_1^* + J_2^*) - n_J J_3^* - g_J J_2^*.$$

Introducing $n_{\rm res} = \frac{\mu^2}{(J_3^*)^3}$, H_0^* can be written:

$$\begin{aligned} H_0^* = &-\frac{\mu^2}{2(J_3^*)^2} + (3n_J - 2n_{\rm res})(J_1^* + J_2^*) \\ &-\frac{\mu^2}{2}\sum_{l=3}^{\infty}\frac{(i+1)}{J_3^{2+i}}2^i(J_1^* + J_2^*)^i - n_J J_3^* - g_J J_2^*. \end{aligned} \quad (18)$$

Because of the definition of n we have $3n_J - 2n_{\rm res} = 0$

The Hamiltonian $H^*(\theta^*, J^*)$ is a two-degrees-of-freedom Hamiltonian where θ_1^* is critical (resonant) and θ_2^* is a long period angle.

In the following expressions, we also expand J_1^* in the neighborhood of $-J_2^*$ through the substitution of the expression

$$J_1^* = -J_2^* + \xi \quad (19)$$

It is worth emphasizing that we do not assume J_1^* and J_2^* small, but only $J_1^* + J_2^*$ small.

The disturbing function is expanded as:

$$\mathcal{R} = \mathcal{R}^0 + \frac{\partial \mathcal{R}}{\partial \xi}|_{\xi=0}(J_1^* + J_2^*) + \cdots$$

where

$$\left.\frac{\partial \mathcal{R}}{\partial \xi}\right|_{\xi=0} = \left.\left[\frac{\partial \mathcal{R}}{\partial a^*}\frac{\partial a^*}{\partial \xi} + \frac{\partial \mathcal{R}}{\partial e^*}\frac{\partial e^*}{\partial \xi}\right]\right|_{\xi=0}. \quad (20)$$

At $\xi=0$ we have (see Equation 7)

$$a^*|_{\xi=0} = \frac{J_3^{*2}}{\mu}$$

$$e^*|_{\xi=0} = \sqrt{1-\left(\frac{J_2^*+J_3^*}{J_3^*}\right)^2}$$

$$\left(\frac{\partial a^*}{\partial \xi}\right)\bigg|_{\xi=0} = -4\frac{J_3^*}{\mu} \tag{21}$$

$$\left(\frac{\partial e^*}{\partial \xi}\right)\bigg|_{\xi=0} = (-1)\left[1-\left(\frac{J_2^*+J_3^*}{J_3^*}\right)^2\right]^{-1/2} \frac{(J_2^*+J_3^*)(2J_2^*-J_3^*)}{J_3^{*3}}$$

It is worth emphasizing that to calculate the partial derivatives with respect to ξ in Equation (21), we use Equation (19) and consider the momentum J_2^* as a constant.

4.2. THE HORI KERNEL

Because of the resonance, $H_0(J^*)$ is not topologically adequate to be used as a Hori kernel for the second averaging. We have to choose a new Hori kernel that reproduces the topological features of the flow in the neighborhood of the resonance. For that sake, orders of magnitude have to be assigned to variables and parameters so that the main terms are all present in the main resonant part of the Hamiltonian, on the same footing, while the remaining non-trivial terms are of higher order. We use, now, the fact that $J_1^*+J_2^*$ is small and assume that:

$$\xi = J_1^* + J_2^* = \mathcal{O}(\sqrt{\varepsilon}).$$

Figure 2 shows that the main critical terms of the disturbing function are

$$\mathcal{R}_{10}^0(J_2^*)\cos\theta_1^* + \mathcal{R}_{01}^0(J_2^*)\cos\theta_2^* + \mathcal{R}_{20}^0(J_2^*)\cos 2\theta_1^*$$

where the functions $\mathcal{R}_{10}^0$, $\mathcal{R}_{01}^0$ and $\mathcal{R}_{20}^0$ are generically defined by

$$\mathcal{R}_{\ell m}^0 = -\left(\sum_{i=0}^{4}\sum_{j=0}^{15}\sum_{k=0}^{15} R_{ijk\ell m0}\frac{(\alpha^*-\alpha_{\text{res}})^i}{a_J}e^{*j}e_J^k\right)_{\xi=0} \tag{22}$$

(In this expression, the eccentricity and semi-major axis are evaluated at $\xi=0$, that is, at $J_3^*=\sqrt{\mu a_{\text{res}}}$ and $J_1^*=-J_2^*$.)

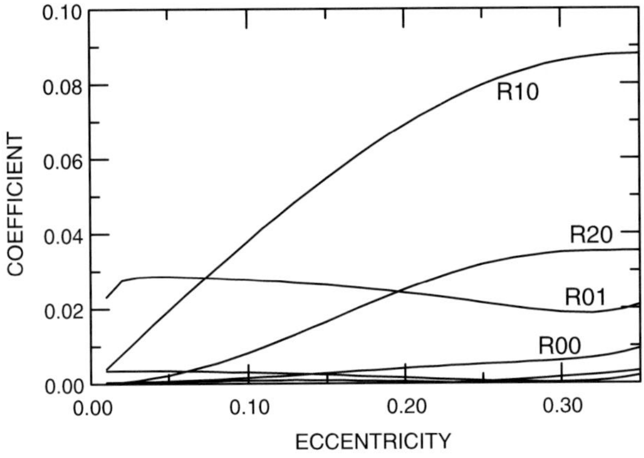

Figure 2. Main coefficients of the disturbing function.

For asteroids with eccentricities in the neighborhood of 0.2, the first term is clearly dominant and we will keep only it in the Hori kernel, letting the other two terms to be considered with the next order perturbations. This choice makes the work considerably simpler, but limits the validity of the theory. In a more precise theory they should also be considered. We note that the routine of calculation would be the same, with only the introduction of the solutions of a more complex Hamiltonian instead of the simple pendulum. This would introduce new technical difficulties but without precluding the construction of the solutions.

Thus, we consider that the main part of the Hamiltonian is now,

$$F_2^* = \frac{1}{2}\nu_{11}(J_1^* + J_2^*)^2 - g_J J_2^* + \varepsilon \mathcal{R}_{10}^0(J_2^*)\cos\theta_1^* \tag{23}$$

where $\nu_{11} = -\frac{12\mu^2}{(J_3^*)^4}$ and the angle θ_2^* is cyclical and thus J_2^* is a constant in the solution of this abridged Hamiltonian. Therefore $\mathcal{R}_{10}^0$ is constant and F_2^* is the Hamiltonian of a simple pendulum whose solution, in the case of small-amplitude librations can be given by (Ferraz-Mello, 2002):

$$\begin{aligned} J_1^* + J_2^* = -8\frac{\omega_1^0}{|\nu_{11}|}\Bigg\{ & [\cos w_1^*]\mathcal{Q} - [2\cos w_1^* - \cos 3w_1^*]\mathcal{Q}^3 \\ & - \left[\frac{17}{2}\cos w_1^* - 5\cos 3w_1^* - \cos 5w_1^*\right]\mathcal{Q}^5\Bigg\} + \mathcal{O}(\mathcal{Q}^7), \end{aligned} \tag{24}$$

where

$$\omega_1^0 = \sqrt{\varepsilon\nu_{11}\mathcal{R}_{10}^0}, \tag{25}$$

and

$$\sin\theta_1^* = 8[\sin w_1^*]\mathcal{Q} - 24[2\sin w_1^* - \sin 3w_1^*]\mathcal{Q}^3$$
$$-8[\frac{25}{2}\sin w_1^* - 3\sin 3w_1^* - 5\sin 5w_1^*]\mathcal{Q}^5 + \mathcal{O}(\mathcal{Q}^7). \quad (26)$$

The quantity $\mathcal{R}_{10}^0$ is a positive quantity. Would $\mathcal{R}_{10}^0$ be negative, the definition of the angle θ_1 should be changed, by adding π, before using the classical pendulum results.

The amplitude $\mathcal{Q}$ is a finite quantity related to the action Λ_1^* of the pendulum through

$$\mathcal{Q} = \sqrt{\frac{\nu_{11}\Lambda_1^*}{32\omega_1^0}}.$$

In what follows, we assume that $\mathcal{Q}$ is a small quantity (*small-amplitude libration hypothesis*).

After introduction of the new variables, Equation (23) becomes

$$F_2^* = \mathcal{F}(\Lambda_1^*, J_2^*) - g_J J_2^*, \quad (27)$$

where

$$\mathcal{F}(\Lambda_1^*, J_2^*) = \varepsilon\mathcal{R}_{10}^0(J_2^*)\left(1 - 32\mathcal{Q}^2 + 64\mathcal{Q}^4\right) + \mathcal{O}(\mathcal{Q}^6) \quad (28)$$

is the Hamiltonian of a pendulum, written in terms of the new action.

4.3. EXTENSION TO THE SECOND DEGREE OF FREEDOM

The variables introduced in the previous section, w_1^*, Λ^* are action-angle variables only while F_2^* is considered as a one-degree-of-freedom Hamiltonian. As the variables of the given system are $\theta_1^*, \theta_2^*, \Lambda_1^*, \Lambda_2^*$, we have to extend the canonical transformation to include the second degree of freedom. For that sake, we use the Henrard-Lemaître transformation (Henrard and Lemaître, 1986)

$$\begin{aligned} \theta_1^* &= \theta_1^*(w_1^*, \Lambda^*) \\ J_1^* &= J_1^*(w_1^*, \Lambda^*) \\ \theta_2^* &= w_2^* - \Xi_2(w_1^*, \Lambda^*) \\ J_2^* &= \Lambda_2^* \end{aligned} \quad (29)$$

where

$$\Xi_2 = \int_0^{w_1} \left(\frac{\partial\theta_1^*}{\partial w_1^*}\frac{\partial J_1^*}{\partial\Lambda_2^*} - \frac{\partial\theta_1^*}{\partial\Lambda_2^*}\frac{\partial J_1^*}{\partial w_1^*}\right) dw_1^*.$$

It is necessary, beforehand, to write θ_1^* in terms of w_1^*, $\mathcal{Q}(\Lambda_1^*)$. We use, for $\sin\theta_1^*$ and $(J_1^*+J_2^*)$ the pendulum solutions given in the previous section. In addition, since we are only considering small-amplitude librations, we may use the approximation

$$\theta_1^* = \arcsin(\sin\theta_1^*) = \sin\theta_1^* + \frac{\sin^3\theta_1^*}{6} + \cdots \tag{30}$$

which is expanded about $\mathcal{Q}=0$. At the order $\mathcal{O}(\mathcal{Q}^7)$, we have

$$\begin{aligned} \theta_1^* &= 8\sin w_1^*\mathcal{Q} + \left[16\sin w_1^* + \frac{8}{3}\sin 3w_1^*\right]\mathcal{Q}^3 \\ &\quad + \left[92\sin w_1^* + 24\sin 3w_1^* + \frac{8}{5}\sin 5w_1^*\right]\mathcal{Q}^5 \end{aligned} \tag{31}$$

and, after some calculations, we obtain

$$\begin{aligned} \Xi_2 &= -8\sin w_1^*\mathcal{Q} - \frac{16}{|\nu_{11}|}\frac{\partial\omega_1^0}{\partial\Lambda_2^*}\sin 2w_1^*\mathcal{Q}^2 - \left[16\sin w_1^* + \frac{8}{3}\sin 3w_1^*\right]\mathcal{Q}^3 \\ &\quad - \frac{16}{|\nu_{11}|}\frac{\partial\omega_1^0}{\partial\Lambda_2}\left[2\sin 2w_1^* + \sin 4w_1^*\right]\mathcal{Q}^4 \\ &\quad - \left[92\sin w_1^* + 24\sin 3w_1^* + \frac{8}{5}\sin 5w_1^*\right]\mathcal{Q}^5 + \mathcal{O}(\mathcal{Q}^6) \end{aligned} \tag{32}$$

and the transformation is completed.

4.3.1. *Some Auxiliary Transformations*

For the sake of writing the Hamiltonian in terms of the new variables, we expand the cosines in powers of $\mathcal{Q}$

$$\begin{aligned} \cos\theta_1^* &= 1 - 16[1-\cos 2w_1^*]\mathcal{Q}^2 - 32[\cos 2w_1^* - \cos 4w_1^*]\mathcal{Q}^4 \\ &\quad - [64 + 112\cos 2w_1^* - 128\cos 4w_1^* - 48\cos 6w_1^*]\mathcal{Q}^6 + \cdots . \end{aligned} \tag{33}$$

Since the action Λ_1^{**} has order $\mathcal{O}(\sqrt{\varepsilon})$, in the application of the Lie Theory, it is necessary to take into account that the Poisson bracket contains two parts of different orders. Indeed, if f, g are two functions with degrees r and s in the variables of the set $\mathcal{S} \equiv (\Lambda_1^{**}, \sqrt{\varepsilon})$, the Poisson bracket is

$$\{f, g\} = \{f, g\}_1 + \{f, g\}_2$$

where

$$\begin{aligned}
\{f, g\}_1 &= \frac{\partial f}{\partial w_1^{**}}\frac{\partial g}{\partial \Lambda_1^{**}} - \frac{\partial f}{\partial \Lambda_1^{**}}\frac{\partial g}{\partial w_1^{**}} = \mathcal{O}(r+s-1) \\
\{f, g\}_2 &= \frac{\partial f}{\partial w_2^{**}}\frac{\partial g}{\partial \Lambda_2^{**}} - \frac{\partial f}{\partial \Lambda_2^{**}}\frac{\partial g}{\partial w_2^{**}} = \mathcal{O}(r+s)
\end{aligned} \tag{41}$$

To solve the second of Equations (40), we adopt

$$F_3^{**} = \frac{1}{2\pi}\int_0^{2\pi} F_3^*(w^{**}, \Lambda^{**})\, dw_1^{**}$$

and obtain the resonant Lie generator

$$W_2^{(2)} = \left(\frac{\partial F_2^{**}}{\partial \Lambda_1^{**}}\right)^{-1} \int [F_3^{**} - F_3^*(w^{**}, \Lambda^{**})]\, dw_1^{**}$$

or

$$\begin{aligned}
\widehat{\omega}_1 W_2^{(2)} = & \left[-\frac{8\varepsilon\omega_1^0}{|\nu_{11}|}\left(\left.\frac{\partial \mathcal{R}_{10}}{\partial \xi}\right|_{\xi=0}\right)\mathcal{Q} \right. \\
& \left. - \left(\frac{64\nu_{111}(\omega_1^0)^3}{|\nu_{11}|^3} - \frac{80\varepsilon\omega_1^0}{|\nu_{11}|}\left.\frac{\partial \mathcal{R}_{10}}{\partial \xi}\right|_{\xi=0}\right)\mathcal{Q}^3 \right] \sin w_1^{**} \\
& - \left(\frac{64\nu_{111}(\omega_1^0)^3}{9|\nu_{11}|^3} + \frac{24\varepsilon\omega_1^0}{|\nu_{11}|}\left.\frac{\partial \mathcal{R}_{10}}{\partial \xi}\right|_{\xi=0}\right)\mathcal{Q}^3 \sin 3w_1^{**} \\
& - \left[4\varepsilon\mathcal{R}_{01}^0\mathcal{Q} - \left(24\varepsilon\mathcal{R}_{01}^0 - \frac{32\varepsilon\mathcal{R}_{01}^0}{|\nu_{11}|}\frac{\partial\omega_1^0}{\partial\Lambda_2^*}\right)\mathcal{Q}^3\right]\sin(w_1^{**} - w_2^{**}) \\
& + \left[4\varepsilon\mathcal{R}_{01}^0\mathcal{Q} - \left(24\varepsilon\mathcal{R}_{01}^0 + \frac{32\varepsilon\mathcal{R}_{01}^0}{|\nu_{11}|}\frac{\partial\omega_1^0}{\partial\Lambda_2^*}\right)\mathcal{Q}^3\right]\sin(w_1^{**} + w_2^{**}) \\
& + \left(\frac{4\varepsilon\mathcal{R}_{1-1}^0}{|\nu_{11}|}\frac{\partial\omega_1^0}{\partial\Lambda_2^*} + 4\varepsilon\mathcal{R}_{01}^0 + \frac{4\varepsilon\mathcal{R}_{01}^0}{|\nu_{11}|}\frac{\partial\omega_1^0}{\partial\Lambda_2^*}\right)\mathcal{Q}^2\sin(2w_1^{**} + w_2^{**})
\end{aligned}$$

$$
\begin{aligned}
&-\left(\frac{4\varepsilon\mathcal{R}^0_{1-1}}{|\nu_{11}|}\frac{\partial\omega^0_1}{\partial\Lambda^*_2}-4\varepsilon\mathcal{R}^0_{01}+\frac{4\varepsilon\mathcal{R}^0_{01}}{|\nu_{11}|}\frac{\partial\omega^0_1}{\partial\Lambda^*_2}\right)\mathcal{Q}^2\sin(2w^{**}_1-w^{**}_2)\\
&+\left(\frac{32\varepsilon\mathcal{R}^0_{01}}{3|\nu_{11}|}\frac{\partial\omega^0_1}{\partial\Lambda^*_2}-4\varepsilon\mathcal{R}^0_{01}\right)\mathcal{Q}^3\sin(3w^{**}_1-w^{**}_2)\\
&+\left(\frac{32\varepsilon\mathcal{R}^0_{01}}{3|\nu_{11}|}\frac{\partial\omega^0_1}{\partial\Lambda^*_2}+4\varepsilon\mathcal{R}^0_{01}\right)\mathcal{Q}^3\sin(3w^{**}_1+w^{**}_2)\\
&+32\varepsilon\mathcal{R}^0_{20}\mathcal{Q}^2\sin 2w^*_1+\mathcal{O}(\mathcal{Q}^4).
\end{aligned}
\tag{42}
$$

$\widehat{\omega}_1$ is the frequency of the angle w^{**}_1:

$$
\widehat{\omega}_1=\frac{\partial F^{**}_2}{\partial\Lambda^{**}_1}=\frac{\partial\mathcal{F}(\Lambda^{**})}{\partial\Lambda^{**}_1}
\tag{43}
$$

From the pendulum equations for small oscillations (Ferraz-Mello, 2002), it is easy to see that,

$$
\widehat{\omega}_1\approx\omega^0_1(\Lambda^{**}_2)(1-4\mathcal{Q}(\Lambda^{**}_1,\Lambda^{**}_2)^2)
$$

The quantities $\nu_{11},\omega^0_1,\mathcal{Q},\mathcal{R}_{ij}$ appearing in the right-hand side of Equation (42) have the same definitions as before, but with Λ^{**} instead of Λ^*.

At this order, the canonical transformation of the semi-mean action-angle elements into mean elements is given by

$$
\begin{aligned}
w^*_1 &= w^{**}_1+\{w^{**}_1,W^{(2)}_2\}_1\\
w^*_2 &= w^{**}_2+\{w^{**}_2,W^{(2)}_2\}_1\\
\Lambda^*_1 &= \Lambda^{**}_1+\{\Lambda^{**}_1,W^{(2)}_2\}_1\\
\Lambda^*_1 &= \Lambda^{**}_1+\{\Lambda^{**}_2,W^{(2)}_2\}_1
\end{aligned}
\tag{44}
$$

and the resulting Hamiltonian for the *mean elements* is:

$$
\begin{aligned}
H^{**}=&\,\mathcal{F}(\Lambda^{**})-g_J\Lambda^{**}_2+\varepsilon\mathcal{R}^0_{00}(\Lambda^{**}_2)+\varepsilon\mathcal{R}^0_{20}(\Lambda^{**}_2)(1-64\mathcal{Q}^2)\\
&+\left[\varepsilon\mathcal{R}^0_{01}(\Lambda^{**}_2)+\varepsilon\mathcal{R}^0_{1-1}(\Lambda^{**}_2)-16\varepsilon\mathcal{R}^0_{01}(\Lambda^{**}_2)\mathcal{Q}^2\right]\cos w^{**}_2.
\end{aligned}
\tag{45}
$$

5. Third Averaging: The Proper Elements

The Hamiltonian obtained in the previous section is a one-degree-of-freedom Hamiltonian (the only variables are, now, w^{**}_2 and Λ^{**}_2), hence integrable.

We may use again a Lie series theory up to first order in the small parameter $\sqrt{\varepsilon}$ to construct the solution of this system. We consider the canonical transformation $(w_2^{**}, \Lambda_2^{**}) \to (w_2^{***}, \Lambda_2^{***})$ generated by the Lie generator $W^{(3)}(w_2^{***}, \Lambda_2^{***})$ and the "perturbation" equations[2]

$$H^{***} = H^{**}(w_2^{***}, \Lambda_2^{***}) + \{F_2^{***}, W^{(3)}\} \tag{46}$$

To solve the latest equation, we adopt the averaging

$$H^{***} = \frac{1}{2\pi}\int_0^{2\pi} H^{**}(w_2^{***}, \Lambda_2^{***})dw_2^{***}$$

and the Lie generator

$$W^{(3)} = \left(\frac{\partial H^{***}}{\partial \Lambda_2^{***}}\right)^{-1} \int [H^{***} - H^{**}(w_2^{***}, \Lambda_2^{***})]dw_2^{***}.$$

We thus obtain

$$H^{***} = \mathcal{F}(\Lambda^{***}) - g_J\Lambda_2^{***} + \varepsilon\mathcal{R}_{00}^0(\Lambda_2^{***}) + \varepsilon\mathcal{R}_{20}^0(\Lambda_2^{***})(1 - 64\mathcal{Q}^2)$$

and

$$W_3^{(3)} = \frac{1}{\widehat{\omega}_2}\left[\varepsilon\mathcal{R}_{10}^0(\Lambda_2^{***}) + \varepsilon\mathcal{R}_{1-1}^0(\Lambda_2^{***}) - 16\varepsilon\mathcal{R}_{10}^0(\Lambda_2^{***})\mathcal{Q}^2\right]\sin w_2^{***} \tag{47}$$

where

$$\widehat{\omega}_2 = \frac{\partial H_0^{**}}{\partial \Lambda_2^{**}}$$

We have thus obtained two quasi integrals Λ_1^{**} and Λ_2^{***}, which are the sought *Dynamical Proper Elements*. Λ_2^{***} is related to Λ_2^{**} through

$$\begin{aligned}\Lambda_2^{***} &= \Lambda_2^{**} + \frac{\partial W_3^{(3)}(w^{**}, \Lambda^{**})}{\partial w_2^{**}}\\ &= \Lambda_2^{**} + \frac{1}{\widehat{\omega}_2}\left[\varepsilon\mathcal{R}_{01}^0(\Lambda_2^{**}) + \varepsilon\mathcal{R}_{1-1}^0(\Lambda_2^{**}) - 16\varepsilon\mathcal{R}_{01}^0(\Lambda_2^{**})\mathcal{Q}^2\right]\cos w_2^{**}\end{aligned}$$

The transformation of the angles is similar.

In the actual application to the asteroids, it maybe considered of interest to construct Keplerian-like semi-major axis and eccentricity as functions of the proper actions Λ_1^{**} and Λ_2^{***}. They can be obtained through the following steps: (i) Equations formally equal to Equation (29) are used to obtain the formal variables J^{***} and θ^{***}; (ii) These variables can be used to determine the equivalent Delaunay's moments; and (iii) Equations formally equal to Equation (7) are used to obtain the *equivalent elliptical proper elements* a^{***} and e^{***}.

[2]For sake of simplicity we omitted in all equations of this section the dependence on the constant Λ_1^{**}.

6. Conclusions

We have presented the mathematical basis for the construction of analytical proper elements for resonant asteroids. This theory becomes possible thanks to two recent theoretical advances: The extension of Lie Series Perturbations Theory to systems including both a resonant and a long-period (i.e. degenerate) angles (Ferraz-Mello, 1997, 2002) and a representation of the disturbing function valid in high eccentricities (Beaugé, 1996; Beaugé and Michtchenko, 2003).

In this low-order theory, the semi-mean elements are calculated up to first order in the mass of Jupiter. The passage to the mean and proper elements is obtained solving the perturbation equations up to order $\mathcal{O}(\varepsilon^{3/2})$ and using an approximation of the libration of the critical angle up to the order of the square of the libration amplitude.

The results show the feasibility of the proposed scheme. However, one should be aware that much work is still necessary to reach he conclusion of this investigation and to have PE for the asteroids in the Hilda group. The extension to the inclined case is the main of the forthcoming steps. The extension of Beaugé's approximation of the disturbing force in the spatial case is done and the actual construction of PE is in progress. However, the influence of the other planets, indispensable for the correct calculation of the proper frequencies, was not yet considered. Only then, it will be possible to apply the theory to the known population of Hildas and to compare to results obtained by numerical means (e.g. Schubart, 1982a,b). It would also be important, then, to compare the whole procedure to the numerical and semi-numerical averaging techniques used by Milani (1993) and by Beaugé and Roig (2001) in the case of the Trojan asteroids.

Acknowledgements

O. Miloni thanks to Dr Pablo Cincotta and Facultad de Ciencias Astronómicas y Geofísicas at La Plata University by technical facilities. This work was supported by a CAPES fellowship and grants from FAPESP, USP and CNPq (Brasil).

Appendix A. The Disturbing Function

The disturbing function is

$$\mathcal{R} = \frac{1}{\Delta} - \frac{\mathbf{r}\mathbf{r}_J}{r_J^3} \tag{48}$$

where

$$\Delta = |\mathbf{r} - \mathbf{r}_J| = \left(r^2 + r_J^2 - 2rr_J \cos S\right)^{1/2}. \tag{49}$$

S is the angle between the asteroid and Jupiter as seen from the Sun.

In this paper we use an improved version of the expansion first suggested by Beaugé (1996). This expansion is valid in large domains of the phase space excluding a region around the singularities associated to collisions between Jupiter and the asteroid.

A.1 BEAUGÉ'S APPROXIMATION. THE PARAMETER δ

The big problem in the expansion of $\mathcal{R}$ comes from the term Δ^{-1}, so let us first deal with it. Introducing the ratio $\rho = r/r_J$, we obtain from Equation (49)

$$\frac{r_J}{\Delta} = \left(1 + \rho^2 - 2\rho \cos S\right)^{-1/2}. \tag{50}$$

Instead of expanding this function in Fourier series of S (Laplace approach) or power series of ρ (Legendre polynomials), we use a best-fit approach. We write

$$\frac{r_J}{\Delta} = (1 + x)^{-1/2}. \tag{51}$$

where

$$x = \rho^2 - 2\rho \cos S \tag{52}$$

and represent the function $(1+x)^{-1/2}$ by a polynomial of order N in x:

$$(1+x)^{-1/2} \simeq \sum_{n=0}^{N} b_n x^n \tag{53}$$

whose coefficients b_n are determined numerically through a linear regression.

The variable $1+x$ is a measure of the proximity of the initial condition to the singularity in $\frac{1}{\Delta}$. It is equal to 0 at the singularity, and increases as we go away from the collision curve. We note that the values of ρ and S are not separately significant; only the distance from the singularity is important.

The numerical fit is performed using values of $x > -1 + \delta$, where δ is a positive parameter close to zero. In Beaugé's approximation, the number of terms necessary to represent Δ^{-1} depends on the magnitude of Δ^{-1} in the domain to be studied: Near the minimum of Δ^{-1}, a few terms are

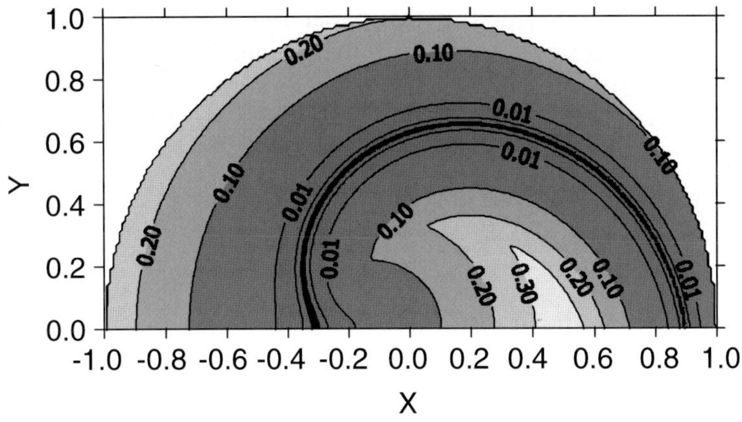

Figure 3. Limits of validity of Beaugé's approximations for asteroids in 3:2 resonance with Jupiter for $e_J = 0$ and different values of δ. The thick black line is the locus of the points where $x_{\min} = -1$ (collision curve). The non-labeled curves adjacent to it correspond to $\delta = 0.001$. Axes: $X = e\cos\theta_1$; $Y = |e\sin\theta_1|$.

enough to have a good representation of $(1+x)^{-1/2}$. This number increases quickly as we approach orbits that may come close to a collision (Beaugé and Michtchenko, 2003). For the Hildas, $\delta > 0.1$ and a good approximation may be obtained with a reasonable number of terms (in this paper, we are using $N = 15$).

Beaugé's approximation no longer requires that the ratio of the distances of the two planets is small. In a mean-motion resonance, the method is valid even for crossing orbits as long as the resonance does not allow the asteroid to come close to Jupiter.

The limits δ in the 3:2 commensurability ($\alpha \simeq 0.763$) are shown in Figure 3 (for $e_J = 0$). The maximum value ($\delta = 0.39$) lies at $e \sim 0.51, \theta_1 = 0$ (on the horizontal axis). This point is very close to the corotation stationary solution of the 3:2 asteroidal resonance ($e_1 = 0.45$ for $e_2 \to 0$; see Ferraz-Mello et al., 1993). The singularities of Δ^{-1} lie on the thick black line (collision curve).

To transform the above approximation into a function having the form needed in a theory, many transformations have to be done. Introducing the explicit expression for x into Equation (53), it becomes

$$\frac{r_J}{\Delta} \simeq \sum_{k=0}^{N} \sum_{j=0}^{n} c_k (-2)^j \binom{k}{j} \rho^{2k-j} \cos^j S \tag{54}$$

where the c_k are constant coefficients, easily obtainable in terms of the b_k.

Since the theory presented in this paper is restricted to coplanar orbits, we introduce the definition of S in the planar case: $S = f - f_J + \Delta\varpi$. Changing from powers of the cosines to multiples of the argument, and introducing the planar expression of S, we can rewrite the last equation as:

$$\frac{a_J}{\Delta} \simeq \sum_{k=0}^{N} \sum_{i=0}^{N-k} 2A_{k,i}\alpha^m \left(\frac{r_1}{a_1}\right)^m \left(\frac{r_2}{a_2}\right)^{-m-1} \cos k(f - f_J + \Delta\varpi) \tag{55}$$

where $m = 2i + k$.

The next step is the transformation of true into mean anomalies. For that sake we use the two-body Fourier expansions:

$$\left(\frac{r}{a}\right)^n \cos(kf) = \sum_{j=0}^{\infty} (X_j^{n,k} + X_{-j}^{n,k}) \cos(j\ell) \tag{56}$$

$$\left(\frac{r}{a}\right)^n \sin(kf) = \sum_{j=0}^{\infty} (X_j^{n,k} - X_{-j}^{n,k}) \sin(j\ell)$$

where the superscript n may be either positive or negative. The coefficients $X_j^{n,k}$ are the Hansen coefficients (see Tisserand, 1889; Kaula, 1962). Hansen coefficients are functions of the eccentricity that may be expanded into power series of the eccentricities:

$$X_j^{n,k} = e^{|k-j|} \sum_{s=0}^{\infty} Y_{s+u_1,s+u_2}^{n,k} e^{2s} \tag{57}$$

($u_1 = \max(0, j-k)$ and $u_2 = \max(0, k-j)$) where the numbers $Y_{s+u_1,s+u_2}^{n,k}$ are the Newcomb operators. Newcomb operators are independent of the eccentricities and, thus, the same for all initial conditions; they obey to some simple recurrence relations allowing them to be easily calculated for all values of the subscripts (see Brouwer and Clemence, 1961).

Introducing Equation (57) into Equation (56), we obtain, after some algebra,

$$\left(\frac{r}{a}\right)^n \cos(kf) = \sum_{i=0}^{\infty} \sum_{m=-\infty}^{\infty} B_{n,k,i,m} e^i \cos(m\ell) \tag{58}$$

$$\left(\frac{r}{a}\right)^n \sin(kf) = \sum_{i=0}^{\infty} \sum_{m=-\infty}^{\infty} C_{n,k,i,m} e^i \sin(m\ell)$$

where $B_{n,k,i,m}$ and $C_{n,k,i,m}$ are constant coefficients expressed as functions of Newcomb operators. These coefficients, first calculated by Leverrier, do not depend on the orbital parameters and may be calculated once for all.

They have some interesting properties. The most important of them is the d'Alembert property: $B_{n,k,i,m} = C_{n,k,i,m} = 0$ when $|m| < i$ or when $|m| - i$ is odd.

The latest expansions are power series in e convergent for $e < 0.6627434\cdots$ (see Wintner, 1941). Introducing now Equation (58), and their analogues for Jupiter, into the expression of the direct part of the disturbing function, and reordering the terms, we get:

$$\frac{1}{\Delta} \simeq \frac{1}{a_J} \sum_{j,k=0}^{\infty} \sum_{m,n=-\infty}^{\infty} \sum_{l=0}^{N} \sum_{i=0}^{N-l} A_{l,i} D_{2i+l,j,k,m,n} \alpha^{2i+l} e^i e_J^j \times \cos(m\ell - n\ell_J + l\Delta\varpi) \tag{59}$$

where the coefficients $D_{2i+l,j,k,m,n}$ are given by:

$$D_{2i+l,j,k,m,n} = \frac{1}{2\gamma_m\gamma_n}(B_{2i+l,l,j,|m|} + \text{sign}(m)C_{2i+l,l,j,|m|}) \times (B_{-2i-l-1,l,k,|n|} + \text{sign}(n)C_{-2i-l-1,l,k,|n|}) \tag{60}$$

and γ_m is a simple bi-valuated function defined as:

$$\gamma_m = \begin{cases} 1/2 & \text{if } m = 0 \\ 1 & \text{if } m > 0 \end{cases} \tag{61}$$

This is the direct part of the disturbing function

A.2 THE INDIRECT PART

The function appearing in the indirect part of $\mathcal{R}$ is

$$\frac{\mathbf{r}\mathbf{r}_J}{r_J^3}. \tag{62}$$

Comparing Equation (48) to Equation (55), we can see that the indirect contribution has exactly the same functional form as one of the terms of the expansion (Equation 55). So, it is possible to add the indirect part directly in the definition of the coefficients used in the previous section.

This can be done very simply modifying the coefficient $A_{k,i}$ with $k = 1$, $i = 0$ in the following way:

$$A_{1,0} \longrightarrow A_{1,0} - 1/2. \tag{63}$$

In this simple manner, we can obtain a single series for the disturbing function of the planetary three-body problem in heliocentric relative coordinates. Recall that the coefficients are constant for all initial conditions, and therefore need only be determined once. (For more details, see Beaugé, 1996; Beaugé and Michtchenko, 2003.)

Each term in $\mathcal{R}$ depends on the mean anomalies ℓ, ℓ_J and on the difference of the perihelion longitudes $\Delta\varpi$. In terms of the resonant canonical variables used in the theory, these arguments may be written as $(m-\ell)\theta_1 + (\ell - n)\theta_2 + (3m-2n)\theta_3$.

References

Beaugé, C.: 1996, 'On a global expansion of the disturbing function in the planar elliptic three-body problem', *Cel. Mech. Dynam. Astron.* **64**, 313–349.

Beaugé, C. and Roig, F.: 2001, 'A Semi-analytical Model for the Motion of the Trojan Asteroids: Proper Elements and Families', *Icarus* **153**, 343–415.

Beaugé, C. and Michtchenko, T.A.: 2003 'Modeling the high-eccentricity planetary three-body problem. Application to the GJ876 planetary system', *Month. Not. Roy. Astron. Soc.* **341**, 760–770.

Brouwer, D.: 1951, Secular variations of the orbital elements of minor planets. *Astron. J.* **56**, 9–32.

Brouwer, D. and Clemence, G.M.: 1961, *Methods of Celestial Mechanics*, Academic Press Inc., New York.

Ferraz-Mello, S.: 1990, 'Averaging Hamiltonians Systems'. In: C. Froeschlé and D. Benest (eds), *Modern Methods in Celestial Mechanics*, Frontières, Gif-sur-Yvette, pp. 151–212.

Ferraz-Mello, S., Tsuchida, M. and Klafke, J.C.: 1993, 'On symmetrical planetary corotations', *Cel. Mech. Dynam. Astron.* **55**, 25–45.

Ferraz-Mello, S.: 1994, 'The convergence domain of the Laplacian expansion of the disturbing function', *Cel. Mech. Dynam. Astron.* **58**, 37–52.

Ferraz-Mello, S.: 1997, 'On Hamiltonian Averaging Theories and Resonance', *Cel. Mech. Dynam. Astron.* **66**, 39–50.

Ferraz-Mello, S.: 2002, 'Ideal Resonance Problem: The Post-Post-Pendulum Approximation', *Cel. Mech. Dynam. Astron.* **83**, 275–289.

Henrard, J. and Lemaître, A.: 1986, 'A perturbation method for problems with two critical arguments', *Celest. Mech.* **39**, 213–238.

Hirayama, K.: 1918, 'Groups of asteroids probably of common origin', *Astronomical J.* **31**, 185–188.

Hirayama, K.: 1923, 'Families of asteroids' *Annales de l'Observatoire astronomique de Tokyo* Append. 11, pp. 55–93.

Kaula, W.M.: 1962, 'Development of the lunar and solar disturbing functions for a close satellite', *Astron. J.* **67**, 300–303.

Lemaître, A. and Morbidelli, A.: 1994, 'Proper elements for highly inclined asteroidal orbits', *Cel. Mech. Dynam. Astron.* **60**, 25–56.

Milani, A. and Knežević, Z.: 1990, 'Secular perturbation theory and computation of asteroid proper elements', *Cel. Mech. Dynam. Astron.* **49**, 347–411.

Milani, A. and Knežević, Z.: 1992, 'Asteroid proper elements and secular resonances', *Icarus* **98**, 211–232.

Milani, A. and Knežević, Z.: 1994, 'Asteroid proper elements and the dynamical structure of the asteroid main belt', *Icarus* **107**, 219–254.

Knežević, Z and Milani, A.: 2000, 'Synthetic Proper Elements for Outer Main Belt Asteroids', *Cel. Mech. Dynam. Astron.* **78**, 17–46.
Schubart, J.: 1968, 'Long-period effects in the motion of Hilda-type planets', *Astron. J.* **73**, 99–103.
Schubart, J.: 1982a, 'Three characteristic parameters of orbits of Hilda-type asteroids', *Astron. Astroph.* **114**, 200–204.
Schubart, J.: 1982b, 'Numerical determination of proper inclinations of Hilda-type asteroids', *Cel. Mech. Dynam. Astron.* **28**, 189–194.
Sessin, W. and Ferraz-Mello, S.: 1984, 'Motion of two planets with periods commensurable in the ratio 2:1. Solutions of the Hori auxiliary system', *Celest. Mech.* **32**, 307–332.
Tisserand, F.: 1889, *'Traité de Mécanique Céleste'*, vol. I, Gauthier-Villars, Paris, chap. XV.
Williams, J. G.: 1969, 'Secular Perturbations in the Solar System', *Ph.D. Thesis* University of California, Los Angeles **60**, 25–56.
Wintner, A.: 1941.: *'The Analytical Foundations of Celestial Mechanics'*, University Press, Princeton.

Celestial Mechanics and Dynamical Astronomy (2005) 92:113–121
DOI 10.1007/s10569-004-3114-5

STABILITY OF CO-ORBITAL MOTION IN EXOPLANETARY SYSTEMS

BÁLINT ÉRDI and ZSOLT SÁNDOR
Department of Astronomy, Eötvös University, Pázmány Péter s.1/A, H-1117 Budapest, Hungary, B.Erdi@astro.elte.hu

(Received: 21 July 2004; accepted: 8 September 2004)

Abstract. The stability of co-orbital motions is investigated in such exoplanetary systems, where the only known giant planet either moves fully in the habitable zone, or leaves it for some part of its orbit. If the regions around the triangular Lagrangian points are stable, they are possible places for smaller Trojan-like planets. We have determined the nonlinear stability regions around the Lagrangian point L_4 of nine exoplanetary systems in the model of the elliptic restricted three-body problem by using the method of the relative Lyapunov indicators. According to our results, all systems could possess small Trojan-like planets. Several features of the stability regions are also discussed. Finally, the size of the stability region around L_4 in the elliptic restricted three-body problem is determined as a function of the mass parameter and eccentricity.

Key words: co-orbital motion, exoplanets, nonlinear stability

1. Introduction

We speak of co-orbital motion when two planets move in nearly the same orbits. In this case they are in a 1:1 mean motion resonance. In the Solar system there are many examples for this type of motion. The most well-known co-orbital objects are Jupiter's Trojan asteroids, sharing their orbits with Jupiter, while also librating around the triangular Lagrangian points L_4 or L_5 of the Sun–Jupiter system. The number of the known Trojans now is above 1600, while their estimated number is one million. Mars and Neptune also have Trojan asteroids, while Earth has a few temporarily captured co-orbital companions.

It can be expected that co-orbital objects exist also in exoplanetary systems. Laughlin and Chambers (2002) outlined a possible formation mechanism of a 1:1 resonant pair in the protoplanetary accretion disc, which is maintained during migration to small semi-major axes as a result of the planets' interaction with the disc. There are a few studies on extrasolar Trojans. Caton et al. (2001) and Davis et al. (2001) searched for Trojan planets in eclipsing binaries by looking for transits about 60° ahead and behind the primary stellar eclipse on the chance that a Trojan planet is found in the L_4 or L_5 points of the binary system. They investigated 18 systems and in a few cases they found anomalies in the light curves near the phases 0.2 and 0.8, however, there has been no confirmation of existing Trojan planets as yet. Laughlin and Chambers (2002) discussed the viability and detectability of extrasolar Trojan planets. They have shown that two planets with masses comparable to Jupiter or Saturn around a solar-mass star can perform stable tadpole-type librations around the vertices of an equilateral triangle. Pairs of Saturn-mass planets can also execute horseshoe orbits around a solar-mass star, but this is not possible for Jupiter-mass pairs. A pair of extrasolar planets both in tadpole-type and horseshoe-type orbits induces a characteristic pattern in the radial velocity component of the central star that could be detected. A third possible configuration for a 1:1 resonance is when one planet is initially in a very eccentric orbit, the other is in a circular orbit and the planets exchange angular momentum and swap eccentricities in a manner that allows them to avoid close encounters (Novak, 2002; Laughlin and Chambers, 2002). Nauenberg (2002) determined numerically the nonlinear stability domain of the triangular Lagrangian solutions in the general three-body problem as a function of the eccentricity and Routh's mass parameter. This study indicates that there is a wide range of Jupiter-size planetary masses (including brown dwarfs) and eccentricities for which such solutions can exist in exoplanetary systems.

In exoplanetary studies an important problem is the existence of terrestrial planets in the habitable zones. If a giant planet moves in the habitable zone the existence of terrestrial planets is questionable. However, as Menou and Tabachnik (2003) noted, habitable terrestrial planets could exist at the stable Lagrangian points of the giant planets. In this paper we study this possibility. We investigate 5 exoplanetary systems, in which the only known giant planet is always in the habitable zone, and determine the region around the Lagrangian point L_4 of the systems in which stable tadpole-type motion is possible. We also study 4 systems, in which the value of the semi-major axis of the giant planet falls between the limits of the habitable zone, but due to the larger eccentricity of the orbit the giant planet leaves for some time the habitable zone.

2. Exotrojan Stability Regions

For the investigation of the nonlinear stability of orbits around the triangular Lagrangian points we used the method of the relative Lyapunov indicators (RLI) (Sándor et al., 2000, 2004). The RLI measures the convergence of the finite-time Lyapunov indicators to the maximal Lyapunov characteristic exponents of two very close orbits. This method is extremely fast in determining the ordered or chaotic nature of individual orbits, as well as distinguishing between stable and unstable regions of the phase space. According to our experiments, gained in different dynamical problems, it is enough to integrate the close orbits for a few hundred times the longest orbital period of the system. In this investigation we made the integration for 10^3 periods of the giant planets.

Table I gives the main parameters of the 5 studied systems, in which the giant planet moves in the habitable zone. The second and third columns give the mass of the central star and the minimum mass of the giant planet in solar mass, the fourth and fifth columns contain the semi-major axis and the eccentricity of the giant planets' orbit, and the last column indicates the region of the habitable zone (HZ).

Table II gives the same parameters for 4 systems, in which the giant planet's orbit is not fully in the habitable zone, due to the values of the semi-major axis and the eccentricity.

We took the other orbital elements of the giant planets zero. Note, that from a dynamical point of view there is no difference between the systems of Tables I and II, the difference is rather conceptual being the possible terrestrial Trojan planets fully in the habitable zone or not.

We studied the nonlinear stability of orbits around the Lagrangian point L_4 of all systems in the model of the elliptic restricted three-body problem by using the method of the RLI. Some of the results are shown in Figures 1–4. Figure 1 displays the stability region around L_4 in the system HD 108874.

TABLE I

Parameters of exoplanetary systems with the giant planet in the habitable zone.

System	m_s [m$_\odot$]	m_p [$m_\odot$]	a [AU]	e	HZ [AU]
HD 17051	1.03	0.00194	0.91	0.24	0.7–1.3
HD 28185	0.99	0.00570	1.03	0.07	0.7–1.3
HD 108874	1.00	0.00178	1.06	0.14	0.7–1.3
HD 177830	1.17	0.00152	1.14	0.10	0.9–1.8
HD 27442	1.20	0.00128	1.18	0.07	0.9–1.8

TABLE II

Parameters of exoplanetary systems with the giant planet partly outside the habitable zone.

System	m_s [m$_\odot$]	m_p [$m_\odot$]	a [AU]	e	HZ [AU]
HD 150706	1.00	0.00100	0.82	0.26	0.7–1.3
HD 114783	0.92	0.00090	1.20	0.10	0.6–1.25
HD 20367	1.05	0.00116	1.25	0.23	0.75–1.40
HD 23079	1.10	0.00261	1.65	0.10	0.85–1.60

Changing the semi-major axis of the orbit of a fictitious planet with negligible mass around the semi-major axis of the giant planet's orbit with a stepsize $\Delta a = 0.002$ AU, and the synodic longitude of the fictitious planet (the difference between the mean orbital longitudes of the fictitious and the giant planets) between 20°–180° with a stepsize $\Delta\lambda = 2°$, and assuming an initial orbital eccentricity for the fictitious planet equal to the eccentricity of the giant planet, we computed the values of the RLI for all orbits originating from these initial conditions. We did the same computations for all systems of Tables I and II. Figure 1 shows the logarithm of the RLI values, corresponding to each initial point, in a black and white scale. Smaller values of the RLI (lighter regions) correspond to more stable orbits, larger RLIs (darker regions) correspond to chaotic orbits. The black background corresponds to escape orbits or to collision orbits with the giant planet. Some structures of the light stability region can be distinguished, among them a ring-like structure near the edge of the stability region. This is more pronounced in Figure 2 corresponding to the case of the system HD 114783. This ring stucture has been also shown by Dvorak et al. (2004), and this might correspond to higher order resonances (13:1, 14:1) between the short and long period components of libration. This is also in agreement with the result obtained by symplectic mappings for coorbital motion by Sándor and Érdi (2003).

The system HD 27442 has a similar stability structure, therefore it is not shown here. From Figures 1 and 2 it can also be seen that the center of the stability region (most light area) is shifted towards the opposite direction of the giant planet, probably due its large orbital eccentricity. It is interesting to compare this result with that obtained in the circular restricted three-body problem. In that model the center of libration is shifted with the increase of

Figure 1. Stability region around L_4 in the system HD 108874.

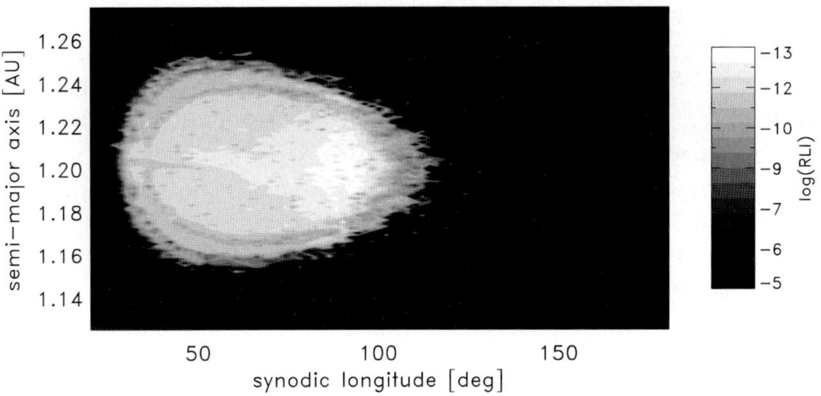

Figure 2. Stability region around L_4 in the system HD 114783.

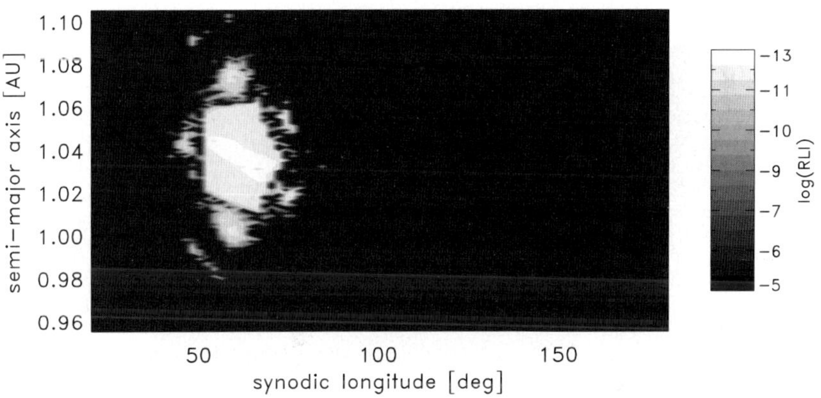

Figure 3. Stability region around L_4 in the system HD 28185.

the orbital eccentricity of the test particle towards the opposite direction of the smaller primary (Namouni and Murray, 2000; Sándor et al. 2002).

Figure 3 shows the stability region around L_4 for the system HD 28185. Here an interesting structure of five stability islands can be noticed, which might be the result of the disruption of a larger stability region due to the larger mass of the giant planet.

Figure 4 displays the case of HD 23079, in which the giant planet moves partly outside the habitable zone. The fine structure of the stability region at the opposite side with respect to the giant planet is well observable. Note that orbits here suffer slow chaotic diffusion as long time integrations show. Note also a bar-like structure of the most stable region in the middle of the figure. This structure has also been described by Dvorak et al. (2004).

In all studied systems there is an extended stability region around the triangular Lagrangian points, where a small planet with negligible mass

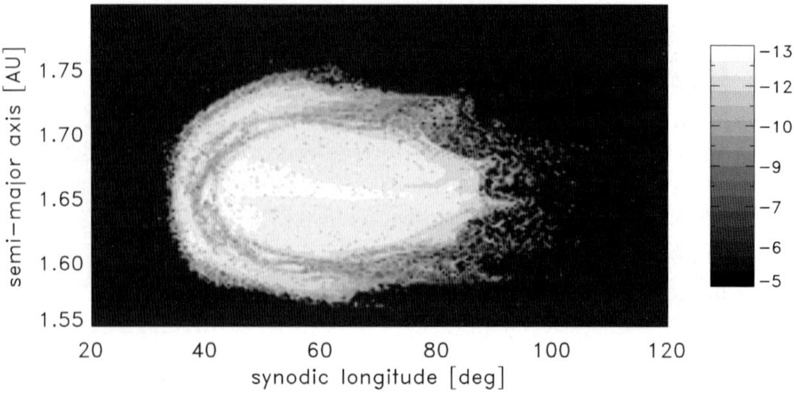

Figure 4. Stability region around L_4 in the system HD 23079.

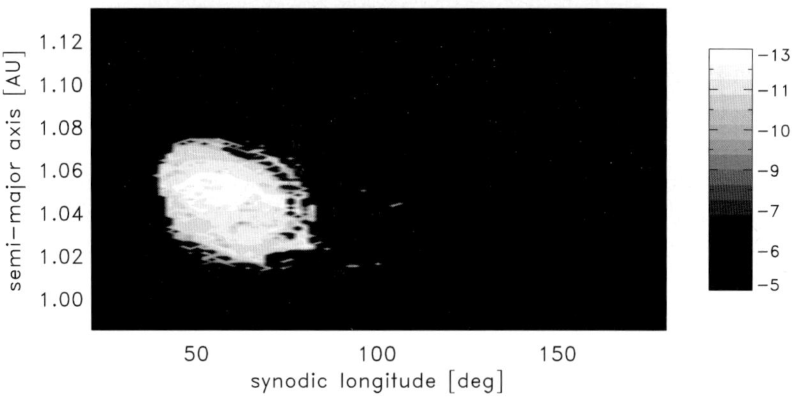

Figure 5. Stability region around L_4 in the perturbed system HD 108874.

could exist. We are going to make computations with larger masses of the Trojan-like planets, and our preliminary results indicate the existence of stability regions around the triangular Lagrangian points in these cases as well.

Certainly, the size of the stability region depends very much on the studied systems. If there is an additional giant planet in the system that can largely modify the structure and the extent of the stability region. This is shown as an example in Figure 5 for the system HD 108874, where we assumed a second giant planet with the same mass as the real one (0.00178 $m_{\odot}$) but in an outer orbit ($a = 1.82$ AU, $e = 0.06$). Comparing Figure 5 with Figure 1 the change is evident.

3. Size of the Stability Region Around L_4

In the model of the elliptic restricted three-body problem, that we have applied to the study of the systems in Tables I and II, the stability of the triangular Lagrangian points depends on the mass parameter and the orbital eccentricity of the primaries. The stability of orbits around L_4 also depends on these parameters. Nonlinear stability of motions around L_4 was studied by Györgyey (1985) for the mass parameter of the Earth–Moon system and for a few values of the eccentricity. Lohinger and Dvorak (1993) made a systematic survey of a wide range of mass parameters and eccentricities. Nauenberg (2002) determined the nonlinear stability domain of the triangular Lagrangian solutions as a function of Routh's mass parameter and the eccentricity.

We also determined the stability region around L_4 in the model of the elliptic restricted three-body problem depending on the mass parameter μ and the eccentricity e of the primaries. We changed the mass parameter in the region 0–0.045 with a stepsize $\Delta\mu = 0.001$, and the eccentricity in the region 0–0.4 with a stepsize $\Delta e = 0.02$. For each pair of μ and e we determined the size of the stability region around L_4 in the same way as we did it for the investigated exoplanetary systems. The size of the stability region is characterized by the number of stable orbits of the test particle. We considered an orbit stable if it did not cross the axis of conjunction of the three bodies, and the value of the RLI remained below 10^{-11} during the whole integration time, which was 10^3 periods of the primaries.

Figure 6 displays the number of stable orbits in a logarithmic black and white scale as the function of the mass parameter and eccentricity of the

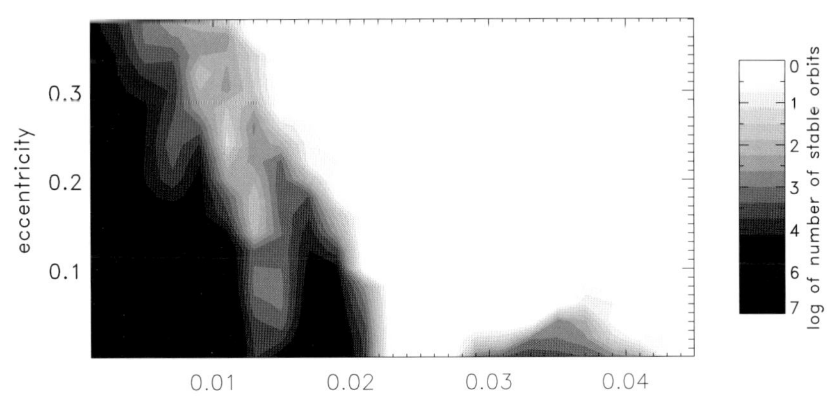

Figure 6. Size of the stability region around L_4 depending on the eccentricity and the mass parameter.

primaries. Figure 6 is in good agreement with the results of Lohinger and Dvorak (1993) and Nauenberg (2002). It can be seen that the stability region is the largest for small values of both e and μ. The investigated 9 exoplanetary systems have stability regions around L_4 with sizes corresponding to the left side of Figure 6. Beside the largest stability region at the left side of Figure 6, there is also an extended stability region near $\mu = 0.02$ for small values of e. This was also found by Lohinger and Dvorak (1993). The unstable regions at about $\mu = 0.015$ and $\mu = 0.024$ correspond to the resonances 3:1 and 2:1 between the librational frequencies. It can be seen in Figure 6 that for a given mass parameter the size of the stability region around L_4 decreases as the eccentricity of the primaries increases, as shown also by Dvorak et al. (2004).

4. Conclusions

We have studied the nonlinear stability of motions around the Lagrangian points L_4 in 9 exoplanetary systems in the model of the elliptic restricted three-body problem by using the method of the relative Lyapunov indicators. Each system has an extended stability region around L_4, whose extent depends on the mass and orbital eccentricity of the giant planet. It is possible that exotrojan planets of small mass exist in these systems. The stability regions have some interesting features, such as a ring-like structure corresponding to higher order resonances, and a bar-like structure of more stable domains in the middle of the stability region. The stability regions are the largest for small values of both the mass parameter and the eccentricity, and for a given mass parameter the size of the stability region decreases with the increase of the eccentricity.

Further works are necessary to investigate the problem for non negligible masses of exotrojan planets and these investigations are underway.

Acknowledgements

The support of the Hungarian National Research Fund, grant numbers OTKA T043739 and OTKA D048424 is acknowledged.

References

Caton, D. B., Davis, S. A., Kluttz, K. A., Stamilio, R. J. and Wohlman, K. D.: 2001, *BAAS*, **33,** 890.

Davis, S. A., Caton, D. B., Kluttz, K. A., Wohlman, K. D., Stamilio, R. J. and Hix, K. B.: 2001, *BAAS*, **33,** 1303.

Dvorak, R., Pilat-Lohinger, E., Schwarz, R. and Freistetter, F.: 2004, *Astron. Astrophys.* **426**, L37–L40.

Györgyey, J.: 1985, *Celest. Mech.*, **36,** 281.
Laughlin, G. and Chambers, J. E.: 2002, *Astron J.* **124,** 592.
Lohinger, E. and Dvorak, R.: 1993. *Astron. Astrophys.*, **280,** 683.
Nauenberg, M.: 2002, Astron J., **124, 2332.**
Novak, G. S.: 2002, *BAAS*, **34.** 939.
Menou, K. and Tabachnik, S.: 2003, *ApJ*, **583,** 473.
Namouni, F. and Murray, C. D.: 2000. *Celest. Mech. Dynam. Astron.,* **76,** 131.
Sándor, Zs., Érdi, B. and Efthymiopoulos, C.: 2000. *Celest. Mech. Dynam. Astron.*, **78,** 113.
Sándor, Zs., Érdi, B. and Murray, C. D.: 2002. *Celest. Mech. Dynam. Astron.*, **84,** 355.
Sándor, Zs. and Érdi, B.: 2003. *Celest. Mech. Dynam. Astron.,* **86,** 301.
Sándor, Zs., Érdi, B., Széll, A. and Funk, B.: 2004, *Celest. Mech. Dynam. Astron.*, **90**, 127.

Celestial Mechanics & Dynamical Astronomy (2005) 92:123–134
DOI 10.1007/s10569-005-1556-z

THE GENERAL SOLUTION OF THE PLANAR LAPLACE PROBLEM

C. MARCHAL
ONERA, BP 72, 92322, Châtillon cedex, France, e-mail: clbmarchal@wanadoo.fr

(Received: 1 July 2004; revised: 9 December 2004; accepted: 1 February 2005)

Abstract. The famous Laplace problem is the three-body, secular, planetary problem. Its plane version has the great theoretical advantage of being integrable (Ferraz-Mello, private correspondence, 2001) and Ferraz-Mello et al. (Chaotic World: from order to Disorder in Gravitational N-Body Systems, Kluwer Academic Publisher, 2004)). Nevertheless it remains a very complex problem with many singularities and many possibilities of collisions. Large eccentricities lead generally to large perturbations especially if the two planets have the same direction of revolution about their star.

Key words: plane secular, three-body problem

1. Introduction

The discovery of many extrasolar planets with a great variety of eccentricities has given a new impetus to the study of the long term evolution and the stability of planetary systems. The simplest non-trivial case is this of a single star with two much less massive planets. The full study of that case remains a dream, even in the plane case, but, if the resonances are negligible, the secular approximation of that plane problem is integrable and a general presentation can be done.

2. Elements of the Problem

We will use the usual notations.
G = Cavendish constant = 6.672×10^{-11} m^3/kg s^2,
M = mass of the star,
m_1 = mass of the inner planet,
m_2 = mass of the outer planet, $m_1 + m_2 \ll M$,
$m' = Mm_1/(M + m_1) \cong m_1$, m' is the "inner reduced mass",
$m'' = (M + m_1)\, m_2/(M + m_1 + m_2) \cong m_2$, m'' is the "outer reduced mass".

with

$$U = \text{mean mutual potential of the two orbits (for G} = 1 \text{ and unit of masses)}$$

$$= (1/4\pi^2) \int_0^{2\pi} dl_1 \int_0^{2\pi} \frac{dl_2}{r_{12}}$$

$$= (1/4\pi^2) \int_0^{2\pi} dE_1(1 - e_1 \cos E_1) \int_0^{2\pi} dE_2(1 - e_2 \cos E_2)/r_{12} \qquad (11)$$

and with

$$E_1 \text{ and } E_2 = \text{eccentric anomalies}, \quad E_1 - e_1 \sin E_1 = l_1, \quad E_2 - e_2 \sin E_2 = l_2. \qquad (12)$$

(Notice that the "$-G\, m_1\, m_2\, U$" term is not the average of the last term of (6), its main part is $-G\, m_1\, m_2/a_{2S}$, fortunately that main part is only a function of the masses and L_{2_S}: it has no effects on the interesting parameters H_{S_1}, H_{S_2}, h_{S_1}, h_{S_2} and can be associated with the $\mathbf{H}_{S_0}$ (L_{S_1}, L_{S_2}) term of (10)).

This elimination assumes that the ratio of the two periods T_1 and T_2 is irrational.

If $T_1/T_2 = p/q$, a second-order term (taking account of l_{S_1} and l_{S_1}) must be added into the expression (10) of $\mathbf{H_S}$, fortunately the main factor of that term is the following K:

$$K = \sup \{r^q e_1^{(q-\varepsilon p)}; r^m e_2^{(q-\varepsilon p)}; r^2 e_2^{(q-2\varepsilon)} e_1^{|p-2|}; r^2 e_2^{(q+2\varepsilon)} e_1^{(p+2)}\}, \qquad (13)$$

with: $1 \leqslant p < q$; p prime to q

$$r = a_1/a_2 \cong (p/q)^{2/3} < 1,$$
$$\varepsilon = \cos i_1 = \pm 1,$$
$$m = \sup \{2; p\}. \qquad (14)$$

For large $(q - \varepsilon p)$ that factor K is generally extremely small.

5. Expression of the Mean Mutual Potential U

Let us call ρ and θ the polar coordinates of an arbitrary point P, either inside or outside of a given Keplerian orbit O (Figure 2).

Let us call a, b, p, e, r, l, E, v the usual orbital elements of that Keplerian orbit:

a = semi-major axis,
e = eccentricity,
$b = a(1 - e^2)^{1/2}$ semi-minor axis,

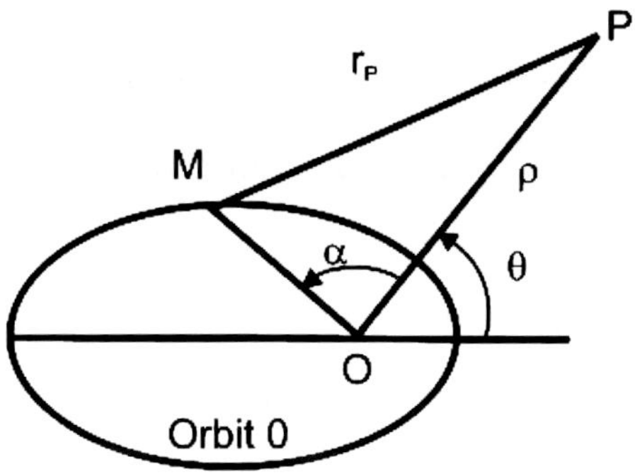

Figure 2. The Keplerian orbit O and the point P.
v = true anomaly = $\theta + \alpha$,
r_p = distance MP.

$p = a(1-e^2)$ = semi-latus rectum,
r = distance OM = $a(1 - e\cos E) = p/(1 + e\cos v)$,
E = eccentric anomaly,
v = true anomaly,
l = mean anomaly = $E - e \sin E$.

The expression of the potential U_K of the Keplerian orbit (with unit G and mass) at the point P is given by

$$U_K = (1/2\pi)\int_0^{2\pi} \mathrm{d}l/\mathrm{r}_p = (1/2\pi)\int_0^{2\pi} \mathrm{d}E(1 - e\cos E)/r_p. \tag{15}$$

We must express U_K in terms of ρ, θ and the orbital elements. We will need the follwing usual expressions (for integers k and m, with $m \geqslant 0$ and with $P_k(\cos\alpha)$ = kth Legendre polynomial).

$$A(k,m) = (1/2\pi)\int_0^{2\pi} \mathrm{d}\alpha\, P_k(\cos\alpha)\cos m\alpha, \tag{16}$$

$$B(k,m) = (1/2\pi)\int_0^{2\pi} \mathrm{d}l\, r^k \cos mv. \tag{17}$$

(These $B(k, m)$ coefficients are related to the "Hansen coefficients" (see Tisserand, 1889a).)

The expression $A(k, m)$ is zero if the difference $(k - m)$ is odd and/or negative, and if not we obtain:

$$n = (k-m)/2 = \text{integer} \geqslant 0, \quad A(k,m) = \mathbf{C}_{2k-2n}^{k-n}\cdot \mathbf{C}_{2n}^{n}/4^k, \tag{18}$$

where $\mathbf{C}_a^b$ is the usual Pascal integer $a!/b!\,(a-b)!$, with $0! = 1$.

This expression (18) gives another formulation of the Legendre polynomial $P_k(\cos\alpha)$:

$$P_k(\cos\alpha) = \sum_{n=0}^{k}\left[\mathbf{C}_{2k-2n}^{k-n}\cdot\mathbf{C}_z 2n^n\cdot\cos(k-2n)\alpha\right]/4^k. \quad (19)$$

Notice that all coefficients of the cos $(k - 2\text{n})\alpha$ are positive and that their sum is $+1$.

For instance

$$P_2(\cos\alpha) = \{6\cos 2\alpha + 4 + 6\cos(-2\alpha)\}/16 = (1 + 3\cos 2\alpha)/4, \quad (20)$$

$$P_3(\cos\alpha) = \{20\cos 3\alpha + 12\cos\alpha + 12\cos(-\alpha) + 20\cos(-3\alpha)\}/64$$
$$= (3\cos\alpha + 5\cos 3\alpha)/8. \quad (21)$$

The expression $B(k, m)$ is more complex, we will especially consider two cases: (1°) The case when $m \leqslant k + 1$, with $A =$ integer part of $(\text{k} + 1 - \text{m})/2$; (and thus $A \geqslant 0$). (2°) The case when $k + 2 + m \leqslant 0$, with $A' =$ integer part of $(-\text{k} - 2 - \text{m})/2$; (and thus $A' \geqslant 0$).

In the case $m \leqslant k + 1$, we obtain:

$$B(k,m) = (1/2\pi)\int_0^{2\pi} \mathrm{d}l r^k \cos mv$$
$$= (-1)^m a^k[\mathbf{C}_{k+m+1}^m/\mathbf{C}_{k+1}^m]\cdot\sum_{j=0}^{A}\left[\mathbf{C}_{m+2j}^j\cdot\mathbf{C}_{k+1}^{m+2j}\cdot(e/2)^{(m+2j)}\right]. \quad (22)$$

For instance

$$B(-1;0) = 1/a, \quad B(+1;0) = a(1 + e^2/2). \quad (23)$$

In the case $k + 2 + m \leqslant 0$ we obtain:

$$B(k,m) = (1/2\pi)\int_0^{2\pi} \mathrm{d}l\, r^k \cos mv = [p^{(3+k)}/b^3]$$
$$\cdot\sum_{j=0}^{A'}\left[\mathbf{C}_{m+2j}^j\cdot\mathbf{C}_{-k-2}^{m+2j}\cdot(e/2)^{(m+2j)}\right]. \quad (24)$$

For instance

$$B(-2;0) = 1/ab, \quad B(-4;0) = (1 + e^2/2)/pb^3. \quad (25)$$

If $m + 2 > (-\text{k}) \geqslant 2$, the quantity $B(k, m)$ is zero, but its expression is mixed and more complex in the remaining cases, i.e. the cases when $m > k + 1 \geqslant 0$; for instance $B(-1; +1) = -e/(a + b)$.

We now need the usual expressions of $1/r_p$ and we will use them in (15).

If ρ is larger than r:

$$1/r_p = 1/\rho + rP_1(\cos\alpha)/\rho^2 + r^2P_2(\cos\alpha)/\rho^3 + \cdots + r^nP_n(\cos\alpha)/\rho^{n+1} + \cdots \quad (26)$$

If ρ is larger than r:

$$1/r_p = 1/r + \rho P_1(\cos\alpha)/r^2 + \rho^2 P_2(\cos\alpha)/r^3 + \cdots + \rho^n P_n(\cos\alpha)/r^{n+1} + \cdots \tag{27}$$

Hence the outer and inner parts U_{out} and U_{in} of the potential U_K:

$$U_{\text{out}}(\rho,\theta) = U_{o0} + 2\sum_{m=1}^{\infty} U_{om}\cos m\theta, \tag{28}$$

with, for positive or zero m:

$$U_{\text{om}} = \sum_{k=0}^{\infty} A(k,m)\cdot B(k,m)/\rho^{(k+1)}, \tag{29}$$

that is, with (18) and (22)

$$U_{\text{om}} = (-1)^m \sum_{q=0}^{\infty}\left[\mathbf{C}_{2q}^{q}\cdot\mathbf{C}_{2m+2q}^{m+q}\cdot\mathbf{C}_{2m+2q+1}^{m}\cdot(a/4)^{(m+2q)}/\mathbf{C}_{m+2q+1}^{m}\cdot\rho^{(m+2q+1)}\right]\cdot\mathbf{P}_{m,q}(e), \tag{30}$$

with, for the polynomial in e, $\mathbf{P}_{m,\,q}(e)$

$$\mathbf{P}_{m,q}(e) = \sum_{j=0}^{q}\mathbf{C}_{m+2j}^{j}\cdot\mathbf{C}_{m+2q+1}^{m+2j}\cdot(e/2)^{(m+2j)} \tag{31}$$

and similarly, with (27), (18), (24) and the polynomials (31):

$$U_{\text{in}}(\rho,\theta) = U_{i_0} + 2\sum_{m=1}^{\infty} U_{im}\cos m\theta, \tag{32}$$

with, for m positive or zero

$$U_{\text{im}} = \sum_{k=0}^{\infty} A(k,m)\cdot B(-k-1,m)\cdot\rho^k, \tag{33}$$

that is, for $m = 0$

$$U_{i_0} = 1/a + \sum_{q=1}^{\infty}\left[\left(\mathbf{C}_{2q}^{q}\right)^2\cdot(\rho/4)^{2q}/b^3 p^{(2q-2)}\right]\cdot\mathbf{P}_{o,q-1}(e) \tag{34}$$

and, for positive m

$$U_{\text{im}} = \sum_{q=1}^{\infty}\left[\mathbf{C}_{2q}^{q}\cdot\mathbf{C}_{2m+2q}^{m+q}\cdot(\rho/4)^{(m+2q)}/b^3 p^{(m+2q-2)}\right]\cdot\mathbf{P}_{m,q-1}(e). \tag{35}$$

These expressions of the outer potential U_{out}, given in (28)–(31), and the inner potential U_{in}, given in (32)–(35), have a well defined domain of convergence. The first is absolutely converging if $\rho > a(1 + e)$ and the second if $\rho < a(1 - e)$. Expressions with a stronger convergence and a larger domain of convergence can be considered (for instance outside the "main circle" of the Keplerian ellipse of interest), but we will not consider them.

It is now easy to obtain a general expression of the mean mutual potential U of the two coplanar non-intersecting Keplerian ellipses of interest (for $G = 1$ and unit of masses). We need either the outer potential U_{out_1} of the inner orbit, or the inner potential U_{in_2} of the outer orbit, and thus

$$U=(1/2\pi)\int_0^{2\pi} U_{\text{out}_1}(r_2,v_2-\omega)\mathrm{d}l_2=(1/2\pi)\int_0^{2\pi} U_{\text{in}_2}(r_1,v_1+\omega)\mathrm{d}l_1. \quad (36)$$

These expressions lead to the following (and for simplicity we will present the mean mutual potential U in a way slightly different from U_{out} in (28) and U_{in} in (32)).

$$U = (1/a_2) + U_{\text{o}} + 2\cdot\sum_{m=1}^{\infty}(-1)^m U_m \cos m\omega, \quad (37)$$

with, for positive or zero m:

$$U_m =(b_2/a_2^2)\sum_{q=1}^{\infty}\left[\mathbf{C}_{2q}^{q}\cdot\mathbf{C}_{2m+2q}^{m+q}\cdot\mathbf{C}_{2m+2q+1}^{m}/\mathbf{C}_{m+2q+1}^{m}\right]\cdot(a_1/4p_2)^{(m+2q)}$$
$$\cdot\,\mathbf{P}_{m,q}(e_1)\cdot\mathbf{P}_{m,q-1}(e_2), \quad (38)$$

with the polynomial $\mathbf{P}_{m,q}\ (e)$ given in (31).

This expresion (37) and (38) of the mean mutual potential U is absolutely converging if $a_1(1 + e_1) < a_2(1 - e_2)$, but it is diverging if $a_1(1 + e_1) > a_2(1 - e_2)$.

If the ratio a_1/a_2 is small, the main terms are the following.

$$\begin{aligned}U = {} & [(1/a_2) + (a_1^2/4b_2^3)\cdot(1+3e_1^2/2)\\ & + (9a_1^4/64b_2^3 p_2{}^2)\cdot(1+5e_1^2+15e_1^4/8)\cdot(1+3e_2^2/2)+\cdots]\\ & - e_1e_2\cos\omega[(15a_1^3/16b_2^3p_2)\cdot(1+3e_1^2/4)+\cdots]\\ & + e_1^2e_2\end{aligned}$$

Notice that all coefficients U_{m} in (37) are positive, and thus, when the only variable is the angle ω, the mean mutual potential U is maximum for $\omega = 180°$. It seems that, for non intersecting coplanar orbits, U is a monotonic and decreasing function cos ω.

Of course, if $a_1(1 + e_1) > a_2(1 - e_2)$, the above expression (37) and (38) has no interest since it is diverging, and we have used a direct numerical computation of the integral (11) with r_{12} given by Figure 1 and the following:

$$\begin{aligned}r_{12}^2 &= (x_2-x_1)^2 + (y_2-y_1)^2,\\ (x_2-x_1) &= a_2(\cos E_2 - e_2) - a_1(\cos E_1 - e_1)\cos\omega + b_1\sin E_1\sin\omega,\\ (y_2-y_1) &= b_2\sin E_2 - b_1\sin E_1\cos\omega - a_1(\cos E_1 - e_1)\sin\omega. \quad (40)\end{aligned}$$

6. Description of Secular Motions

The remaining problem is the following.

We consider two planets with coplanar, non-intersecting elliptical orbits about their star. The secular, second-order approximation leads to constant Delaunay's parameters L_1 and L_2 (and thus to constant semi-major axes a_1 and a_2, and constant mean angular motions n_1 and n_2), it leads also to ignorable mean anomalies l_1 and l_2, with time-derivative very close to the mean angular motions n_1 and n_2.

Hence the main remaining questions are that of the secular evolution of the remaining Delaunays parameters H_1, H_2, h_1 and h_2 ($H_1 = L_1 (1 - e_1^2)^{1/2} \cos i_1$, with $i_1 =$ either 0° or 180°; $H_2 = L_2 (1 - e_2^2)^{1/2}$, with $i_2 = 0°$; h_1 and h_2 are the longitude of pericenters with: $\omega = h_1 - h_2$).

The main elements of the secular evolution are:

(A) The integral of angular momentum c is the sum $H_1 + H_2$

$$\mathbf{c} = H_1 + H_2 = \text{constant}. \tag{41}$$

(B) The Hamiltonian $\mathbf{H}_S$ of Equation (10) is a second integral of secular motion, and, because of its expression, the mean mutual potential U is also an integral of the secular motion.

(C) For large values of the ratio a_2/a_1, the main terms of $\mathrm{d}h_1/\mathrm{d}t$ and $\mathrm{d}h_2/\mathrm{d}t$ are the following:

$$\begin{aligned} \mathrm{d}h_1/\mathrm{d}t &= [3n_1 m_2 \cos i_1 a_1^2 b_1/4Mb_2^3] + O\{n_1 m_2 a_1^4/Mb_2^3 p_2\}, \\ \mathrm{d}h_2/\mathrm{d}t &= [(6 + 9e_1^2) n_2 m_1 a_1^2/8Mp_2^2] + O\{n_2 m_1 a_1^3/Mp_2^3\}, \\ \mathrm{d}\omega/\mathrm{d}t &= (\mathrm{d}h_1/\mathrm{d}t) - (\mathrm{d}h_2/\mathrm{d}t). \end{aligned} \tag{42}$$

Hence a good presentation of the secular motions is to draw the $U =$ constant curves on an x, y plane with:

$$\begin{aligned} x &= H_1/L_1 - (1 - e_1^2)^{1/2} \cos i_1, \quad -1 \leqslant x \leqslant +1, \\ y &= H_2/L_2 = (1 - e_2^2)^{1/2} \quad 0 \leqslant y \leqslant +1. \end{aligned} \tag{43}$$

The $U =$ constant curves are symmetrical with respect to the y axis, and, in the x, y plane, the motion will remain on the straight line defined by:

$$L_1 x + L_2 y = \mathbf{c}. \tag{44}$$

Notice that this straight line has always a negative slope, and this detail will give the main difference between planets rotating in the same direction ($i_1 = 0°$) and planets rotating in opposite directions ($i_1 = 180°$); the latter ones will have much less perturbations.

The Figure 3 presents the $U = 0.2066$ curves for $a_2/a_1 = 5$ and for several values of the angle ω.

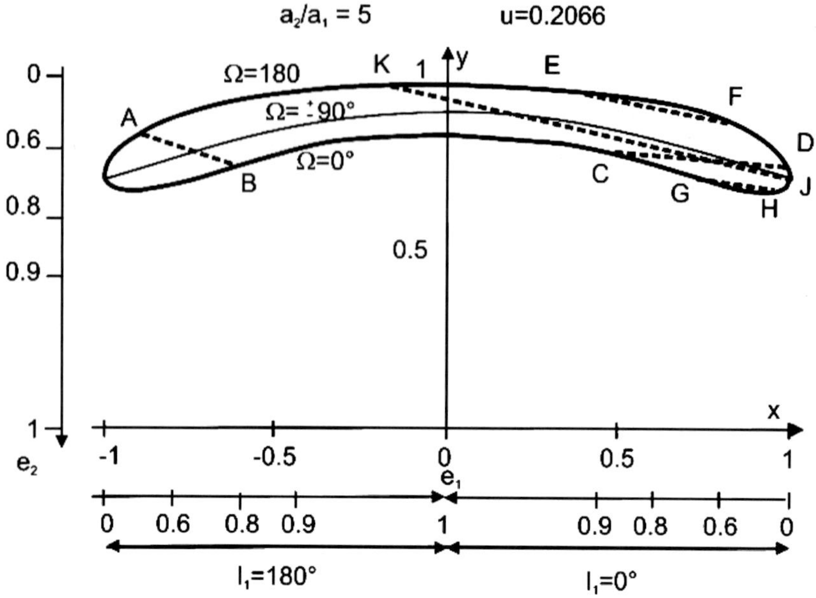

Figure 3. The curves $U = 0.2066$ for $a_1 = 1$; $a_2 = 5$ and several values of the angle ω.

On Figure 3, we can recognize many types of motion, according to the straight line $L_1\, x + L_2\, y = \mathbf{c}$.

Along the straight line AB, we find a periodic solution with a decreasing circulation of ω, along CD a periodic solution with an increasing circulation ω, along EF a periodic solution with a libration of ω about 180°, along GH a periodic solution with a libration of ω about 0°.

Straight lines tangent to the curve drawn for $\omega = 180°$ corresponds to stable solutions with a constant ω, while straight lines tangent to the concave part of the curve drawn for $\omega = 0°$ corresponds either to unstable solutions with a constant ω, or to solutions that are asymptotic to these unstable solutions.

Notice the straight line JK. If we start at J the conditions are $e_1 = 0$, $i_1 = 0°$ and $e_2 = 0.7$, while the slope corresponds to a ratio $m_2/m_1 = 2$. But the point K is very far from J and the motion cross the y axis and thus reaches the eccentricity $e_1 = 1$: the inner planet will fall into its star!

In the solar system a libration of the angle ω of the planets Jupiter and Uranus has been noticed during the nineteenth century by the astronomers Stockwell and Tisserand (Tisserand, 1889b). That angle remains forever between 180° − 71°33′ and 180° + 71°33′.

Figure 4 presents some $U =$ constant curves for large values of the ratio a_2/a_1. The perturbations remains slow and small, especially for motions in

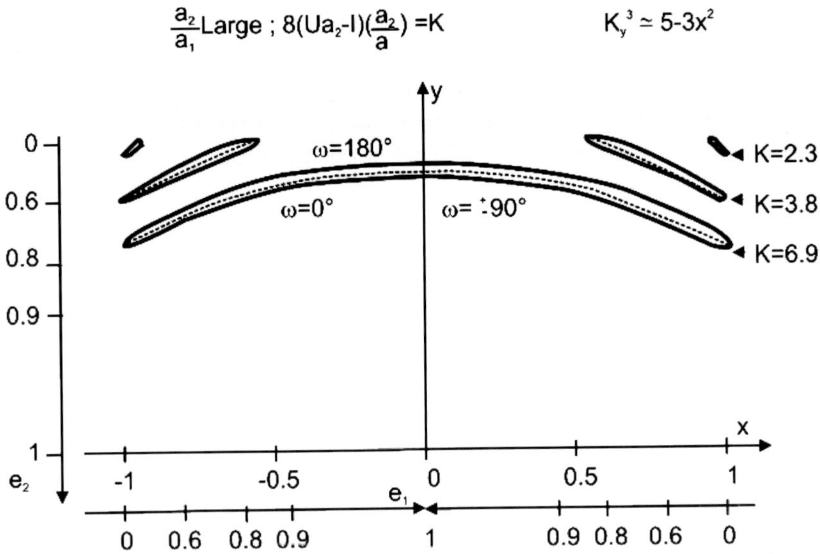

Figure 4. Some $U =$ constant curves for large values of the ratio a_2/a_1.

the opposite directions (left half of the figure) and for motions with small eccentricities (the two upper corners of the figure).

Figure 5 presents two $U =$ constant curves for $a_2/a_1 = 3$. Only the case of two very small eccentricities leads to small perturbations.

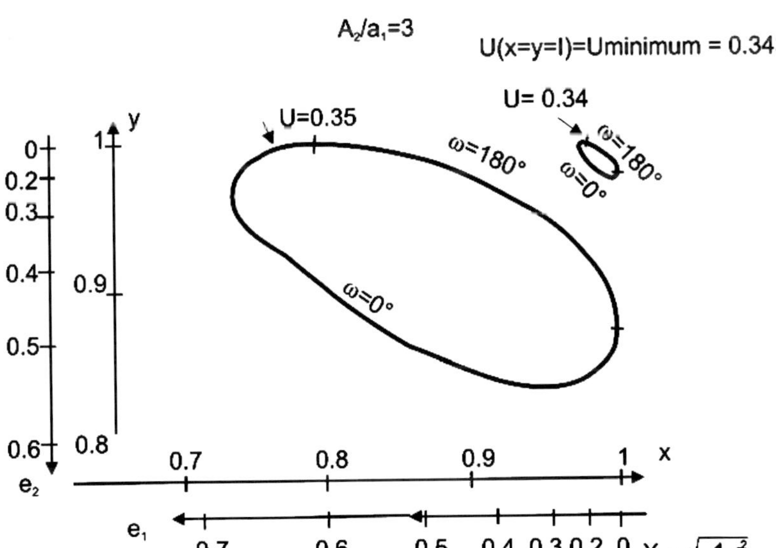

Figure 5. Two $U =$ constant curves for $a_1 = 1$; $a_2 = 3$.

7. Conclusions

In the non-resonant cases the secular version of the three-body, planar, planetary problem is integrable and shows that the perturbations of planetary orbits can be very large, even for very small planetary masses.

The small perturbations occur especially for quasi-circular orbits and/or widely separated orbits.

Planets revolving in opposite directions have, in general, less perturbations than planetes revolving in the same direction.

It remains to enlarge this analysis and especially to study the cases leading to planetary collisions or to very close planetary approaches

References

Ferraz-Mello, S.: 2001, *Private correspondance.*

Ferraz-Mello, S., Michtchenko T. A. and Beauge, C.: 2004, 'Regular motions in planetary systems' In B.A. Steves (ed.), '*Chaotic Worlds: From Order to Disorder in Gravitational N-Body Systems*', Kluwer Academic Publisher.

Tisserand, F.: 1889a, *Formules de Hansen pour le développement de certaines fonctions des coordonées du mouvement elliptique.* Traité de Mécanique céleste. Tome 1, pages 249–261. Gautier-Villars ed.

Tisserand, F.: 1889b, *Rapport sur les calculs de Monsieur Stockwell dans le tome XVIII des «Smithsonian contributions to knowledge».* Traité de Mécanique céleste. Tome 1 page 427. Gautier-Villars ed.

Von Zeipel, H.: 1916–1917, *Recherches sur le mouvement des petites planètes.* Ark. Matematik Astonomi och Fysik **11**(1), **12**(9), **13**(3).

Celestial Mechanics and Dynamical Astronomy (2005) 92:135–156
DOI 10.1007/s10569-004-1333-4

DYNAMICS OF 2/1 RESONANT EXTRASOLAR SYSTEMS APPLICATION TO *HD82943* AND *GLIESE876*

DIONYSSIA PSYCHOYOS and JOHN D. HADJIDEMETRIOU
Department of Physics, University of Thessaloniki, 54124 Thessaloniki, Greece,
e-mail: hadjidem@auth.gr

(Received: 8 June 2004; revised: 5 July 2004; accepted: 12 July 2004)

Abstract. A complete study is made of the resonant motion of two planets revolving around a star, in the model of the general planar three body problem. The resonant motion corresponds to periodic motion of the two planets, in a rotating frame, and the position and stability properties of the periodic orbits determine the topology of the phase space and consequently play an important role in the evolution of the system. Several families of symmetric periodic orbits are computed numerically, for the 2/1 resonance, and for the masses of some observed extrasolar planetary systems. In this way we obtain a global view of all the possible stable configurations of a system of two planets. These define the regions of the phase space where a resonant extrasolar system could be trapped, if it had followed in the past a migration process.

The factors that affect the stability of a resonant system are studied. For the same resonance and the same planetary masses, a large value of the eccentricities may stabilize the system, even in the case where the two planetary orbits intersect. The phase of the two planets (position at perihelion or aphelion when the star and the two planets are aligned) plays an important role, and the change of the phase, other things being the same, may destabilize the system. Also, the ratio of the planetary masses, for the same total mass of the two planets, plays an important role and the system, at some resonances and some phases, is destabilized when this ratio changes.

The above results are applied to the observed extrasolar planetary systems *HD 82943*, *Gliese 876* and also to some preliminary results of *HD 160691*. It is shown that the observed configurations are close to stable periodic motion.

Key words: periodic orbits, resonances, extrasolar systems, HD 82943, GJ 876

1. Introduction

The position and the stability properties of the periodic orbits (or, equivalently, of the fixed points of the Poincaré map) play an important role in the study of the dynamical evolution of a planetary system, because they determine the topology of the phase space. In particular, the mean motion resonances of a planetary system correspond to a periodic motion, in a rotating frame. This is the reason why the resonances play an important role in the study of the long term evolution of a planetary system.

There are several papers on the dynamics of the 2/1 resonant planetary motion and on the mechanisms that stabilize the system, or generate chaotic motion and instability: Gozdjiewski and Maciejewski (2001), Kinoshita and

Nakai (2001), Laughlin and Chambers (2001), Lee and Peale (2001, 2002), Ferraz-Mello (2002), Ferraz-Mello et al. (2003), Ji et al. (2002, 2003a,b), Mahlhotra (2002a, b), Bois et al. (2003), Gozdjiewski et al. (2004), Beaugé et al. (2004). In these papers different methods have been applied (averaging method, direct numerical integrations of orbits, or various numerical methods which provide indicators for the exponential growth of nearby orbits), for a range of orbital parameters. In this way the regions where stable motion exists have been detected, in the orbital elements space.

In the present paper a complete study is made of the resonant motion of two planets revolving around a star, which will be called the *sun*, in the model of the general planar three body problem, by computing all the basic families of resonant periodic orbits. The families of periodic orbits are very useful in the study of the stability and the evolution of an extrasolar planetary system. This is so, because it is close to a stable periodic orbit that a stable system could exist. In addition, the motion close to a periodic motion is the motion with the smallest variation of the orbital elements, a condition which may play an important role in the appearance of life. Finally, the regions of phase space close to a stable periodic motion are the regions where a planetary system may be finally trapped, if it had followed a migration process in the past, before it settled down to its present position. The study of this process is however beyond the object of the present paper.

We consider symmetric periodic orbits, which in this case are the most important resonances, which means that the perihelia of the two planets are either in the same direction or in opposite directions (aligned or antialigned), when the two planets are in the same line with the sun. In this symmetric case, the line of apsides of the two planets precesses slowly, in such a way that $\Delta\omega = 0$ or 2π. Several families of periodic orbits are computed numerically, for the 2/1, and for the masses of the systems *HD 82943*, *HD 160961* and *Gliese 876*.

The factors that affect the stability of a resonant system are studied. For the same resonance and the same planetary masses, the value of the planetary eccentricities is in some cases important and a large value of the eccentricities may stabilize the system, which for smaller eccentricities is unstable. The phase of the two planets (position at perihelion or aphelion when the star and the two planets are aligned) plays an important role, and the change of the phase, other things being the same, may destabilize the system. Also, the ratio of the planetary masses, for the same total mass of the two planets, plays an important role and the system, at some resonances and some phases, is destabilized when this ratio changes. The stability analysis of a resonant planetary system by the method of periodic orbits, that we present in this paper, allows us to obtain a global view of the dynamics of all the observed planetary systems at the 2/1 resonance. In this way the stability of all these systems can bet treated in a unified way.

The above results are applied for the study of the observed extrasolar planetary systems *HD 82943*, *Gliese 876*, *HD 160691*, at the 2/1 resonance and all the possible configurations which lead to stable motion are found. The elements of the above systems, as obtained from the observations, given in the web site *http://www.obspm.fr/encycl/catalog.html* maintained by Jean Schneider. All the masses are multiplied by $\sin i$, where i is the inclination of the planetary orbit and are therefore the minimum masses. In the present study we considered $\sin i = 1$. In the case of HD160961 there is some ambiguity on the value of the elements (Gozdjiewski et al., 2003), but we used the elements published in the above mentioned site, in order to show that even for very high eccentricities, the system may be stable for a suitable phase.

In all the following the central star will be called the *sun*, the inner planet will be called P_1 and the outer planet P_2.

2. The Dynamical Model

2.1. THE EQUATIONS OF MOTION

The model we used in the study of periodic motion of the planetary system is the general three body problem, for planar motion. As we shall see in the following, the gravitational interaction between the two planets is important, even for small planetary masses.

The center of mass of the planetary system is considered as fixed in an inertial frame, and the study is made in a rotating frame of reference xOy, whose x-axis is the line *sun*-P_1, the origin O is the center of mass of these two bodies and the y-axis is perpendicular to the x-axis (Figure 1). In this rotating frame P_1 moves on the x-axis and P_2 on the xOy plane. The coordinates are the position x_1 of P_1, x_2, y_2 of P_2 and the angle θ between the x-axis and a fixed direction in the inertial frame. The coordinates x_1, x_2, y_2, define the position of the system in the rotating frame and the angle θ defines the orientation of the rotating frame, so these four coordinates determine the position of the system in the inertial frame. This is a system of four degrees of freedom, and the Lagrangian of the system is (Hadjidemetriou, 1975)

$$\begin{aligned}\mathcal{L} =& \frac{1}{2}(m_1 + m_0)\{q(\dot{x}_1^2 + x_1^2\dot{\theta}^2) + \frac{m_2}{m}[\dot{x}_2^2 + y_2^2 + \dot{\theta}^2(x^2 + y^2) \\ &+ 2\dot{\theta}(x_2\dot{y}_2 - \dot{x}_2 y_2)]\} - V,\end{aligned} \tag{1}$$

where

$$V = -\frac{Gm_0m_1}{r_{01}} - \frac{Gm_0m_2}{r_{02}} - \frac{Gm_1m_2}{r_{12}}, \tag{2}$$

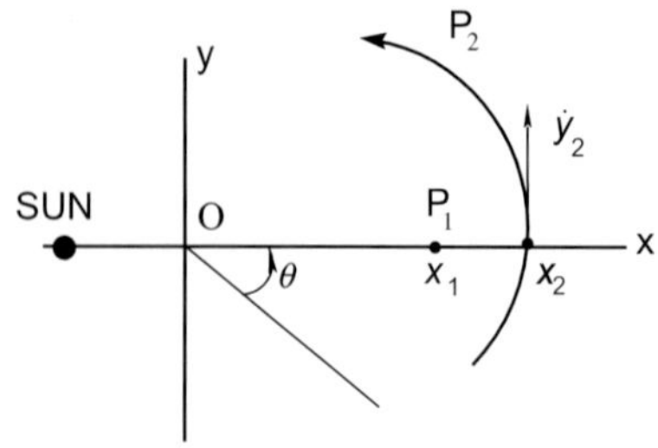

Figure 1. The rotating frame xOy.

and

$$m = m_0 + m_1 + m_2, \quad q = m_1/m_0. \tag{3}$$

G is the gravitational constant and r_{01}, r_{02} and r_{12} are the distances between the sun and P_1, the sun and P_2 and P_1P_2, respectively. We note that the angle θ is ignorable, so besides the energy (or Jacobi) integral there also exists the angular momentum integral, $L = \partial\mathcal{L}/\partial\dot{\theta} =$ constant,

$$L = (m_0 + m_1)\Big\{\dot{\theta}\Big[qx_1^2 + \frac{m_2}{m}(x_2^2 + y_2^2)\Big] + \frac{m_2}{m}(x_2\dot{y}_2 - \dot{x}_2y_2)\Big\}. \tag{4}$$

By making use of this latter integral, we can eliminate the angle θ and thus reduce the system to a system of three degrees of freedom. This can be achieved by constructing the Routhian function, which is the new Lagrangian of the reduced three degrees of freedom system (Pars, 1965; Hadjidemetriou, 1975). The value of the angular momentum appears as a fixed parameter in the differential equations of motion in the rotating frame.

2.2. PERIODIC ORBITS

The differential equations of motion in the rotating frame xOy are invariant under the transformation

$$x_1 \to x_1, \quad x_2 \to x_2, \quad y_2 \to -y_2, \quad t \to -t,$$

which implies that if the planet P_2 starts perpendicularly from the x-axis ($y_2 = 0$, $\dot{x}_2 = 0$) and at that time $\dot{x}_1 = 0$, and after some time $t = T/2$ the planet P_2 crosses again the x-axis perpendicularly and at that time it is $\dot{x}_1 = 0$, the orbit is periodic with period equal to T, symmetric with respect to the x-axis. We remark that the second perpendicular crossing of P_2 from the x-axis may take place after several non perpendicular crossings.

From the above we see that the non zero initial conditions of a symmetric periodic orbit, in the rotating frame, are

$$x_{10}, \quad x_{20}, \quad \dot{y}_{20}. \tag{5}$$

This implies that a family of symmetric periodic orbits is represented by a smooth curve in the three dimensional space x_{10}, x_{20}, $\dot{y}_{20}$.

The periodic orbits that we will construct are in this rotating frame, which means that the *relative position* of the planetary system is repeated in the inertial frame. In order to avoid duplication of the results we fix the units of mass, length and time. This is achieved by taking the total mass of the system as the unit of mass, the gravitational constant equal to unity and also by keeping a fixed value of the angular momentum L for all the orbits of a family of periodic orbits. So, the normalizing conditions are

$$m_0 + m_1 + m_2 = 1, \quad G = 1, \quad L = \text{constant}.$$

In practice, we made the integration of the planetary system in the inertial frame (where the center of mass is fixed) and the reduction to three degrees of freedom in the rotating frame was made by a coordinate transformation. The method of integration was based on Taylor series expansion, and the accuracy was 10^{-14}.

We have computed all the basic families of periodic orbits of two planets in planar motion, in the 2/1 mean motion resonance. Along these orbits the resonance (or the ratio of the semi major axes) of the two planets is almost constant. The planetary orbits are perturbed ellipses and their eccentricity varies along the family, starting from zero values. Most of these families bifurcate from the family of circular orbits of the two planets, along which the orbits of the planets are almost circular, and the ratio T_1/T_2 of their periods starts from very large values and decreases along the family. At the 2/1 resonance a gap appears, for non zero planetary masses, and two distinct families of resonant elliptic orbits start from this gap (Hadjidemetriou, 2002; Hadjidemetriou and Psychoyos, 2003).

The periodic orbits that we computed are symmetric with respect to the rotating x-axis, which means that at $t = 0$, when the two planets are on the same line with the sun, the line of apsides are on this line and the position of the perihelia are either in the same direction or in opposite directions. Consequently, we have at $t = 0$ eight different phases, that are equivalent in pairs, and are given in Table I and Figure 2. All the possible initial phases of a periodic orbit, and the equivalent phase at $t = T/2$, are summarized in Table I.

Each of the symmetric families of periodic orbits that we present in the following sections belongs to a certain type, as described in Table I. In order to distinguish between the different families we will use the terminology:

- Family 1: Type 1 ($\omega_1 = \omega_2$), perihelia in the same direction.
- Family 2: Type 4 ($\omega_1 = \omega_2 + \pi$), perihelia in opposite directions.
- Family 3: Types 2 and 3. Starts as type 2 and ends as type 3, because the eccentricity of P_1 crosses zero and the planet goes from perihelion to aphelion.

As we shall see in the following, there are two basic families of periodic orbits, family 1 and family 2 (family 3 appears only in one case and is

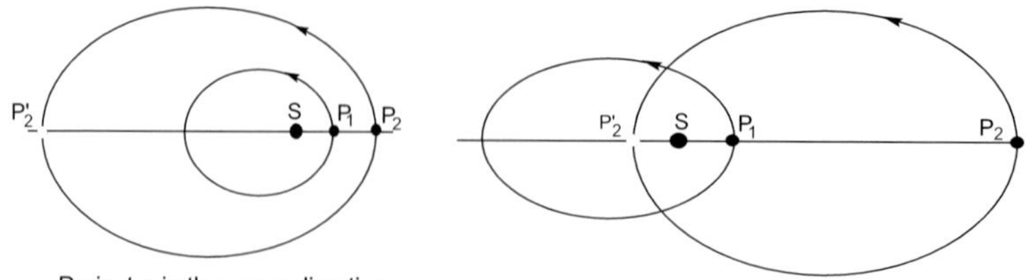

Figure 2. The four equivalent pairs of configurations at the 2/1 resonance.

TABLE I

All possible phases at $t = 0$ and $t = T/2$ for the 2/1 resonance

Type 1:	Sun – P_1(per) – P_2(per)	→	P_2 (ap) – Sun – P_1 (per)
Type 2:	Sun – P_1 (ap) – P_2 (ap)	→	P_2 (per) – Sun – P_1 (ap)
Type 3:	Sun – P_1 (per) – P_2 (ap)	→	P_2 (per) – Sun – P_1 (per)
Type 4:	Sun – P_1 (ap) – P_2 (per)	→	P_2 (ap) – Sun – P_1 (ap)

unstable). In family 1 the perihelia of the two planetary orbits are in the same direction and the two planets are both at perihelion at $t = 0$. In family 2, the perihelia are in opposite directions and at $t = 0$ P_1 is at aphelion and P_2 at perihelion (see Figure 2).

In the following we will present families of periodic orbits for the masses of the extrasolar planetary systems HD 82943, Gliese 876 and HD 160961. In order to have a better physical insight, we will not present the families of periodic orbits in the space of initial conditions, but in the space e_1e_2 of the planetary eccentricities. In order to avoid artificial discontinuities, we use the convention $e > 0$ if the planet is at aphelion and $e < 0$ if it is at perihelion.

3. The System HD 82943

3.1. FAMILIES OF PERIODIC ORBITS FOR THE MASSES OF THE SYSTEM HD 82943

We computed all the basic families of periodic orbits at the 2/1 resonance, for the masses of HD 82943, normalized to $m_0 + m_1 + m_2 = 1$. The basic results have been presented in Hadjidemetriou and Psychoyos (2003), but for reasons of completeness, and for comparison with further results on the 2/1 resonance, we will repeat them here. The basic families are given in Figure 3. The normalized masses (corresponding to the masses given by Israelian et al., 2001) are $m_0 = 0.9978$, $m_1 = 0.0008$, $m_2 = 0.0014$. Note that $m_1 < m_2$. There are three different families: *Family 1*, corresponding to the initial phase *sun – perihelion – perihelion*, *family 2*, corresponding to the initial phase *sun – aphelion – perihelion* and *family 3*, corresponding to the initial phase *sun –*

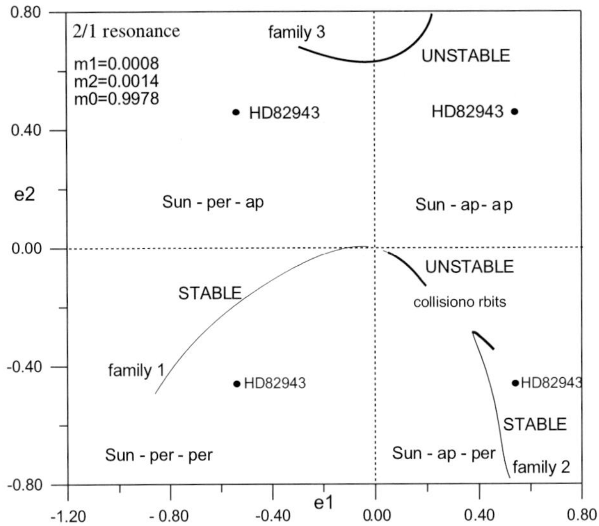

Figure 3. The symmetric families of periodic orbits at the 2/1 resonance, for the masses of HD 82943. The unstable parts are presented by a thicker line. The four possible positions of HD 82943 are also shown.

aphelion – aphelion → sun – aphelion – perihelion. In *family 1* the perihelia are in the same direction and in *family 2* they are in opposite directions. In *family 3* the perihelia are in the same direction at one end, but as e_1 decreases and passes from the value $e_1 = 0$, the perihelia shift to the opposite direction. All orbits of the family 1 are stable. The family 2 is unstable for small eccentricities. A gap appears on this family, due to close encounters between the two planets, and after the collision area the orbits on this family become stable, although the planetary orbits intersect. All orbits of family 3 are unstable. We remark that along the family 2, although all orbits have the same phase, the orbits with small planetary eccentricities are unstable, but the system is stabilized when the eccentricities are large. Four typical orbits on these families are presented in Figure 4.

The stability we mention above is the linear stability. In order to study the non linear stability, we considered two types of perturbations, that preserve the resonance: We shift the position of P_2 along its orbit to a new, non symmetric position, and also we rotate the orbit of P_2 by a certain angle. The evolution of the perturbed orbits were studied by computing the Poincaré map on the surface of section $y_2 = 0$. We found that in all cases the linearly stable orbits have a stable region, where bounded motion exists, with bounded variation of the orbital elements. The linearly unstable orbits become chaotic and in many cases one planet escapes. The details of the computations are in Hadjidemetriou and Psychoyos (2003).

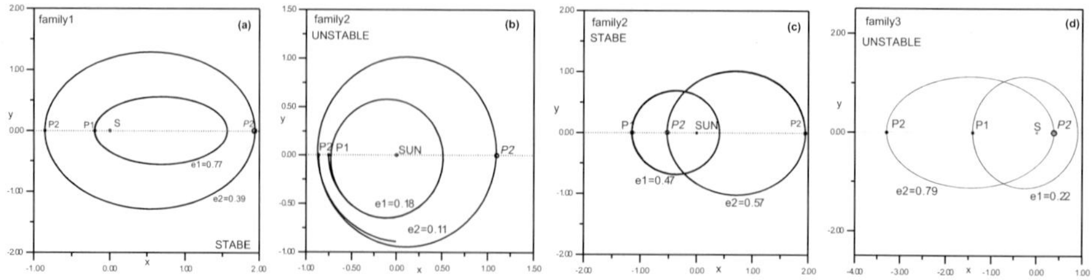

Figure 4. Four typical orbits on the families 1, 2 and 3. (a) family 1, $e_1 = 0.77, e_2 = 0.39$, stable. (b) family 2, $e_1 = 0.18, e_2 = 0.11$, unstable. (c) family 2, $e_1 = 0.47, e_2 = 0.57$, stable. (d) family 3, $e_1 = 0.22, e_2 = 0.79$, unstable.

3.2. THE PLANETARY SYSTEM HD 82943 WITH VARIABLE RATIO OF THE PLANETARY MASSES

In order to study the effect of the mass ratio of the planets on the stability of the system, we considered a planetary system with the same total mass of the planets as HD 82943, but with the masses reversed: $m_1 \rightarrow m_2$ and $m_2 \rightarrow m_1$. Now $m_1 = 0.0014, m_2 = 0.0008$, $m_1 > m_2$. The new families of periodic orbits are shown in Figure 5a. We note that the family 3 of Figure 3 no longer exists and a large part of the family 1 is now unstable. The stability of the family 2 is not affected by the inversion of the masses.

In Figure 5b we present the families 1 both for the true and the inverse masses, and we select two orbits, orbit 1 and orbit 2 on these two families. Orbit 1 is stable and orbit 2 is unstable. They both have the same eccentricity for P_2, $e_2 = 0.30$, and the eccentricities of P_1 are $e_1 = 0.06$ and 0.09 for the orbit 1 and orbit 2, respectively. In order to study the long term stability we computed the evolution of these two orbits by shifting the position of P_2 on its orbit by about 45°, by the Poincaré map on the surface of section $y_2 = 0$. The results are in Figure 6, for the evolution of the eccentricities and the semi major axes. We note that in both cases the system remains bounded, but there is an important qualitative difference between the linearly stable and unstable orbits. The variation of the eccentricities of the stable orbit 1 is very small, while it is large for the unstable orbit 2. The variation of the semimajor axes is small in both cases, but still there is a difference between the stable and the unstable orbit.

As we showed above, the whole family 1, corresponding to the phase *sun – perihelion – perihelion*, is stable when the mass of the inner planet is smaller than the mass of the outer planet, but the system is destabilized if the mass of the inner planet is larger than the mass of the outer planet. So the mass ratio m_1/m_2 plays an important role on the stability, for this phase.

We remark that a periodic orbit has two non unit pairs of eigenvalues (and one more pair which is always equal to unity, because of the existence of the energy integral). In the present case, where we have a nearly integrable

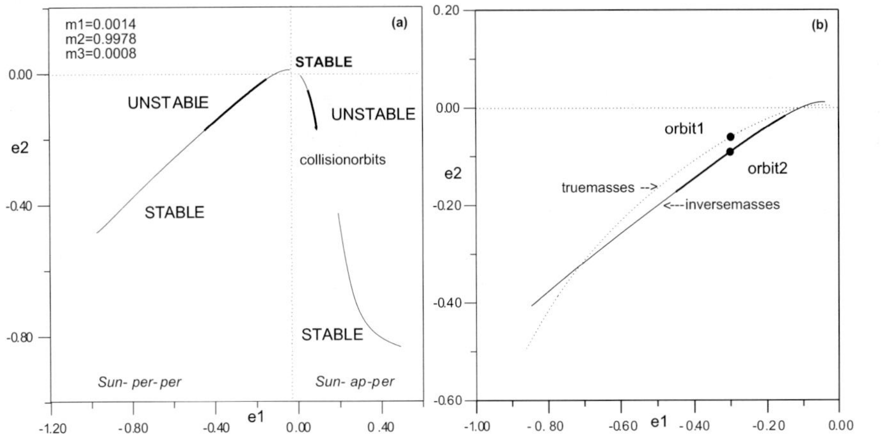

Figure 5. Families of periodic orbits for the inverse masses of HD 82943. (a) The families 1 and 2 with inverse masses. (b) The families 1 both for the true and the inverse masses, and the orbits 1 and 2 on these families, respectively. Note the unstable region on family 1.

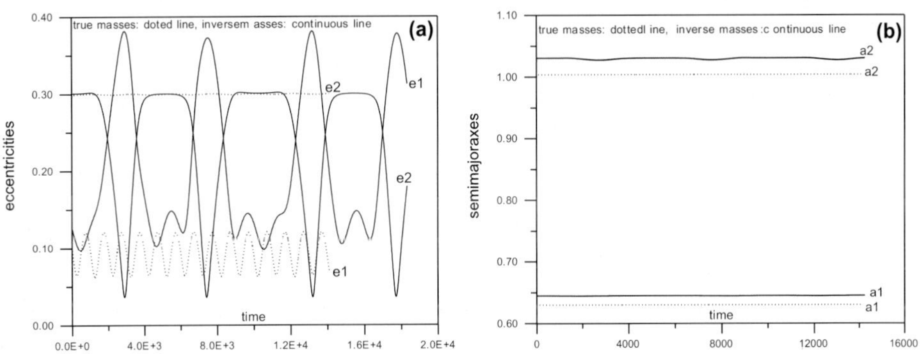

Figure 6. The evolution of the orbits 1 and 2 of the Figure 5b, when the position of the planet P_2 is shifted along its orbit by 45°. Dotted lines refer to the true masses and continuous lines to the inverse masses. (a) The evolution of the eccentricities. (b) The evolution of the semi major axes.

dynamical system (two weakly coupled Keplerian orbits), there exist two stability indices, which are close to the value -2 (corresponding to the two non unit pairs of eigenvalues). The orbit is stable if both stability indices are larger than -2 (and smaller than 2).

It turns out that along the family 1 the first stability index is in all cases larger than -2 (and close to this value), but the second stability index, which we will call b, may cross the value -2, $b < -2$, for a region of the family 1, and the system is destabilized. This is the mechanism how the system is destabilized at the 2/1 resonance.

In order to find the exact value of m_1/m_2 where instability appears for the first time on a region of the family 1, as m_1/m_2 varies (and $m_1 + m_2 = \text{constant}$),

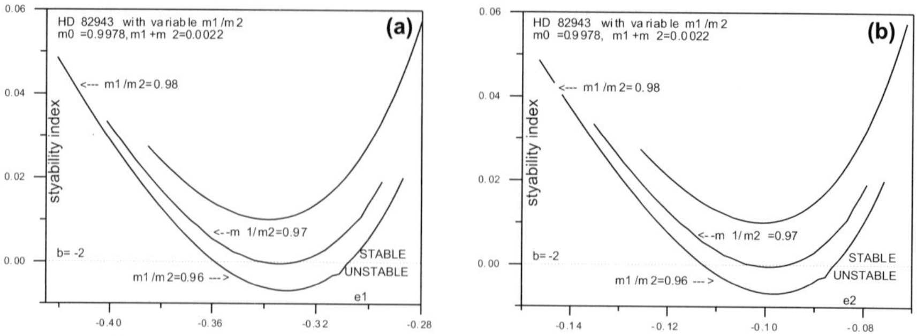

Figure 7. The stability along the *family 1* for three different mass ratios: (a) The stability vs. e_1. (b) The stability vs. e_2. Instability is generated when $m_1/m_2 < 0.97$.

we computed several families for different mass ratios and the results are shown in Figure 7a and b. In this figure we present the stability index b along the *family 1* for three different mass ratios, close to the transition value, which is equal to $m_1/m_2 = 0.97$. The x-axis in panel (a) is the eccentricity e_1, and in panel (b) the eccentricity e_2, which play the role of a parameter along the family. For the y-axis we used the value $10^3(2 + b)$, instead of the stability index b, so the transition value is zero, corresponding to $b = -2$.

From all the above we see that the system is stable, at the phase *sun – perihelion – perihelion*, if the mass ratio of the planets is $m_1/m_2 < 0.97$, provided that the mass of the sun is kept equal to $m_0 = 0.9978$ in normalized units.

In the case $m_1/m_2 > 0.97$ an unstable region appears on the family 1 (see Figure 5a), which increases as the ratio m_1/m_2 increases. At both ends of this unstable region there are critical points, with one stability index equal to $b = -2$, and we have a bifurcation of a new resonant 2/1 family of *non-symmetric* periodic orbits, from each critical point. It turns out that these two non symmetric families are connected, so in fact there is one non symmetric family of periodic orbits that starts from the first critical point and ends to the second critical point (Voyatzis and Hadjidemetriou, 2005). This is in agreement with the work of Beaugé et al. (2004), who found non symmetric periodic orbits in the 2/1 resonance.

3.3. THE EVOLUTION OF THE SYSTEM HD 82943

The elements of the system HD 82943 are (Israelinian et al., 2001): $m_0 = 1.05$ M_{SUN}, $m_1 \sin i$ =0.88 MJ, $m_2 \sin i = 1.63$ MJ, $a_1 = 0.73$ AU, $a_2 = 1.16$ AU, $T_1 = 221.6 \pm 2.7$ d, $T_2 = 444.6 \pm 8.8$ d, $e_1 = 0.54 \pm 0.05$, $e_2 = 0.41 \pm 0.08$, $\omega_1 = -$, $\omega_2 = 117.8 \pm 3.4$. This is a system very close to the 2/1 resonance.

In our numerical computations we considered four different cases, all with the same masses, semimajor axes and eccentricities, of the system HD 82943, given above, corresponding to the four different phases presented in Figure 1. These four positions are shown in Figure 2, where the families of periodic orbits are presented. It turned out that two phases, *sun – perihelion – perihelion* and *sun – aphelion – perihelion*, which are close to stable periodic orbits, are stable and the other two phases are unstable. The study was made by considering the Poincaré map on the surface of section $y_2 = 0$. This map is in the four dimensional phase space $x_1, \dot{x}_1, x_2, \dot{x}_2$ and in the above two stable phases the motion is clearly on a 4-torus. In the other two phases, close encounters between the two planets take place (line $x_1 = x_2$ in the Poincaré maps of Figure 8c and d) and the system is destabilized in a rather short time. Projections of the Poincaré map for all the above four cases are shown in Figure 8. In the first two cases, where we have bounded motion, the variation of the eccentricities is quite large, contrary to the semimajor axes whose variation has a small amplitude (Hadjidemetriou and Psychoyos, 2003).

Ji et al. (2003, submitted for publication) found, by a numerical exploration of the evolution of HD 82943 that this system is stabilized if, in addition to being in the 2/1 resonance, it is also in an apsidal resonance, with the axes of the two planets antialigned. This means, in fact, that the orbit should be close to a periodic orbit. This result coincides with our result, as presented in Figure 8b (phase: *sun – aphelion – perihelion*). Note that this phase is stable, despite the fact the the two planetary orbits intersect.

3.4. NEW ORBITAL VALUES FOR HD 82943

New values for the system HD 82943 were given recently by Mayor et al. (in preparation). The orbital elements and the values of the masses are quite different from those published before. The new values are: $m_0 = 1.05\ M_{\mathrm{SUN}}$, $m_1 \sin i = 1.85$ MJ, $m_2 \sin i = 1.84$ MJ, $a_1 = 0.75$ AU, $a_2 = 1.18$ AU, $T_1 = 219.4 \pm 0.2$ d, $T_2 = 435.1 \pm 1.4$ d, $e_1 = 0.38 \pm 0.01$, $e_2 = 0.18 \pm 0.04$, $\omega_1 = 124 \pm 3$, $\omega_2 = 237 \pm 13$. This is a system very close to the 2/1 resonance and although it is stated that the values of the eccentricities may be different from those published, we repeated the study of the evolution of this system, as in Section 3.2, using the new values. In Figure 9 we present the families of periodic orbits for the new masses of HD 82943 and we also show the position of the system corresponding to these new elements. We note that the phase *sun – perihelion – perihelion* is very close to the stable region of the family 1, but not far from the unstable region.

In Figure 10a–d we show the evolution of the system for the above four configurations, using the new elements. We note that the only stable phase is *sun – perihelion – perihelion*. Note that the phase *sun – aphelion – perihelion* is unstable, contrary to the case of Figure 8b (for the same phase), for the old

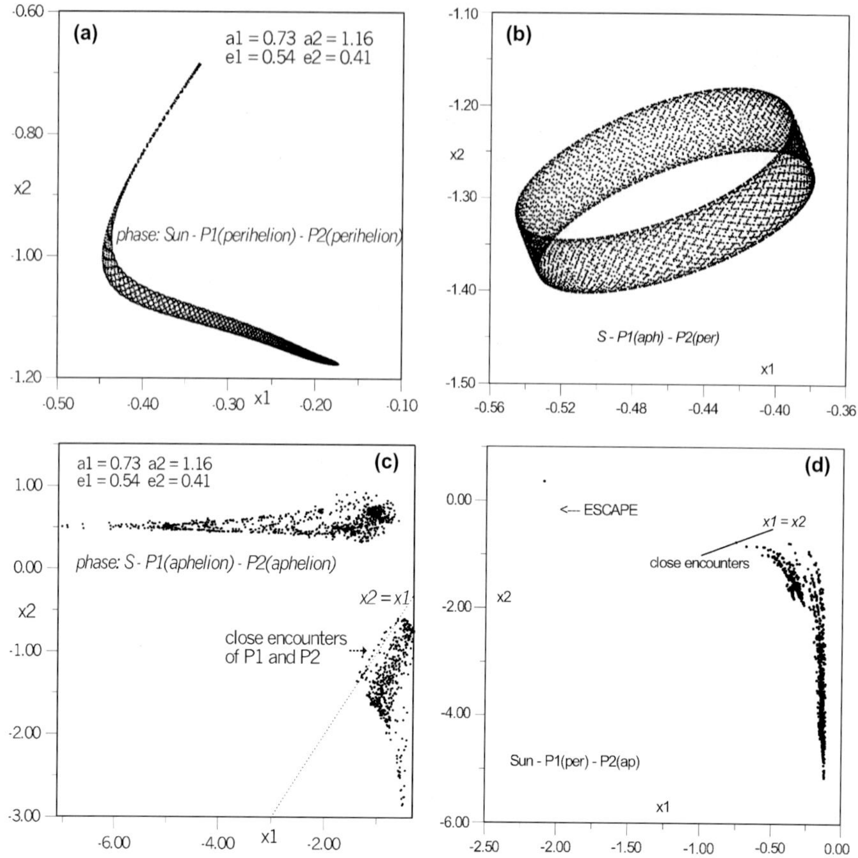

Figure 8. The evolution of HD 82943 for the four different possible phases. Projection of the Poincaré map on coordinate planes. (a) *sun – perihelion – perihelion*, (b) *sun – aphelion – perihelion*, (c) *sun – aphelion – aphelion*, (d) *sun – perihelion – aphelion*. In cases a and b the motion is bounded, on a 4-torus.

elements of HD 82943, which is stable. This is so because now this phase is close to the close encounter zone, because the values of the eccentricities are in the new case smaller, and the system is destabilized.

4. The System Gliese 876

4.1. FAMILIES OF PERIODIC ORBITS FOR THE MASSES OF GLIESE 876

We repeated the study of the families of periodic orbits for the masses of the system Gliese 876, as for the system HD 82943. In this case we also have

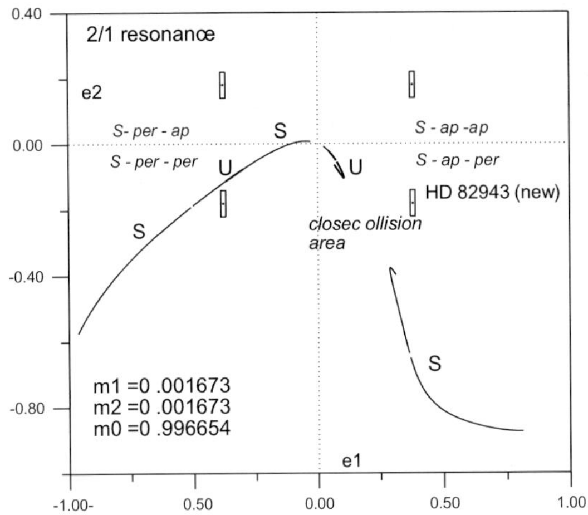

Figure 9. The symmetric families of periodic orbits at the 2/1 resonance, for the *new* masses of HD 82943. The unstable parts are presented by a thicker line. The four possible positions of HD 82943 for the new elements are also shown, together with the error bars. Compared with the figure 3, we note that a small unstable region appears on the family 1 (sun – perihelion – perihelion).

$m_1 < m_2$, but now the normalized planetary masses are larger, compared to the normalized masses of the system HD 82943. The normalized masses of Gliese 876 (corresponding to the masses given by Marcy et al., 2001) are $m_0 = 0.98275, m_1 = 0.00166, m_2 = 0.00559$. This means that the gravitational interaction between the two planets is stronger and in particular, for the phase *sun – aphelion – perihelion* and small planetary eccentricities, it dominates. As a consequence, the part of the family 2 corresponding to small eccentricities practically does not exist at all and we have in this region a much larger gap. We remind that we also had a gap for the masses of HD 82943 (Figure 3). In fact, a 2/1 resonant motion for this phase and small eccentricities cannot exist at all, because the two planets are trapped into a 1/1 resonance and the two planets revolve around the sun as a close binary, as we will show in the next section. (Such orbits do exist however, if the planetary masses are smaller, as we verified by numerical computations). In Figure 11 we present the families 1 and 2 of periodic orbits, for the masses of Gliese 876. In this figure we also indicate the position of Gliese 876, for the observed elements (*sun – perihelion – perihelion*) and for the three other possible phases.

From this figure we can find all the possible stable configurations that the system Gliese 876 could obtain. A different approach to this problem was made by Gozdjiewski et al. (2002), who made a complete stability investigation using the MEGNO technique, and they found estimations of the 2/1 mean motion resonance widths.

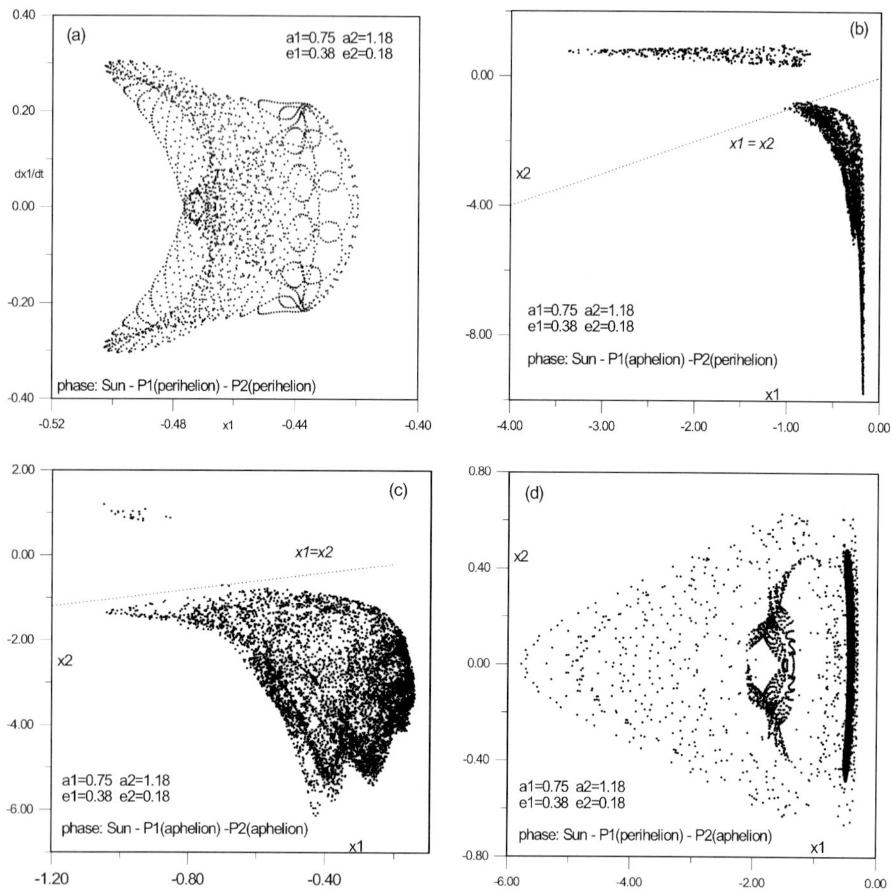

Figure 10. The evolution of HD 82943, new, for the four different possible phases. Projection of the Poincaré map on coordinate planes. (a) *sun – perihelion – perihelion*, (b) *sun – aphelion – perihelion*, (c) *sun – aphelion – aphelion*, (d) *sun – perihelion – aphelion*. The motion is bounded, on a 4-torus, only in the case (a). In the case (d) the motion is initially on a 4-torus (black region to the right), but later chaotic motion develops. In the cases (b) and (c) close encounters between P_1 and P_2 take place (points close to the line $x_1 = x_2$) and the system is destabilized.

4.2. THE EVOLUTION OF GLIESE 876

The observed system Gliese 876 corresponds to the phase where the line of apsides of both planets are almost on the same line and the perihelia in the same direction. The elements of this system are (Marcy et al., 2001; Laughlin and Chambers, 2001; Rivera and Lissauer, 2001): $m_0 = 0.32\ M_{\rm SUN}$, $m_1 \sin i = 1.89$ MJ, $m_2 \sin i = 0.56$ MJ, $a_1 = 0.21$ AU, $a_2 = 0.13$ AU, $T_1 = 61.02$ d, $T_2 = 30.1$ d, $e_1 = 0.10$, $e_2 = 0.27$, $\omega_1 = 333$, $\omega_2 = 330$. This position is indicated in Figure 11, for $e_1 = -0.10, e_2 = -0.27$. We also show

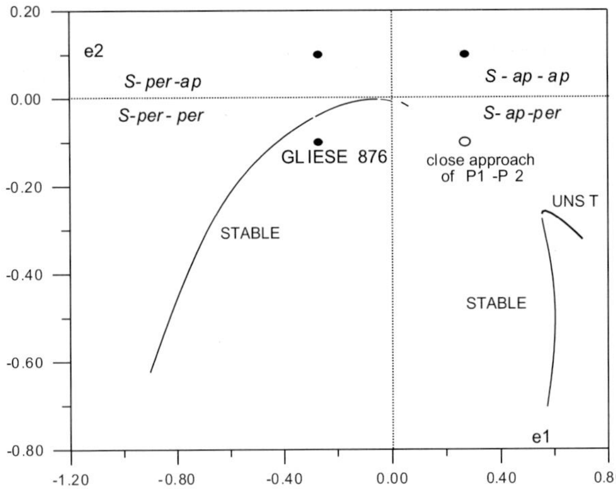

Figure 11. The families 1 and 2 of periodic orbits, for the masses of Gliese 876. Orbits with the phase *sun – aphelion – perihelion*, for small eccentricities do not exist, due to the strong gravitational interaction between the planets.

in this figure the other three possible phases with the same eccentricities. The phase $e_1 = +0.10, e_2 = -0.27$ (*sun – aphelion – perihelion*) is inside the close approach region, and is indicated in Figure 11 by an empty circle.

The evolution of the true system is studied by computing the Poincaré map on the surface of section $y_2 = 0$. In Figure 12a we show a projection of the Poincaré map on the plane $x_1, \dot{x}_1$ and in Figure 12b and c we present the evolution of the eccentricities and semimajor axes. We note that the system moves on a well defined 4-torus (a projection is in Figure 12a) and the amplitude of the variation of the eccentricities and semimajor axes is small.

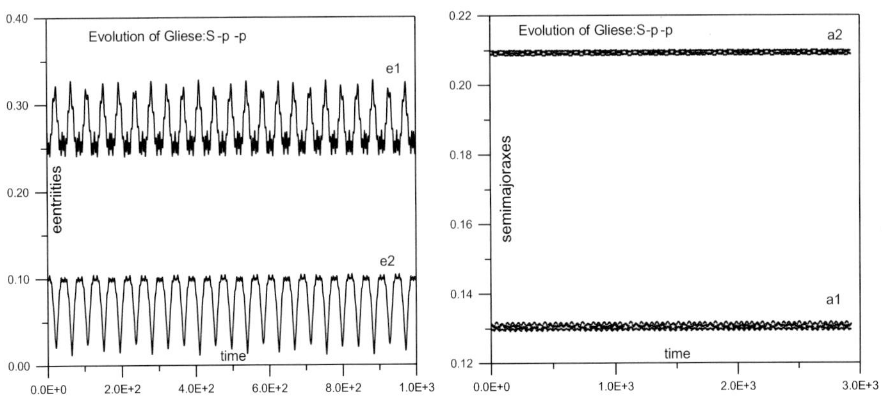

Figure 12. The Poincaré map of the true system Gliese 876, *sun – per – per*. (a) Projection on the $x_1\dot{x}_1$ plane. (b) Evolution of the eccentricities. (c) Evolution of the semimajor axes.

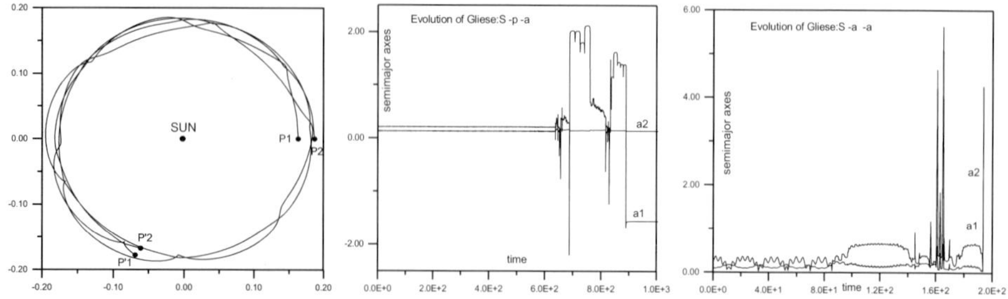

Figure 13. (a) The orbit of Gliese 876 for the phase *sun – aphelion – perihelion* inside the close approach zone. Due to the strong gravitational interaction between P_1 and P_2 the two planets are trapped into a close binary which revolves around the sun. (b) The evolution of the semimajor axes of Gliese 876 for the unstable phase *sun – aphelion – aphelion*. (c) The evolution of the semimajor axes of Gliese 876 for the unstable phase *sun – perihelion – aphelion*.

In Figure 13a we show the orbit of the system when initially the two planets are placed at the phase *sun – aphelion – perihelion*. The gravitational interaction between the two planets dominates the attraction from the sun and the system is trapped to a 1/1 resonance, forming a close binary which revolves around the sun. Note that the same phase gives a stable configuration for the system HD 82943 (Figure 8b), because the planetary eccentricities are larger, and this fact stabilizes the system, because the close encounters between the planets are avoided.

In Figure 13b and c we show the evolution of the eccentricities and semimajor axes for the phases *sun – aphelion – aphelion* and *sun – perihelion – aphelion*. In both cases the system is destabilized, due to close encounters between the two planets, as is verified from the projection of the Poincaré map on the x_1x_2 plane. Note that in both cases the system is at first trapped on a torus, but soon chaotic motion develops and the system is destabilized.

A stability analysis of the system Gliese 876 was made by Gozdziewski et al., 2002. They used the MEGNO technique and proved that the system is stable if, in addition to being at the 2/1 resonance, it is also in a $\omega_1 - \omega_2 \simeq 0$ secular resonance. This means that the system should be close to a periodic orbit, and this coincides with our results, as shown in Figure 12, which corresponds to the stable phase *sun – perihelion – perihelion*.

5. The System HD 160961

5.1. FAMILIES OF PERIODIC ORBITS FOR THE MASSES OF HD 160961

In this case we also computed the basic 2/1 resonant families of periodic orbits, corresponding to the masses of the system HD 160961, which is a system very close to the 2/1 resonance. The normalized masses (corre-

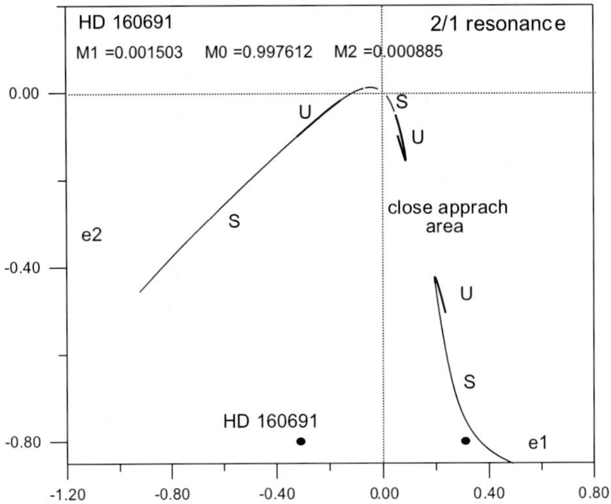

Figure 14. The families 1 and 2 of periodic orbits, for the masses of HD 160961. The position of the real system HD 160961 for two different phases are also shown.

sponding to the masses given by Jones et al., 2002) are $m_0 = 0.997612, m_1 = 0.001503, m_2 = 0.000885$. Two families, *family 1* and *family 2* are presented in Figure 14. Note that $m_1 > m_2$ and consequently an unstable region appears in family 1. In this figure we also present the position of the real system HD 160961, for two different phases: *sun – perihelion – perihelion* and *sun – aphelion – perihelion*. Two more possible phases (*sun – aphelion – perihelion* and *sun – aphelion – aphelion*) are not shown. Only one phase, namely *sun – aphelion – perihelion* is close to a periodic orbit, and it is the only stable configuration, as we will show in the following.

5.2. THE EVOLUTION OF THE SYSTEM HD 160961

The elements of the system HD 160961 are given (Jones et al., 2002): $m_0 = 1.08$, M_{SUN}, $m_1 \sin i = 1.7$ MJ, $m_2 \sin i = 1$ MJ(?), $a_1 = 1.48$ AU, $a_2 = 2.3$ AU(?), $T_1 = 637.3$ d, $T_2 = 1300$ d(?), $e_1 = 0.31$, $e_2 = 0.8$(?), Omega (deg):, $\omega_1 = 320$, $\omega_2 = 99$(?). Although the above values are not the only possible fits to the observational data and possibly they are not correct (Gozdjiewski et al., 2003), we computed the Poincaré map with these elements, for all four different phases, mentioned in the previous section. These are computed for the symmetric case. The only stable configuration is the one corresponding to the phase *sun – aphelion – perihelion* (perihelia in opposite directions) and its evolution is given in Figures 15 and 16. The motion is clearly on a 4-torus (Figure 16a) and the evolution of the eccentricities of the two planets is almost periodic. The same is true for the semimajor axes (not shown). We also note

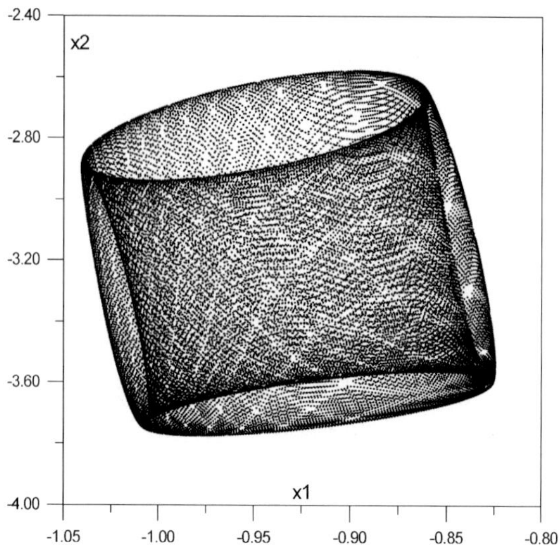

Figure 15. The Poincaré map of the system HD 160961, for the phase *sun – aphelion – perihelion*. Projection on the x_1x_2 plane. The motion is bounded, on a 4-torus.

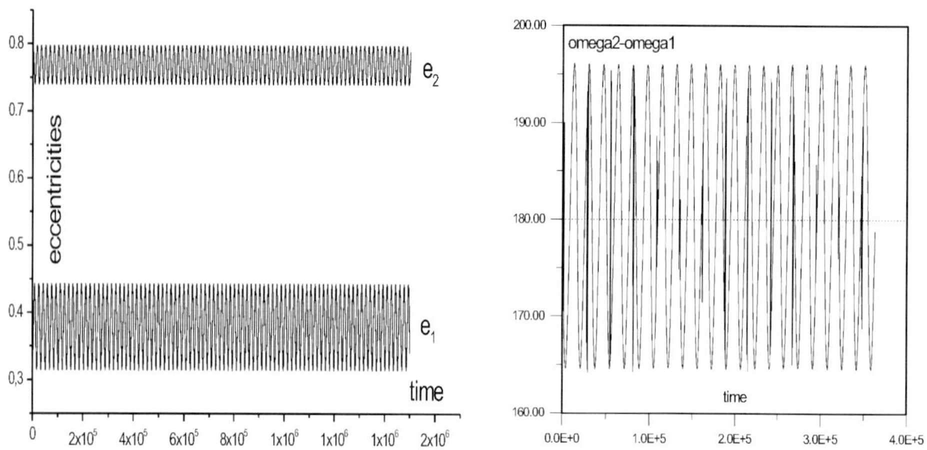

Figure 16. The evolution of the eccentricities and the angle $\omega_2 - \omega_1$, which librates around the value π, corresponding to the system HD 160961, for the phase of Figure 15.

that the angle $\omega_2 - \omega_1$ librates around 180 (Figure 16b). Note that this configuration is stable, despite the fact that the two planetary orbits intersect.

Starting from the above configuration, we extended the study of the evolution of the system, by changing the angle $\omega_2 - \omega_1$. We found that the system remains bounded up to $\omega_2 - \omega_1 = 45°$. Beyond this value, the system is chaotic.

In Figure 17 we present the evolution of the semimajor axes and the angle $\omega_2 - \omega_1$ of the system HD 160961 for the phase *sun – perihelion – aphelion*. The motion is clearly chaotic.

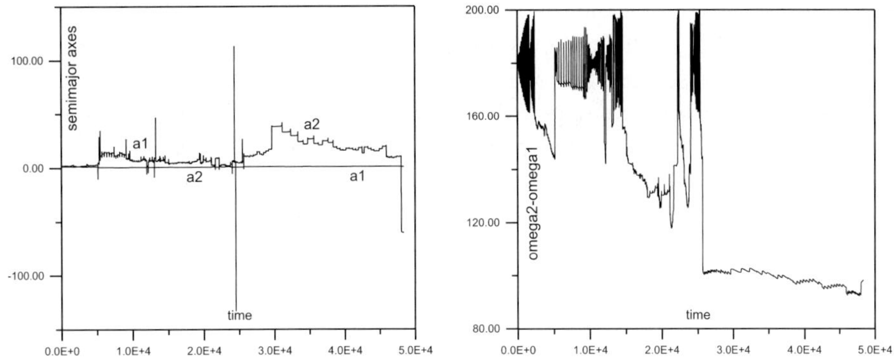

Figure 17. The evolution of the semimajor axes and the angle $\omega_2 - \omega_1$ corresponding to the system HD 160961, for the phase *sun – perihelion – aphelion*.

In Figure 18 we present the evolution of the semimajor axes of the system HD 160961, for the phases *sun – aphelion – aphelion* and *sun – perihelion – perihelion*. In this case also the motion is chaotic.

Bois et al. (2003), studied the stability of HD 160961 by the MEGNO technique. They found that the system is stable if it is in a 2/1 resonance, combined with an apsidal secular resonance, corresponding to the phase *aphelion – sun – aphelion* $\rightarrow$ *sun – aphelion – perihelion*. Their results are in complete agreement with the results that we obtained, as shown in Figures 15 and 16.

6. Conclusions

Several techniques have been used to study the dynamical evolution and the stability of an extrasolar planetary system. In the present work we present a

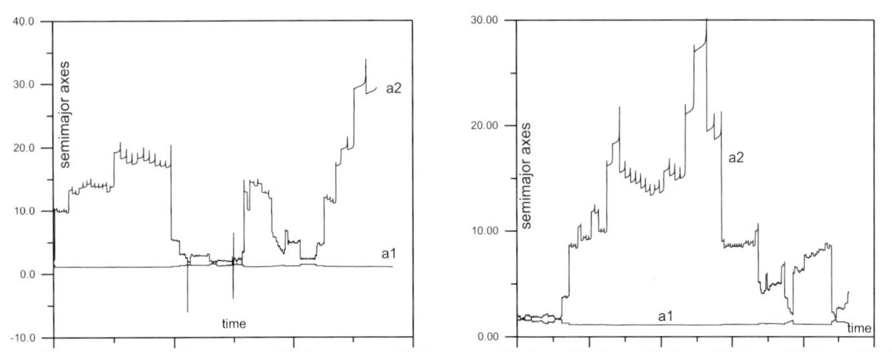

Figure 18. The evolution of the semimajor axes corresponding to the system HD 160961, for the phases *sun – aphelion – aphelion* (panel (a)) and *sun – perihelion – perihelion* (panel (b)).

method to obtain a global view of all the stable and unstable resonant configurations of a planetary system, by making a complete study of all the basic families of resonant periodic orbits.

The periodic orbits (or, equivalently, the corresponding fixed points in a Poincaré map on a surface of section) determine the topology of the phase space. In particular, close to a stable periodic orbit there exists a region where the orbit librates around the exact periodic orbit, and consequently stable, bounded, motion could be expected to exist in nature. On the contrary, the motion close to an unstable periodic orbit is chaotic, and in some cases one planet escapes. Consequently, a planetary system could not exist in nature at this region of the phase space. So, if we know the families of periodic orbits, we know in what regions of the phase space, or, equivalently, in the space of the orbital elements, a planetary system could exist in nature. These are the regions where a planetary system could be trapped in its present form, if it had followed a migration process in the past. The stable regions can also serve as a guide to select the best fits of elements in the observation of a new planetary system. These latter topics however, are beyond the scope of the present paper.

The periodic orbits that we study are in the model of the general planar three body problem and are periodic in a *rotating* frame. This means that the *relative* configuration is repeated in the inertial frame. The two planets revolve around the sun in elliptic orbits, which are perturbed because of their mutual gravitational interaction, and are in mean motion resonance. In addition, since the most important families are symmetric with respect to the rotating x-axis (Section 2), they are also in an apsidal secular resonance, which means that either $\omega_1 - \omega_2 = 0$ or $\omega_1 - \omega_2 = \pi$. This means that the apsidal lines are either aligned or antialigned.

The present study is restricted to the 2/1 resonance, and we found, in a global way, all the factors that stabilize a resonant planetary system. In this way, the study of the dynamics of all the observed 2/1 resonant planetary systems, HD 82943, Gliese 876 and HD 160961, can be made in a unified way.

We found that the phase of the two planets, that is, their position at perihelion or aphelion, when they are in the same line with the sun, plays an important role for the stability. The most stable phase is *sun – perihelion – perihelion*, for $m_1 < m_2$, which is equivalent to the phase *aphelion – sun – perihelion*. The perihelia are in this case aligned.

The value of the eccentricities of the planetary orbits is also an important stabilizing parameter, especially in the phase *sun – aphelion – perihelion*, which implies that the two perihelia are antialigned. For small eccentricities the system is chaotic, but if the eccentricities are large, the close encounters are avoided, and the system is ordered and stays bounded. This is clearly seen by comparing Figures 8b and 10b. Thus, the increase of the eccentricities plays a stabilizing role. We remark that in this latter phase the two planetary orbits intersect, but the resonance generates a phase protection mechanism, which does not allow the two planets to come close to each other.

The other two possible phases, *sun – perihelion – aphelion* and *sun – aphelion – aphelion* are always unstable. In some cases (Figure 13b and c) the system sticks on a 4-torus for a long time interval, but finally chaotic motion develops and the system is destabilized. The mechanism of generation of instability is the close approach between the two planets, as is seen in Figures 8c and d and 10c and d.

Another factor that plays an important role on the stability of a planetary system is the ratio m_1/m_2 of the planetary masses. We found that in the phase *sun – perihelion – perihelion* the system is stable if the mass of the inner planet is smaller than the mass of the outer planet and becomes unstable if $m_1 > m_2$. The change of the mass ratio however, does not affect the stability in the phase *sun – aphelion – perihelion*.

The above results, applied to the observed systems HD 82943, HD 160961 and Gliese 876, showed that the phase *sun – perihelion – perihelion* is stable in all cases. The real systems can be considered as perturbed periodic orbits corresponding to this resonance. The phase *sun – aphelion – perihelion* is stable only in the case HD 82943, old data, because the eccentricities are large. In all other systems, including HD 82943 with the new data, the system is unstable, because the eccentricities are small.

References

Beauge, C. and Michtchenko, T.: 2003, 'Modelling the high-eccentricity planetary three-body problem. Application to the GJ876 planetary system', *MNRAS* **341**, 760–770.

Beaugé, C., Ferraz Mello, S. and Michtchenko, T. A.: 2003, Extrasolar Planets in Mean-Motion Resonance: Apses Alignment and Asymmetric Stationary Solutions. *Astrophys. J.* **593**, 1124–1133.

Bois, E., Kiseleva-Eggleton, L., Rambaux, N. and Pilat-Lohinger, E.: 2003, 'Conditions of dynamical stability for the HD 160961 planetary system', *ApJ.* **598**, 1312–1320.

Callegari, N., Michtchenko, T. and Ferraz-Mello, S.: 2002, 'Dynamics of two planets in the 2:1 and 3:2 mean motion resonances', In: *34th DPS Meeting, BAAS*, 34, 30.01.

Ferraz-Mello, S.: 2002, 'Tidal acceleration, rotation and apses allignment in resonant extra-solar planetary systems', In: *34th DPS Meeting, BAAS*, **34**, 30.01.

Ferraz-Mello, S., Beaugé, C. and Michtchenko, T.: 2003, 'Evolution of migrating planet pairs in Resonance', *Celest. Mech. Dynam. Astron.* **87**, 99–112.

Ford, E. B., Havlickova, M. and Rasio, F. A.: 2001, 'Dynamical instabilities in extrasolar planetary systems containing two giant planets', *Icarus* **150**, 303–313.

Gozdziewski, K. and Maciejewski, A.: 2001, 'Dynamical analysis of the orbital parameters of the HD 82943 planetary system', *ApJ. Lett.* **563**, L81–L85.

Gozdziewski, K., Bois, E. and Maciejewski, A.: 2002, 'Global dynamics of the Gliese 876 planetary system', *MNRAS* **332**, 839–855.

Gozdziewski, K., Konacki, M. and Maciejewski, A.: 2003, 'Where is the second planet in the HD 160691 planetary system?', *ApJ.* **594**, 1019–1032.

Hadjidemetriou, J. D.: 1975, 'The continuation of periodic orbits from the restricted to the general three-body problem', *Cel. Mech.* **12**, 155–174.

Hadjidemetriou, J. D.: 2002, 'Resonant periodic motion and the stability of extrasolar planetary systems', *Cel. Mech. Dynam. Astron.* **83**, 141–154.

Hadjidemetriou, J. D. and Psychoyos, D.: 2003, 'Dynamics of extrasolar planetary systems: 2/1 resonant motion', In: G. Contopoulos and N. Voglis (eds.), *Lecture Notes in Physics: Galaxies and Chaos*, Vol. 626, Springer-Verlag, pp. 412–432.

Israelian, G., Santos, N., Mayor, M. and Rebolo, R.: 2001, 'Evidence for planet engulfment by the star HD 82943', *Nature* **411**, 163–166.

Ji, J., Li, G. and Liu, L.: 2002, 'The dynamical simulations of the planets orbiting GJ 876', *ApJ.* **572**, 1041–1047.

Ji, J., Liu, L., Kinoshita, H., Zhou, J., Nakai, H. and Li, G.: (2003a), 'The Librating Companions in HD 37124, HD 12661, HD 82943, 47 Uma and GJ 876: alignment or anti-alignment?', *ApJ. Lett.* **591**, L57–L60.

Ji, J., Kinoshita, H., Liu, L., Li, G. and Nakai, H.: 2003b, The antialignment of the HD 82943 system, Cel. Mech. Dynam. Astron.

Jones, H., Butler, P., Marcy, G., Tinney, C., Penny, A., McCarthy, C. and Carter, B.: 2002, 'Extra-solar planets around HD 196050, HD 216437 and HD 160691', *MNRAS* **337**, 1170–1178.

Kinoshita, H. and Nakai, H.: 2001a, 'Stability of the GJ 876 Planetary System', *PASJ* **53**, L.25–L.26.

Kiseleva-Eggleton, L., Bois, E., Rambaux, N. and Dvorak, R.: 2002, 'Global dynamics and stability limits for planetary systems around HD 12661, HD 38529, HD 37124 and HD 160691', *ApJ. Lett.* **578**, L145–L148.

Laughlin, G. and Chambers, J.: 2001, 'Short-term dynamical interactions among extrasolar planets', *Ap.J.* **551**, L109–L113.

Lee, M. and Peale, S.: 2001, 'Evolution of the GJ876 planets into the 2:1 orbital resonance', *BAAS* **34**, 3.03.

Lee, M. and Peale, S.: 2002, 'Dynamics and origin of the 2:1 orbital resonance of the GJ 876 planets', *Ap.J.* **567**, 596–609.

Malhotra, R.: 2002a, 'A dynamical mechanism for establishing apsidal resonance', *ApJ. Lett.* **575**, L33–L36.

Malhotra, R.: 2002b, 'Eccentricity excitation and apsidal alignment in exo-planetary systems', In: *34th DPS Meeting, BAAS,* **34**, 42.05.

Marcy, G., Butler, P., Fischer, D., Vogt, S., Lissauer, J. and Rivera, E.: 2001, 'A pair of resonant planets orbiting GJ 876', *ApJ.* **556**, 296–301.

Mayor, M., Udry, S., Naef, D., Pepe, F., Queloz, D., Santos, N. C. and Burnet, M.: 2004, *The CORALIE survey for southern extra-solar planets. XII. Orbital solutions for 16 extra-solar planets discovered with CORALIE*, A& A **415**, 391–402.

Pars, L. S.: 1965, 'A Treatise on Analytical Dynamics', *Heinemmann*, p. 159.

Rivera, E. J. and Lissauer, J. J.: 2001, 'Dynamical models of the resonant pair of planets orbiting the star GJ 876', *Ap. J.* **558**, 392–402.

Voyatzis, G. and Hadjidemetriou, J. D.: 2005, 'Symmetric and asymmetric librations in planetary and satellite systems at the 2/1 resonance', submitted to *Celest. Mech. Dynam. Astron.* (submitted)

Celestial Mechanics and Dynamical Astronomy (2005) 92:157–188
DOI 10.1007/s10569-004-0029-0

GLOBAL DYNAMICS IN SELF-CONSISTENT MODELS OF ELLIPTICAL GALAXIES

C. KALAPOTHARAKOS[1,2] and N. VOGLIS[1]
[1]*Academy of Athens, Research Center for Astronomy, 4 Soranou Efesiou, GR-11527, Athens, Greece, e-mail: ckalapot@cc.uoa.gr; nvogl@cc.uoa.gr*
[2]*Department of Physics, Section of Astrophysics, University of Athens, Panepistimiopoli, Athens GR 15783, Greece*

(Received: 7 September 2004; revised: 9 December 2004; accepted: 31 December 2004)

Abstract. We compare two different N-body models simulating elliptical galaxies. Namely, the first model is a non-rotating triaxial N-body equilibrium model with smooth center, called SC model. The second model, called CM model, is derived from the SC by inserting a central mass in it, so that all possible differences between the two models are due to the effect of the central mass. The central mass is assumed to be mainly due to a massive central black hole of mass about 1% of the total mass of the galaxy. By using the fundamental frequency analysis, the two systems are thoroughly investigated as regards the types of orbits described either by test particles, or by the real particles of the systems at all the energy levels. A comparison between the orbits of test particles and the orbits of real particles at various energy levels is made on the rotation number plane. We find that extensive stable regions of phase space, detected by test particles remain empty, i.e. these regions are not occupied by real particles, while many real particles move in unstable regions of phase space describing chaotic orbits. We run self-consistently the two models for more than a Hubble time. During this run, in spite of the noise due to small variations of the potential, the SC model maintains (within a small uncertainly) the number of particles moving on orbits of each particular type. In contrast, the CM model is unstable, due to the large amount of mass in chaotic motion caused by the central mass. This system undergoes a secular evolution towards an equilibrium state. During this evolution it is gradually self-organized by converting chaotic orbits to ordered orbits mainly of the short axis tube type approaching an oblate spheroidal equilibrium. This is clearly demonstrated in terms of the fundamental frequencies of the orbits on the rotation number plane and the time evolution of the triaxiality index.

Key words: chaos, frequency maps, N-body simulations, galaxies: elliptical, galaxies: evolution

1. Introduction

Self-consistent galactic models realized by N-body simulations are efficient tools in studying the structure and evolution of galaxies. We use these tools in investigating the orbital structure, the evolution and the stability of models simulating elliptical galaxies.

If we assume that elliptical galaxies are non-rotating triaxial equilibrium configurations with a smooth center, i.e. the density near the center is flat, the potential in this region tends to be harmonic. The most favored type of ordered orbits in this case is the type of box orbits, i.e. orbits composed of three oscillations along the corresponding principal axes of the anisotropic system. They satisfy three integrals of motion, which are approximately the energies along the three axes of the box. The pericenters of these orbits can be arbitrarily close to the center.

In real galaxies, in principle, the central region may not be smooth. In many cases the density near the center, instead of being flat, presents a cuspy density profile (Ferrarese et al., 1994; Lauer et al., 1995; Gebhardt et al., 1996; Faber et al., 1997). Furthermore, the dominant opinion today is that the center is occupied by a black hole. Recent investigations provide accumulated evidence that massive central black holes in galaxies must be quite common (Kormendy and Richstone, 1995; Kormendy et al., 1997, 1998; van der Marel et al., 1997, 1998; Magorrian et al., 1998; Cretton and van den Bosch 1999; Gebhardt et al., 2000). The massive central black holes may be surrounded by other types of mass, e.g. gas, clouds, stars, etc.

The dynamical effect of a central force field (produced by a massive central black hole and its direct environment) on the orbits of stars in a galaxy is important. Orbits passing near the center are appreciably deflected by the central force field that dominates locally. As a consequence, the integrals of motion of ordered orbits are destroyed when these orbits approach the center. Thus, the orbits become chaotic (Gerhard and Binney 1985; Merritt and Fridman 1996).

This effect may have important consequences on the stability of galaxies. This field of research is very active in recent years (e.g. Merritt and Fridman, 1996; Merritt and Valluri, 1996; Fridman and Merritt, 1997; Valluri and Merritt, 1988; Merritt and Quinlan, 1998; Siopis, 1999; Siopis and Kandrup, 2000; Kandrup and Sideris, 2002; Poon and Merritt, 2002, 2004; Kandrup and Siopis, 2003; Kalapotharakos et al 2004; Voglis and Kalapotharakos, 2005)

In order to investigate further this problem, we concentrate our study on a comparison between the following two models:

(i) a triaxial N-body model with smooth center in virial equilibrium, hereafter called Smooth Center (SC) model and
(ii) a model, called Central Mass (CM) model, that is created from the SC model by inserting a central mass. The central mass is assumed to be mainly due to a black hole of size comparable to the largest masses of black holes in the centers of galaxies estimated from observational data.

The comparison regards the various types of orbits described by the individual particles in each system, as they are mapped on the space of their fundamental frequencies. It is obvious that all the differences between the two models must be attributed to the presence of the central mass.

As we will see below, this comparison reveals important features of the dynamical role a central black hole can play in real galaxies, such as the destabilization of orbits of certain types (e.g. box orbits) or enhancement of the stability of other types (e.g. short axis tube orbits). We show that redistribution of different types of orbits can take place, so that the system develops a secular evolution towards a new equilibrium configuration.

In Section 2 we describe the main features of the two models (Sections 2.1 and 2.2, respectively). In Section 3 we give a short description of the method developed earlier (Voglis et al., 2002) to distinguish the mass in chaotic motion from the mass in ordered motion in N-body models and explain how this method is applied in our problem. We use also the frequency analysis method to make a similar distinction and we compare the results of the two methods. In Section 4, we use the frequency analysis method to identify all the types of orbits of test particles in the SC model and compare them with the orbits of the real particles of the system on the rotation number plane i.e. the plane of the two fundamental frequencies of oscillation of the orbits along the middle and the longest axis, respectively, divided by the fundamental frequency along the shortest axis. The types of orbits that appear at various energy levels are examined. In Section 5 the same method is applied at a particular snapshot of the CM model to identify the various types of orbits on the rotation number plane and compare with the results of Section 4. In section 6 the secular evolution of the CM model is examined by representing the orbits on the rotation number plane at different snapshots of the system. Our conclusions are summarized in Section 7.

2. Description of the Models

2.1. THE SC MODEL

The derivation of the smooth center model is described in detail in previous papers (Contopoulos et al., 2002; Voglis et al., 2002). In brief, this N-body model is the outcome of dissipationless collapse and relaxation derived from quiet cosmological initial conditions. In the relaxed equilibrium configuration this model is characterized by a triaxial bar. Let the principal axes, shortest, intermediate and longest axes be respectively along the x,y,z cartesian axes. The maximum ellipticity of the equidensity contours (on the x–z plane) in the inner parts is of E7 type, while it tends to an E5 type in the outer parts (on the same plane).

The number of particles used is $N \simeq 1.5 \times 10^5$. The evolution is followed by a N-body code based on the conservative technique Allen et al., 1990 In this code the Poisson equation is solved in terms of an orthogonal set of basis functions (Spherical Bessel functions and Spherical Harmonics). The potential provided by this technique is smooth. Particles do not see their neighbors. They move under the global field created by their distribution at any time. In the central region this potential is roughly harmonic. In its typical version the code gives the global potential expanded in 120 terms. In spherical coordinates (r, θ, ϕ) the radial expansion is extended up to 20 terms and the angular expansions (θ, ϕ) up to quadrupole terms. Among the 120 terms, 20 are monopole terms (depending only on r), 20 quadrupole terms (depending on r and θ) and 80 triaxial terms (depending on the three coordinates). This potential can be written as

$$\begin{aligned} V(r,\theta,\phi) = {} & \underbrace{\sum_{l=0}^{19} B_{l00} V_{l00}(r)}_{\text{20 monopole terms}} + \underbrace{\sum_{l=0}^{19} B_{l20} V_{l20}(r,\theta)}_{\text{20 quadrupole terms}} \\ & \underbrace{+ \sum_{l=0}^{19} B_{l21} V_{l21}(r,\theta)\cos\phi + \sum_{l=0}^{19} C_{l21} V_{l21}(r,\theta)\sin\phi + \sum_{l=0}^{19} B_{l22} V_{l20}(r,\theta)\cos 2\phi + \sum_{l=0}^{19} C_{l22} V_{l20}(r,\theta)\sin 2\phi}_{\text{80 triaxial terms}} \end{aligned} \tag{1}$$

where $B_{l00}, B_{l20}, B_{l21}, C_{l21}, B_{l22}, C_{l22}$ (with $l = 0, \ldots, 19$) are the coefficients of the expansion.

The adopted scaling units are as follows. The unit of energy is defined so that the value of the potential at the center (deepest value) of the SC model is equal to $V(0) = -100$. The length unit is the half mass radius R_h of this system. The time unit is the half mass crossing time defined as $T_{\rm hmct} = (2R_h^3/GM_g)^{1/2}$ in terms of $R_{\rm h}$ and the mass $M_{\rm g}$ of the galaxy.

It is useful to express some important time scales in this unit. The period $T_{\rm cp}$ of a circular orbit of radius R_h is $T_{\rm cp} = 2\pi T_{\rm hmct}$. The relation between the radial period $T_{\rm r}$ of an orbit (i.e. the time needed for a star to go from the pericenter to the apocenter and back to the pericenter of its orbit) and the circular period $T_{\rm cp}$ of the same energy can be written as $T_{\rm cp} = fT_{\rm r}$ where the factor f is $f = 1$ in the Kelperian potential, and $f = 2$ in the harmonic potential. In the potential of our N-body system f ranges between these two values. For orbits of the N-body systems with energies near the value of the potential at the half mass radius, the values of f are close to 2.

The radial period of the orbits with energy equal to the mean value of the potential at the half mass radius, denoted by $T_{\rm hmr}$, is $T_{\rm hmr} \approx 3T_{\rm hmct}$. A star in circular motion, in a typical galaxy, describes about 50 cycles in a Hubble time $t_{\rm Hub}$. Therefore, a Hubble time can be written as

$$t_{\rm Hub} \approx 50T_{\rm cp} = 50fT_{\rm hmr} \approx 300T_{\rm hmct}. \tag{2}$$

In a self-consistent run the coefficients of the potential (1) are re-evaluated at regular small time steps $\Delta t = 0.025T_{\rm hmct}$. The variations of the coefficients in this model are quite small. They have only a small noise depending on the number N of particles, i.e. of the order of $1/\sqrt{N} \lesssim 1\%$.

If the coefficients of the potential are fixed at a given snapshot we can write an autonomous 3D Hamiltonian

$$H = \frac{\dot{r}^2}{2} + \frac{L_\phi^2}{2r^2 \sin\theta^2} + \frac{L_\theta^2}{2r^2} + V(r, \theta, \phi). \tag{3}$$

Using this Hamiltonian we can study the phase space of the system. We can find the various types of orbits using test particles and compare with the orbits of the real particles of the system in this Hamiltonian.

Any possible changes in the types of orbits, due to the variations of the coefficients in time, can be checked if we fix the values of the coefficients at a different snapshot and repeat the study of the Hamiltonian (3). No serious changes are detected in the phase space of this model even after run times that considerably exceed the Hubble time.

2.2. THE CM MODEL

We choose a particular snapshot of the SC model (after a run time of $t_{\rm SC} = 100T_{\rm hmct}$ in the relaxed configuration). At this snapshot the time is reset to $t = 0$ and a central mass of a given size is inserted abruptly to create the CM model. So all the differences between the two models can be attributed to the presence of the central mass of this size.

We consider the case when the time of growth of the central black hole is much less than the galaxy's life time. In this case the main dynamical effect of the black hole depends mainly on its final size. For this reason we simplify the model by neglecting the transient period of growth of the black hole and we insert abruptly the central mass.

For the central mass $M_{\rm cm}$, we have adopted the central potential

$$V_{\rm cm}(r) = \frac{GM_{\rm cm}}{a}\left[\arctan\left(\frac{r}{a}\right) - \frac{\pi}{2}\right], \tag{4}$$

where a is a softening length given by

$$a = 0.05 \frac{M_{\rm cm}}{M_{\rm g}} R_{\rm g}, \tag{5}$$

where R_g is the radius of the galaxy. Notice that the force derived from this potential is

$$F_{\rm cm}(r) = -\frac{\rm G\ M_{cm}}{r^2 + a^2}, \tag{6}$$

i.e. it is of a Keplerian nature only for orbits with pericenters much larger than a. The force at the center of this model tends to a finite constant. Stars with pericenters below the softening length a are not deflected strongly by the CM. This softening length does not significantly alter the global behavior of the system, since the number of orbits with pericenters less than a is small.

Notice that the density profile derived from the above potential is

$$\rho_{\rm cm}(r) = \frac{\rm G\ M_{cm} a^2}{2\pi r(r^2 + a^2)^2}, \tag{7}$$

i.e. this model for $r < a$ gives an r^{-1} cuspy density profile.

The relative size m of the central mass with respect to the mass of the galaxy, i.e.

$$m = M_{\rm cm}/M_{\rm g} \tag{8}$$

is an important parameter. In the CM model we use the value of $m = 0.01$. Such a value is of the same order of magnitude as the largest black hole masses estimated by observational data (Magorrian et al., 1998; Merritt and Ferrarese, 2001).

In this model, particles move under the superposition of the potential (4) and the potential due to their distribution in space at any time given by (1) in which the coefficients are re-evaluated at every small time step ($\Delta t = 0.025 T_{\rm hmct}$).

Notice that after the introduction of the central mass the radial profile of the density of particles near the center does not remain so flat as before. It develops a weekly cuspy profile.

In order to study the types of orbits which are consistent with a given snapshot of the total potential we fix the coefficients in (1) at this particular snapshot and we run the particles in the autonomous Hamiltonian

$$H = \frac{\dot{r}^2}{2} + \frac{L_\phi^2}{2r^2 \sin^2\theta} + \frac{L_\theta^2}{2r^2} + V(r, \theta, \phi) + V_{\rm cm}(r). \tag{9}$$

As we will see in Section 6 this model presents a secular evolution that can be studied by applying (9) at many successive snapshots of the potential.

3. Distinction of Particles in Ordered Motion from Particles in Chaotic Motion

As a consequence of the non-integrability of the potential (1) a considerable part of the orbits in the two systems are chaotic.

In Voglis et al. (2002) we have developed a method to identify particles of N-body models moving in chaotic orbits provided that their mean logarithmic divergence rate exceeds a certain threshold.

In brief this method of distinguishing the chaotic from ordered orbits is based on the combination of two different tools. First, is the calculation of the mean exponential divergence in a given period of time, between the orbit of a real particle in the system and a neighboring orbit. The mean exponential divergence is measured in units of the inverse radial period T_{rj} of the particular orbit j. We call it Specific Finite Time Lyapunov Characteristic Number (S-FT-LCN) and we use simply the symbol L_{j} for it. The orbit j is integrated in a 3D autonomous Hamiltonian, as the Hamiltonians (3) or (9) when they are applied to a particular snapshot of the potential (1).

Second, we use the Alignment Index $\mathrm{AI_j}$ of the orbit j, i.e. the magnitude of the sum or the difference of two initially arbitrary deviation vectors of this orbit, normalized to unity at every Δt.

If the orbit is chaotic, the two deviation vectors tend exponentially to be parallel or antiparallel (depending on their initial orientation). In this case, the alignment index $\mathrm{AI_j}$, expressed by the difference of the deviation vectors if they are parallel, or by the sum if they are antiparallel, i.e. the smaller value between the two, tends exponentially to zero (Voglis et al., 1998, 1999; Skokos 2001).

If the orbit is ordered, or if it is so weakly chaotic that the chaotic character can not appear during the available integration time, the two deviation vectors oscillate around each other. Thus, the corresponding values of $\mathrm{AI_j}$ do not tend exponentially to zero, but instead they maintain finite values, in principle close to unity. (It is very improbable to become less than 10^{-3}).

The run time t_{j} of the orbit j is $t_{\mathrm{j}} = N_{\mathrm{rp}} T_{\mathrm{rj}}$, where N_{rp} is a number of radial periods common for all the orbits. In the case of ordered orbits, or chaotic orbits that temporarily behave as ordered, the values of L_{j} decrease on the average as N_{rp}^{-1} and they are almost independent of the orbit. As N_{rp} increases, the L_{j} of chaotic orbits stops decreasing and it is saturated at a roughly constant value $L_{\mathrm{j}} > N_{\mathrm{rp}}^{-1}$. Chaotic orbits are characterized by such constant values of L_{j}. The detection limit of chaotic orbits is determined by the maximum number of radial periods $N_{\mathrm{rp-max}}$. For the adopted value $N_{\mathrm{rp-max}} = 1200$ the minimum L_{j} of the detected chaotic orbits is the threshold of $\simeq 10^{-2.8}$.

The index L_j gives the rate of exponential divergence per radial period. The advantage of using L_j is the fact that the values of L_j measure the chaotic character independently of the characteristic time scale of the orbit. If the value of L_j is smaller than the threshold the orbit is either ordered, or close to ordered, obeying two integrals of motion (at least approximately) beyond its binding energy. In this sense the index L_j is a measure of the departure from integrability, or in other words, L_j is a measure of the complexity of the phase space of the orbit.

The Lyapunov time derived from L_j, i.e. the time L_j^{-1} is expressed in radial periods T_{rj} and corresponds to different real time scales depending on the size of T_{rj}. Such a Lyapunov time, L_j^{-1}, cannot be directly compared with a fixed time scale, for example, the Hubble time. Furthermore, the threshold value of this Lyapunov time, derived from the threshold value of $L_j = 10^{-2.8}$, is smaller for orbits of short radial periods (innermost orbits, i.e. orbits of low binding energies) than for orbits of long radial periods.

In Voglis et al. (2002), we have also used the CU-FT-LCN, of simply L_{cuj}, i.e., the Finite Time LCN measured in a common time unit for all the orbits, equal to the inverse radial period T_{hmr} of an orbit with energy equal to the mean value of the potential at the half mass radius. The index L_{cuj} gives the total exponential divergence in time T_{hmr}, independently of the number of radial periods during this time. This quantity measures the combined action of two effects, i.e the chaotic character of the orbit and the characteristic time scale of the orbit. For the innermost orbits (where $T_{rj} < T_{hmr}$), L_{cuj} takes larger values than the corresponding values of L_j, while for the outermost orbits it takes smaller values than L_j. If the value of L_{cuj} is smaller than the adopted threshold there is no guarantee that the orbit obeys other integrals of motion beyond its binding energy. Very small values of L_{cuj} can be derived for long period orbits, even if they wander in a large chaotic sea. Therefore, L_{cuj} is not an objective measure of the departure from integrability, because its values are biased by the dynamical time scales of the orbits. For this reason we prefer using L_j instead L_{cuj} for distinguishing the orbits.

However, L_{cuj} is also a useful quantity because the Lyapunov times L_{cuj}^{-1}, expressed in common units T_{hmr}, can be directly compared to the Hubble time. Furthermore, a particular threshold value of L_{cuj} defines a common threshold value of the Lyapunov times for all the orbits independently of their characteristic time scales.

As a consequence of the above properties of the two indices, the number of chaotic orbits, that can be detected by using a threshold value in L_{cuj}, is not exactly the same with those detected by using the same threshold value in L_j. The threshold of L_{cuj} gives more chaotic orbits in the inner parts, while the threshold of L_j gives more chaotic orbits in the outer parts. However, provided that the adopted thresholds are small enough, so that the minimum

Lyapunov times considerably exceed the Hubble time, the difference between the two sets of the detected chaotic orbits regards only very weakly chaotic orbits that might equally well be considered as ordered orbits.

Large values of L_{j} are correlated with large variability per radial period of the actions of the orbits and hence with larger rates of chaotic diffusion (Lichtenberg and Lieberman 1992). Chaotic diffusion, however, is important in a given time provided that this time is longer than the Lyapunov time. The values of L_j can give us the ability of chaotic diffusion in a given number of radial periods, but not in a fixed period of time. Thus, after obtaining the separation of chaotic orbits in terms of L_j, we find the chaotic orbits that can in principle develop chaotic diffusion in a Hubble time, by calculating their L_{cuj} in terms of their L_j as

$$L_{\mathrm{cuj}} = L_{\mathrm{j}} \frac{T_{\mathrm{hmr}}}{T_{\mathrm{rj}}}. \tag{10}$$

As mentioned above, the Lyapunov time L_{cuj}^{-1}, expressed in units of T_{hmr}, can be directly compared with the Hubble time $t_{\mathrm{Hub}} \simeq 100 T_{\mathrm{hmr}}$. Therefore, a necessary condition for a remarkable chaotic diffusion in one Hubble time is $L_{\mathrm{cuj}}^{-1} < t_{\mathrm{Hub}}$, or $L_{\mathrm{cuj}} \gtrsim 10^{-2} T_{\mathrm{hmr}}^{-1}$.

The threshold $L_{\mathrm{j}} = 10^{-2.8}$ is adopted for two reasons. First, because the rate of appearance of new chaotic orbits slows down remarkably by increasing further $N_{\mathrm{rp-max}}$ (Kalapotharakos et al., 2004). Second, all the orbits with $L_{\mathrm{cuj}} > 10^{-2}$ have values of L_{j} well above the threshold $10^{-2.8}$, so they are certainly included in the set of chaotic orbits. This is seen in Figure 1, where all the detected chaotic orbits of the SC model are plotted on the $\log L_{\mathrm{j}} - \log L_{\mathrm{cuj}}$ plane.

Chaotic orbits in a 3D Hamiltonian system can be either fully chaotic, or partially chaotic. The fully chaotic orbits are characterized by two positive Lyapunov numbers, the first (or the maximal) LCN corresponding to L_{j} and a second LCN, smaller than L_{j}. The partially chaotic orbits are characterized by only one positive LCN corresponding to L_{j} and they obey one more integral of motion beyond their binding energy. Due to this integral, partially chaotic orbits are confined in space, and are not allowed to diffuse to all directions.

In our distinction between ordered and chaotic orbits in terms of L_{j}, partially and fully chaotic orbits are bunched together. Most of the chaotic orbits with large values of L_{j} are fully chaotic, e.g. orbits with $\log L_{\mathrm{j}} \gtrsim -2$ in Figure 1. Those with smaller values of L_{j} can be characterized as partially chaotic because their second Lyapunov number is either very small or zero. They obey an approximate integral of motion so they are confined to particular regions of phase space.

Muzzio and Mosquera (2004), Muzzio et al. (2005), in their study of the mass in chaotic motion in self-consistent models of stellar systems, have used a

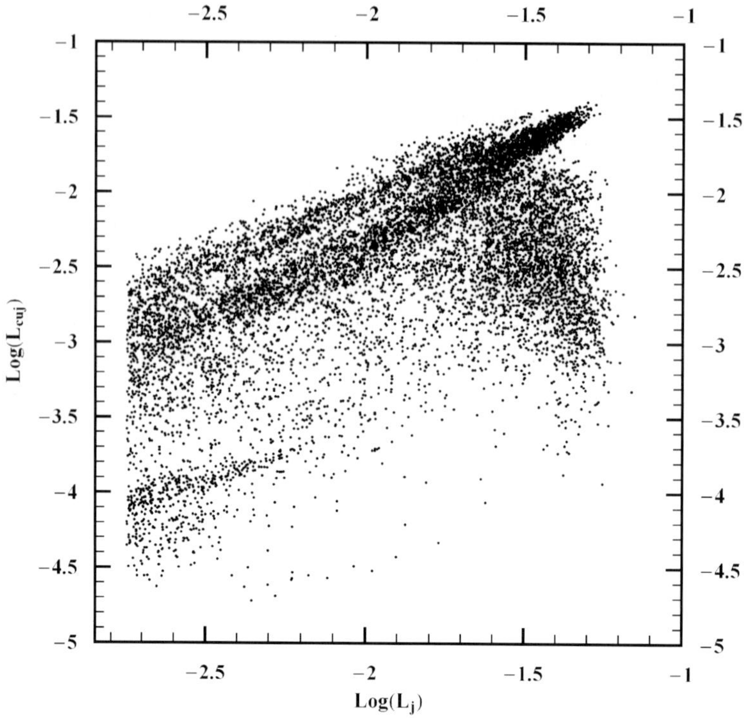

Figure 1. The chaotic orbits in the SC model detected by L_j above the threshold of $10^{-2.8}$ are plotted on the plane of $\log Lj - \log L_{\mathrm{cuj}}$. Very small values of L_{cuj} are only due to the orbits of long radial periods. The orbits with $\log L_{\mathrm{cuj}} > -2$ that can develop chaotic diffusion in a Hubble are well above the threshold of $\log L_j = -2.8$. Partially chaotic orbits are found mainly among the orbits with low values of log L_j.

common unit for the evaluation of the Lyapunov numbers (as in our L_{cuj}). They evaluate both the first and the second FT-LCNs of the orbits above a threshold. They show that the orbits of their systems that are characterized as partially chaotic, but also those of the fully chaotic orbits with low values of the first FT-LCN, have different spatial distributions compared with the fully chaotic orbits with high values of the first FT-LCN. Fully chaotic orbits with low FT-LCN and partially chaotic orbits are confined in a prolate-triaxial distribution, while fully chaotic orbits with high FT-LCN show an almost spherical distribution. Fully chaotic orbits with low FT-LCN behave as partially chaotic for a long time. Therefore, a distinction between the chaotic orbits having a confined spatial distribution and those having an almost spherical distribution can be obtained by using the values of the first FT-LCN (small or large, respectively). It is not quite necessary to identify separately the partially and the fully chaotic orbits, although such a distinction can give more details about the spatial distribution of the orbits with low FT-LCN.

As it is well known, an alternative method to distinguish between ordered and chaotic motion is the frequency analysis method introduced by Laskar (1990). (See also Laskar et al., 1992; Laskar, 1993a, 1993b; Papaphillipou and Laskar, 1996, 1998). This method exploits the fact that the frequencies of ordered orbits remain constant in time (within the available accuracy). The frequencies of chaotic orbits, on the other hand, develop a considerable variability. The distinction is obtained by evaluating accurately the fundamental frequencies of the orbits.

For this purpose we use the FMFT code by Sidlichovsky and Nesvorny (1997) to find the fundamental frequencies ν_x, ν_y, ν_z of the oscillations of the orbits along the x, y, z axes. Notice that these frequencies are measured in units of the inverse radial period of each particular orbit, as in the case of the L_{j}.

In order to check the variability of the fundamental frequencies of the orbits in a particular snapshot of the potential (1), we select two spans for every orbit with the same number of radial periods $\Delta N_{\mathrm{rp}} = 300$, common for all the orbits. The run time of the orbit j is:

span(1): from $t = 0$ to $t = 300T_{\mathrm{rj}}$

span(2): from $t = 900T_{\mathrm{rj}}$ to $t = 1200T_{\mathrm{rj}}$.

The fundamental frequencies along the shortest axis x (the most unstable direction in the system) are always larger and more sensitive than the fundamental frequencies along the other two axes. For this reason we use the quantity

$$\delta\omega_j = \frac{|\nu_{xj}^{(1)} - \nu_{xj}^{(2)}|}{\nu_{xj}^{(1)}} \tag{11}$$

as a measure of the variability of frequencies. The upper index gives the corresponding span from which the frequency is evaluated.

In terms of the three indices, L_{j}, AI_j, $\delta\omega_{\mathrm{j}}$, we can obtain a very clear distinction between ordered and chaotic orbits. This can be seen in Figure 2a,b, where all the orbits of the two systems are plotted in the three-dimensional space $(\log AI_j, \log L_{\mathrm{j}}, \log \delta\omega_j)$. For most of the orbits ($\approx 95\%$) the index $\log AI_j$ works as a switch taking values either in the region from -3 to 0 for ordered orbits, or very small values, less than -10 for chaotic orbits. The intermediate values of $\log AI_j$ correspond to a relatively small number ($\approx 5\%$) of weakly chaotic orbits of the lane joining the two groups.

Using the above method we have found that in the SC model about 32% of the total mass moves in chaotic orbits and about 68% in ordered orbits. As we have checked, these fractions remain almost the same at various snapshots selected at very different times even larger than the Hubble time.

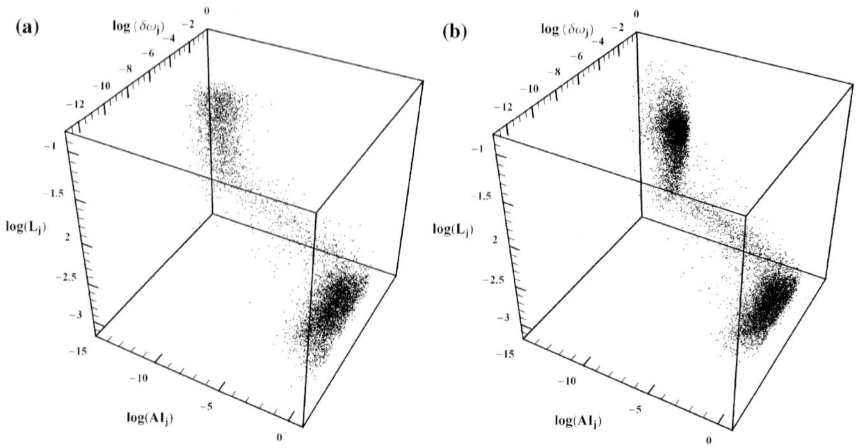

Figure 2. Distinction between ordered and chaotic orbits using the three indices $\log \mathrm{AI_j}, \log L_\mathrm{j}, \log \delta\omega_\mathrm{j}$ for the two models SC (a) and CM (b). The sensitivity of $\log AI_j$ is considerable larger than the sensitivity of the other two indices.

The identities of the particles in the groups are also well preserved, in spite of the noise due to the variations of the coefficients of the potential. Only a fraction of 1–2% of the total number of particles jump from one group to the other and vice versa.

It is worth noticing that Muzzio et al. (2005), in their smooth center model of an elliptical galaxy, have found a fraction of mass in chaotic motion of about 53%, i.e. considerably larger than in our study. Partially, this diffrence is due to an overestimation (with respect to our estimation) of chaos (especially in the low energy orbits) because of the use of a smaller threshold in their FT-LCN similar to L_{cuj}. Another source of this difference is the fact that our SC model is more flat near the center than their model. However, their results are in agreement with our conclusion that, even in models of stellar systems with smooth centers, the fraction of mass in chaotic motion is not negligible. It is at least of the order of a few tens per cent, although in most of the chaotic orbits the corresponding Lyapunov numbers may be relatively small.

Applying our method to various snapshots of the CM model we find that at the snapshot of $t = 0$ the fraction of mass in chaotic motion is $\approx 80\%$. At the snapshot of $t = 150$ the chaotic mass is reduced to $\approx 58\%$, while at the snapshot of $t = 300$ this fraction falls down to $\approx 22\%$ and remains permanently (i.e. for a time much longer than a Hubble time) on this level, indicating that the system has reached a well established equilibrium. The reduction of the fraction of mass in chaotic motion from one snapshot to another is due to a process of self-organization occurring in the CM model that is discussed in Section 6.

4. The Frequency Map of the SC Model

We define the rotation numbers ν_y/ν_x and ν_z/ν_x and we mark the frequencies on the rotation number plane (e.g. as in Papaphillipou and Laskar, 1998; Wachlin and Ferraz-Mello, 1998; Valluri and Merritt, 1998; Holley-Bocklmann et al., 2002). Such diagrams are usually called frequency maps. Typical examples of orbits found in the SC model are shown in Figure 3a–f. The letters a,b,... label the rows of this figure. Every row gives the three projections of the same orbit on the planes $x-y, x-z, y-z$, respectively. Following a standard terminology (de Zeeuw, 1985; Statler, 1987), the types of ordered orbits found in this model can be classified in terms of their rotation numbers as follows:

(a) Box orbits. They are combinations of three oscillations along the x, y, z axes. They can pass arbitrarily close to the center. Their rotation numbers ν_y/ν_x and ν_z/ν_x are irrational. The major axis of most of the box orbits is in average along the longest axis z of the system but it performs two librational motions around the axis z with constant irrational frequencies either on the $y-z$ or on the $x-z$ plane, filling in this way a region having the shape of a parallelepiped, with curved surfaces, (Figure 3a). This type of orbits is compatible only with triaxial equidensity surfaces.

(b) Inner Long Axis Tube (ILAT) orbits. Such orbits fill tube-like regions with maximum size along the longest axis z (Figure 3b). The hole of the tube appears along the z axis due to the component of the angular momentum of the orbit along this axis. For a number of ILAT orbits this hole may be small, so that they can pass quite close to the center. In this case the ILAT orbits resemble the box orbits. For this reason we call them box-like orbits. The ILAT orbits are characterized by the resonant value of the rotation number $\nu_y/\nu_x = 1$. Due to this resonance the major axis of the ILAT orbits describes a precession around the z axis. This type of orbits is compatible with triaxial and prolate equidensity surfaces.

(c) Outer Long Axis Tube (OLAT) orbits. This is another type of resonant orbits with the same rotation number $\nu_y/\nu_x = 1$ as the ILAT orbits and with the hole of the tube being again along the longest axis z (Figure 3c). The main difference of the OLAT from the ILAT orbits is that the major axis of the OLAT orbits oscillates up and down the $x-y$ plane instead of precessing around the z axis. Furthermore, the angular momentum along the z axis and the radius of the hole are larger compared with the corresponding quantities of the ILAT orbits. Thus, they do not approach so close to the center. This type of orbits is compatible with triaxial and prolate equidensity surfaces.

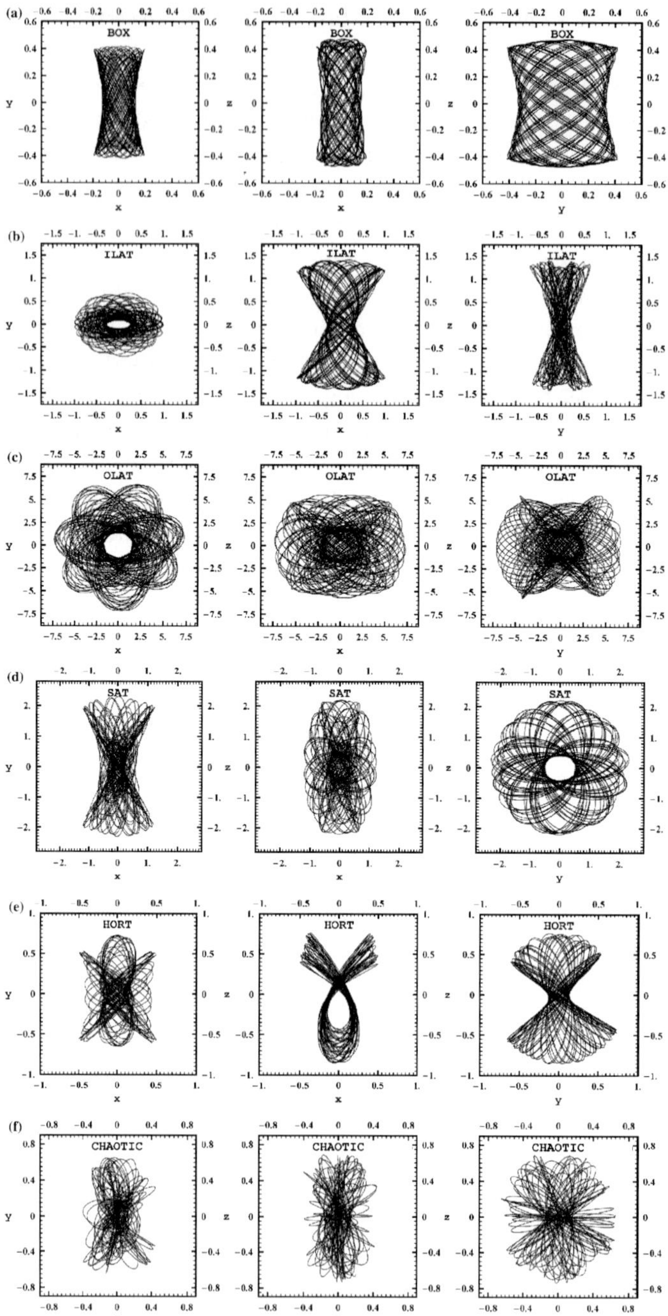

Figure 3. Examples of the various types of orbits of the particles in the SC model. Each row gives a type of orbit projected on the three planes x–y, x–z, y–z, respectively. The types of orbits are box (a), ILAT (b), OLAT (c), SAT (d), HORT (e), chaotic (f).

(d) *Short Axis Tube (SAT) orbits.* They form tubes surrounding the shortest axis x (having their hole along this axis, Figure 3d). They are characterized by equal rotation numbers, i.e. $\nu_y/\nu_x = \nu_z/\nu_x$. Their major axis oscillates up and down the plane of the intermediate-longest axes ($y-z$ plane). Due to their angular momentum along the shortest axis they can not approach the center. They support the flatness of the system along the shortest axis. This type of orbits is compatible with triaxial and oblate equidensity surfaces.

(e) *Higher Order Resonant Tube (HORT) orbits.* These are tube-like ordered orbits corresponding to various resonances of higher order. For example the ratio of frequencies of the orbit in Figure 3e is approximately $\nu_z : \nu_x = 2 : 3$. In general, HORT orbits do no approach very near the center. This type of orbits is compatible only with triaxial equidensity surfaces.

(f) *Chaotic orbits* have been discussed in the previous section. Their rotation numbers vary irregularly in time. An example of chaotic orbit is shown in Figure 3f. This type of orbits is compatible with all the kinds of equidensity surface.

As it is well known the resonant orbits satisfy the condition of Diofantos

$$n_x\nu_x + n_y\nu_y + n_z\nu_z = 0, \qquad (12)$$

where n_x, n_y and n_z are integers. If one of the three frequencies is irrational to the others the above equation is satisfied if the corresponding coefficient is zero. This is the case of the ILAT, OLAT and SAT orbits. For ILAT and OLAT orbits the vector (n_x, n_y, n_z) that satisfies (12) is $(1, -1, 0)$, while for the SAT orbits this vector is $(0, 1, -1)$.

For the HORT orbit shown in Figure 3e $(n_x, n_y, n_z) = (-2, 0, 3)$. There are of course HORT orbits with non-zero integers in all the three components of the vector (n_x, n_y, n_z) satisfying (12). For box orbits equation (12) is satisfied only for $(n_x, n_y, n_z) = (0, 0, 0)$.

As we will see by the end of this section a good number of the detected chaotic orbits follow the geometry of the above types of ordered orbits and they satisfy resonance conditions with a good accuracy for not negligible periods of time.

At high energy levels all the types of orbits are present but this is not the case at low energy levels. In order to study how the various types of orbits appear along different energy levels at a given snapshot of the potential, we run test particles scanning all the available phase space of the corresponding Hamiltonian at an energy level h. We find the fundamental frequencies of their orbits and construct the frequency map on the rotation number plane. This is repeated for a series of energy levels. The same process is repeated for the orbits of the real particles of the system with energies inside a small window (of width ± 1) around h.

In Figure 4$a, \alpha, b, \beta, c, \gamma, d, \delta, e, \varepsilon, f, \zeta$ the rotation number planes of the orbits of test particles and the orbits of real particles are shown for comparison in pairs at energy levels

$$h = -80, -70, -60, -50, -40, -10,$$

respectively, as they result from the snapshot of the potential (1) at $t = 0$. The figures of the left column (labelled by Latin letters) give the orbits of test particles (covering all the available phase space at the corresponding energy). The figures of the right column (labelled by Greek letters) give the orbits of the real particles of the system (with energies inside a small window around the same value of energy).

We first describe Figure 4f ($h = -10$) where all the types of orbits are present as indicated in this figure.

The ILAT and OLAT orbits are located on two separated segments along a vertical line $v_y/v_x = 1$. The segment of ILAT is for $v_z/v_x \lesssim 0.7$. The almost empty part of this line near the value $v_z/v_x = 0.7$ corresponds to a zone of instability that separates these two families of orbits.

In Figure 4f the SAT orbits are located along the diagonal $v_y/v_x = v_z/v_x$. The box and the HORT orbits occupy a wide range on this plane with $v_z/v_x \lesssim 0.7$ and $v_y/v_x < 1$. The box orbits form an almost continuous distribution in this area, interrupted by straight lines corresponding to HORT orbits. The points that are irregularly dispersed in between the above types of orbits correspond to chaotic orbits.

As we can see in Figure 4ζ the real particles of the system at the same energy level occupy all the types of orbits of test particles and they are distributed in a rather similar way. However, this is not always the case especially for low energies as we will see below.

In the deepest parts of the potential only box and HORT orbits are allowed. For example, for $h = -80$ (Figure 4a) only box orbits appear (and a small number of HORT orbits that are not clearly seen in this figure). The same happens also for smaller energy levels. It is remarkable that on the rotation number plane the area of box orbits at low energy levels has a shape of a curvilinear triangle as in Figure 4a. The three corners of the triangle correspond to harmonic oscillations along the longest axis z (top corner), the intermediate axis y (left corner) and the shortest axis x (right corner), with small amplitude of oscillations along the other axes in each case.

In Figure 4α the real particles of the system occupy preferably the left side of the triangle than the right side, which is almost empty. This is because of the self-consistency. The existence of stable orbits of test particles does not necessarily imply that these orbits can be occupied by real particles of the system. The orbits of real particles are further restricted by the self-consis-

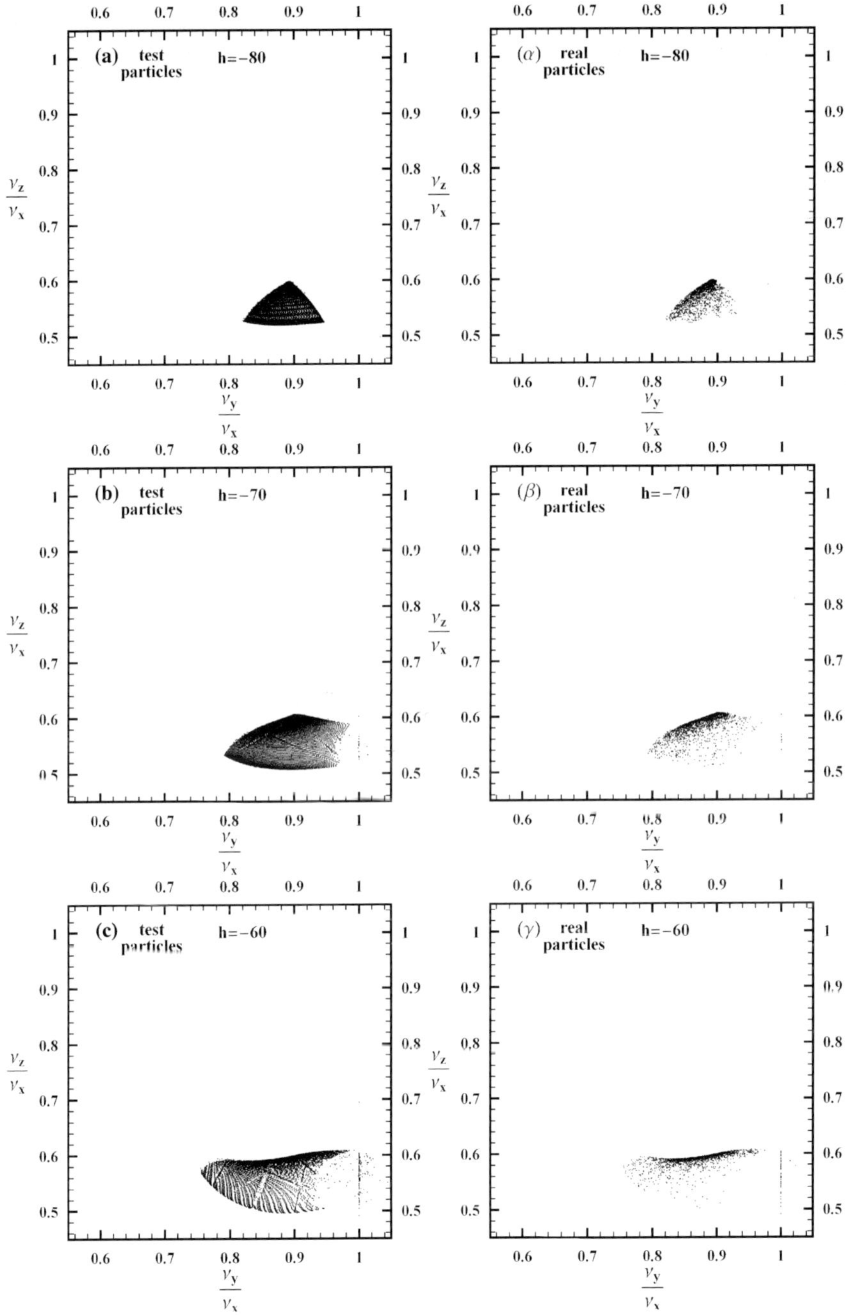

Figure 4. The rotation number plane at the energy levels $h = -80, -70, -60, -50, -40, -10$ for the snapshot at $t = 0$ of the potential (1) of the SC model. In the figures of the left column (a), (b), (c), (d), (e), (f) orbits of test particles are plotted scanning all the available phase space at every energy level. In the right column α, β, γ, δ, ϵ, ζ the orbits of the real particles of the system with energies in a small window ± 1 around the corresponding energy are plotted for comparison. The real particles of the system do not necessarily occupy all the available phase space, because they are further restricted by self-consistency.

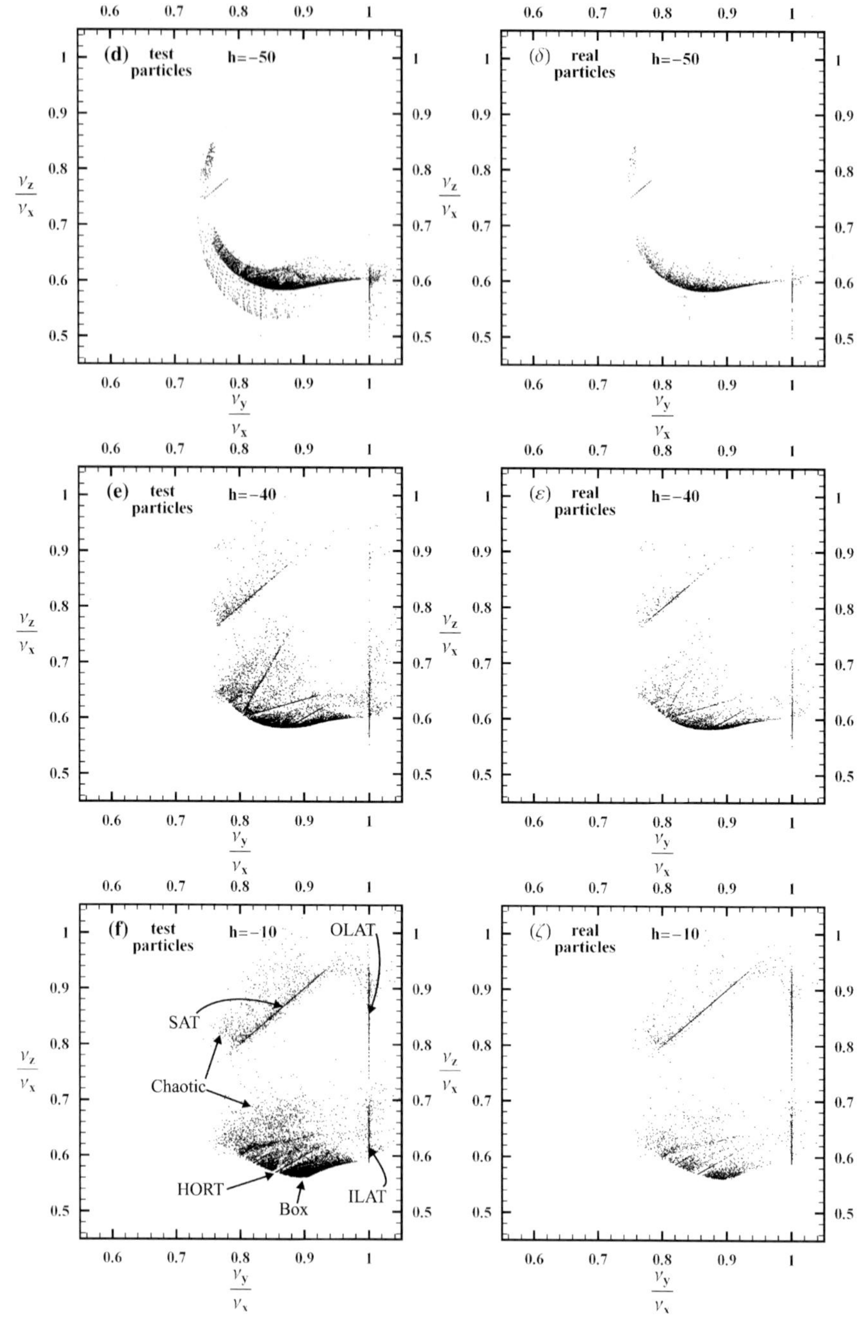

Figure 4. (Continued)

tency to follow a distribution that supports the equidensity surfaces having their axis along y larger than their axis along x, otherwise the system could not be in equilibium.

In Figure 4b (for $h = -70$) the area of box orbits is expanded covering a wider range of values of v_y/v_x. The upper angle of the curvilinear triangle flattens. This flattening corresponds to the increase of the amplitude of the libration of the major axis of the box orbits either on the $y-z$ plane (expansion to the left), or on the $x-z$ plane (expansion to the right). The right corner of the triangle has been destroyed because the oscillation along the shortest axis has already been unstable. A weakly populated chaotic layer appears in this area separating the box orbits from the ILAT orbits, that start appearing at about this level of energy (a few points along the line $v_y/v_x = 1$). The left corner still exists, indicating that the corresponding oscillations (along the y axis) are still stable. This corner is expanded to smaller values of v_y/v_x.

As shown in Figure 4β the orbits of real particles again do not cover all the area of stable orbits of test particles. They are preferably located near the left-upper limit of this area because of self-consistency reasons, as explained above.

At the energy level $h = -60$ (Figure 4c) the area of box orbits is further expanded along smaller values of v_y/v_x, due to further increasing of the libration of the major axis on the $y-z$ plane. The area of box orbits takes the shape of a lane rather than a triangle. The upper limits of v_z/v_x form a curved line with an inverse curvature than the previous curvature (from convex it becomes concave). This concavity becomes larger and larger as the energy increases. The box orbits of real particles lie preferably along this line (Figure 4γ).

For $h = -50$ (Figure 4d) the left end of the lane turns abruptly upwards. The box orbits at this end become unstable and the SAT orbits appear as a new family (short straight line with slope 1.0 in Figure 4d). As the energy increases the area of the SAT orbits increases. The orbits of real particles follow a roughly similar distribution (Figure 4δ).

At the energy level of about $h = -40$ (Figure 4e,ε) the family of OLAT orbits appears (a few points along the line $v_y/v_x = 1$ with $v_z/v_x \gtrsim 0.75$. For higher energy levels all the types of orbits appear as in Figure 4f,ζ.

As described in Section 3, we have found that the ordered and the chaotic components are respectively, $\approx 68\%$ and $\approx 32\%$ in the SC model. Furthermore, these fractions as well as the identities of particles in each type of motion are almost permanent.

All the ordered orbits (of all the energy levels) of the real particles in this system are plotted on the rotation number plane in Figure 5a, while the chaotic orbits are plotted in Figure 5b.

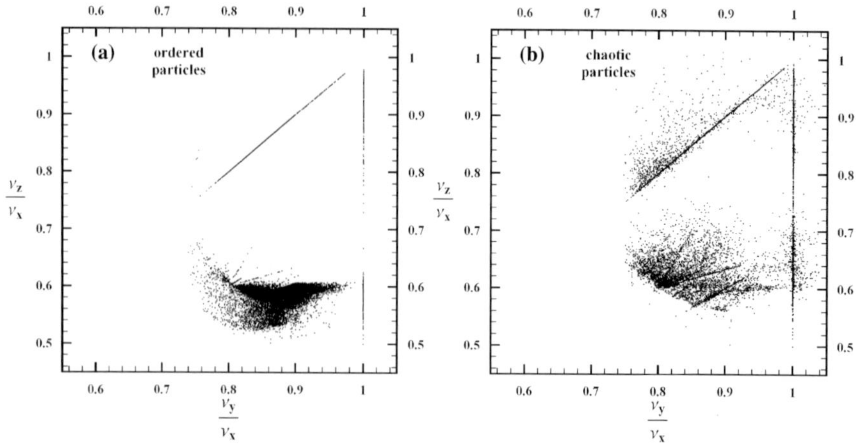

Figure 5. (a) All the ordered orbits of the SC model plotted on the rotation number plane independently of their energy. (b) As in (a) but for the chaotic orbits. A number of chaotic orbits are temporarily trapped near the resonant lines.

In Figure 5a we see clearly the groups of SAT, ILAT, OLAT and HORT orbits on sharp straight lines. The box orbits form a not very sharp area due to the superposition of the box orbits of different energy levels. Note that the majority of the ordered orbits, i.e. $\sim 91\%$ of the ordered component, moves in box (and HORT) orbits.

In Figure 5b we see that many chaotic orbits are located along resonant lines. These orbits diffuse mainly along the resonant lines. As long as they are trapped on a resonant line they maintain a very small or zero second FT-LCN, i.e. they are partially chaotic. We collect those of the chaotic orbits of Figure 5b that are projected upon the most important resonant lines and we plot them separately in Figure 6a. We plot also their distribution along the $\log L_j$ axis in Figure 6b by a dashed line together with the distribution of all the chaotic orbits (solid line). We see that among the more weakly chaotic orbits (i.e. $\log L_j \lesssim -2$) there is an almost constant difference between the values of the two curves at any given L_j. This means that there are many weakly or partially chaotic orbits located outside the resonant lines. These are mainly orbits resembling the box type. Among the more strongly chaotic orbits the ratio of the values of the dashed curve to the values of the solid curve decreases to zero as $\log L_j$ increases. The probability of more strongly chaotic orbits to be trapped along resonant lines is small.

As we have checked, due to the small variations of the potential, some of the resonant orbits can escape the resonant lines, but other chaotic orbits are trapped on the same resonant lines, so that the system preserves a dynamical equilibrium, in which the number of orbits in every type remains remarkably constant in time.

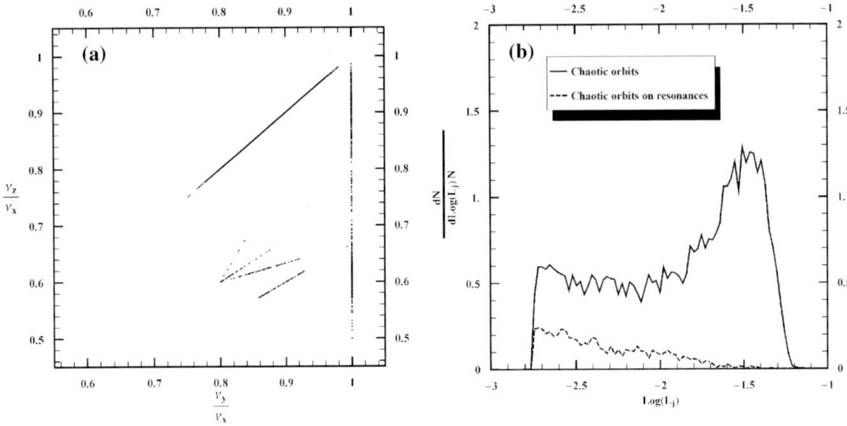

Figure 6. (a) A separate plot of those chaotic orbits in Figure 5b that occupy the most important resonant lines in the SC model. (b) Distribution of all the chaotic orbits (solid line) along the log L_j axis and the distribution of the orbits in (a) (dashed line). For small values of L_j there is a roughly constant difference between the two curves, due mainly to weakly chaotic orbits resembling box orbits. For large values of L_j the relative occurrence of chaotic resonant orbits decreases considerably.

5. The Frequency Map of the CM Model

A similar analysis is performed in the CM model using the Hamiltonian (9) at a snapshot of the self-consistent potential corresponding to a run time $t = 150$ after the central mass is inserted.

At this snapshot, as mentioned in Section 3, we found that the fraction of mass in ordered motion ($L_j < 10^{-2.8}$) and in chaotic motion ($L_j > 10^{-2.8}$) are respectively, $\approx 42\%$ and $\approx 58\%$. All the particles of these two components, plotted on the rotation number plane, are shown in Figure 7a,b, respectively.

In Figure 7a we see that the particles of the ordered component belong to resonant orbits only, forming quite sharp straight lines. No box orbits appear at all. Also the ILAT type is almost absent. The majority ($\sim 95\%$) of the ordered orbits in this figure belong to SAT orbits, but there are some OLAT and HORT orbits as well.

On the other hand, the chaotic component, shown in Figure 7b, contains a large number of orbits in the area of box orbits. These orbits come from the box orbits in SC model that became chaotic by passing near the cental mass.

As in Figure 5b, the resonant lines in Figure 7b are occupied by a good number of weakly or partially chaotic orbits, diffusing mainly along the same resonant line. Due to the secular evolution of the potential these orbits escape from the resonant lines, but other chaotic orbits can be trapped along res-

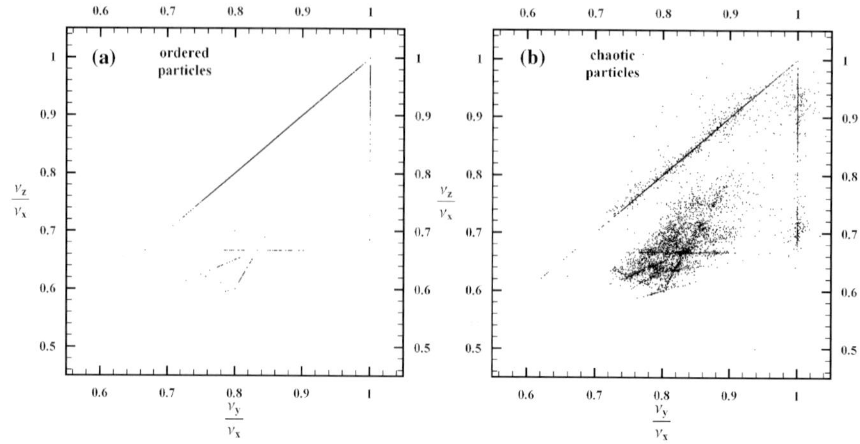

Figure 7. As in Figure 5a,b but for the CM model. Ordered motion (a) occurs in resonant orbits only. Orbits in the region of box and ILAT orbits are chaotic because of the central mass and they are shown in (b).

onant lines. As we will see in Section 6, trapping along the line ($v_z/v_y = 1$) of SAT orbits is much favored by the new shape of the potential, so most of the material is organized in SAT orbits.

In Figure 8a,b, which is similar to Figure 6a,b, but for the CM model, we see that almost all the weakly chaotic orbits ($\log L_j < -2$) are located on

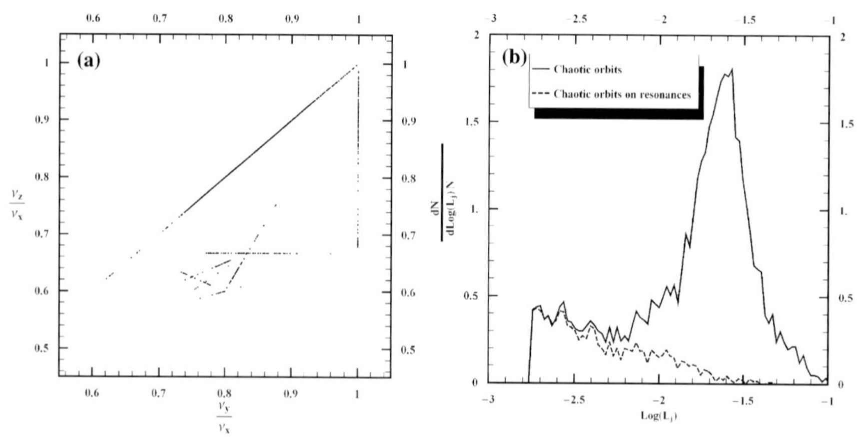

Figure 8. (a) A separate plot of those of the chaotic orbits in Figure 7b that occupy the most important resonant lines. (b) As in Figure 6b but for the CM model at $t = 150$. For small values of L_j the difference between the two curves is quite small, unlike Figure 6b, because those of the weakly chaotic orbits that resemble box orbits (mentioned in Figure 6b) are converted either to more strongly chaotic orbits, or to resonant chaotic orbits. For large values of L_j the relative occurrence of chaotic resonant orbits decreases considerably, as in Figure 6b.

resonant lines, unlike Figure 6b. This is because those of the weakly chaotic orbits in the SC model that were outside the resonant lines (resembling box orbits) have been converted partly to resonant orbits and partly to more strongly chaotic orbits in the CM model.

On the other hand, in Figure 8b, like in Figure 6b, more strongly chaotic orbits have a very small probability of being trapped along resonant lines.

An idea of how the various types of orbits are distributed at different energy levels at the snapshot $t = 150$ of the CM model is given in Figure 9a,α,b,β. The left column (Figure 9a,b) refers to test particles, while the right column (Figure 9α,β) refers to real particles.

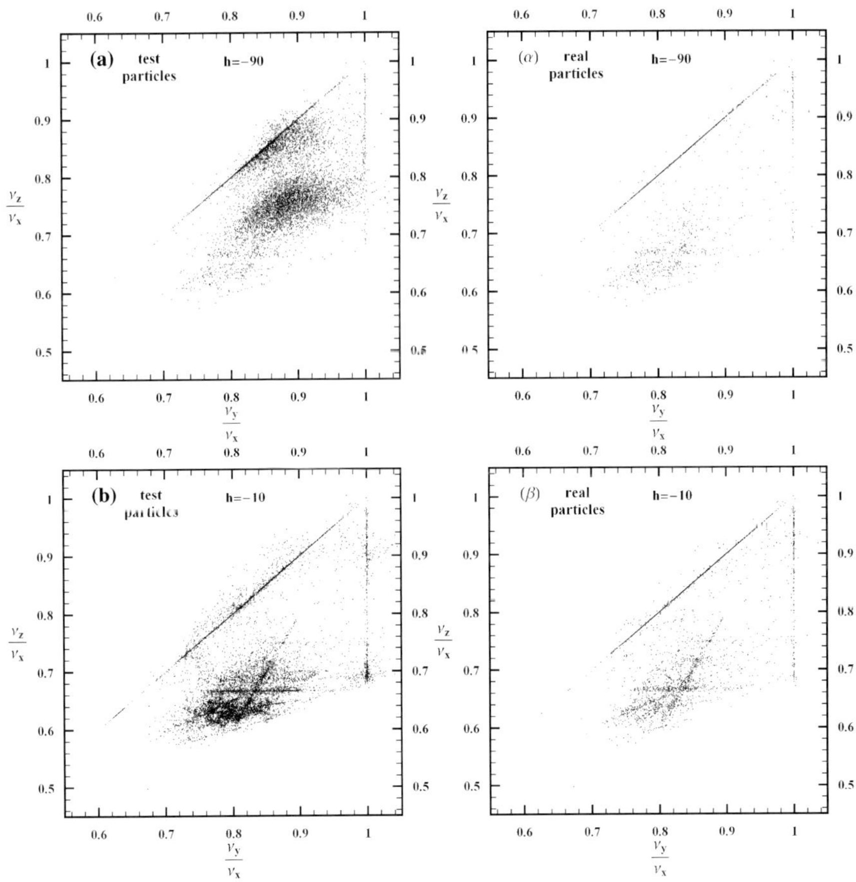

Figure 9. As in Figure 4, but for the snapshot at $t = 150$ of the CM model for two energy levels. (a),(α) are for the level $h = -90$. In this model SAT orbits appear from the lowest energy levels. In low energies the orbits of real particles do not follow the same distribution as the orbits of test particles. (b), (β) are for a high level $h = -10$. For high energies the distribution of real particles is not very different than the distribution of test particles.

At the level of $h = -90$ (Figure 9a,α) the distribution of the orbits of real particles, outside the resonant lines, is not the same with the distribution of the orbits of test particles. Note that the system is not in equilibrium at this snapshot.

In these figures the majority of the orbits are chaotic, but there are also many ordered orbits mainly of SAT type. As we have seen (Section 4), in the SC model, at low energy levels, only ordered box (and a few HORT) orbits appear. At low energy levels ($h \lesssim -50$) of the SC model there are no SAT orbits at all. In contrast, in the CM model SAT orbits appear already from the lowest energy levels. This is another effect due to the presence of the central mass. The central mass favors ordered motion of particles in SAT orbits even at very low energy levels.

At higher energy levels of the CM model a relative increase of the ordered resonant orbits of all the types occurs. This can be seen on the rotation number plane as, for example, in Figure 9b,β for $h = -10$.

6. Time Evolution and Self-organization of the CM Model

In this section we examine the evolutionary features of the CM model.

The introduction of the central mass destabilizes the well established initial equilibrium structure of the SC model. As we have seen above, almost all the box and the box-like (ILAT) orbits are converted into chaotic orbits. As a consequence, at the snapshot $t = 0$ of the CM model, the fraction of mass in chaotic motion is found to be about $\approx 80\%$, in contrast with the fraction $\approx 32\%$ of mass in chaotic motion found in the SC model. This serious increase of the mass in chaotic motion is accompanied by a serious change in the distribution of the Lyapunov numbers. This distribution shows a peak in the region of large values of L_j, due mainly to box orbits that became chaotic.

As mentioned in Section 3, the diffusion of the orbit j is effective in a Hubble time, if $L_{cuj} > 10^{-2}$. In the SC model, the mass in chaotic motion with $L_{cuj} \gtrsim 10^{-2}$ is less than 8% and it is almost spherically distributed, unable to cause any considerable secular evolution in the system.

Notice that almost spherically distributed chaotic orbits cannot have serious consequences on producing secular evolution in the system by chaotic diffusion. Chaotic diffusion is important for producing secular evolution provided that the corresponding chaotic mass has initially an anisotropic spatial distribution. A more detailed examination of this effect is given in Voglis and Kalapotharakos (2005).

In the CM model, the mass in chaotic motion with $L_{cuj} > 10^{-2}$ is $\approx 55\%$ at the snapshot $t = 0$ and it has an anisotropic distribution consistent with the configuration of the SC model. As this mass diffuses, it causes serious changes on the equidensity surfaces and hence on the self-consistent equipotential surfaces of the system. Thus, the system becomes unstable and undergoes secular evolution towards an equilibrium state. During this evolution many chaotic orbits are trapped, by the new shapes of the equipotential surfaces, in different regions of the phase space, where they are converted to ordered orbits. The fraction of mass in ordered motion increases in time, while the mass in chaotic motion decreases. The system is self-organized. This process goes on until the remaining mass in chaotic motion cannot considerably affect the equipotential surfaces any more and the system achieves an equilibrium configuration.

The secular evolution of the CM model can be clearly demonstrated on the rotation number plane. We plot all the particles of the system on the rotation number plane using the Hamiltonian (9) for a series of successive snapshots of the potential taken from the self-consistent run. In this way we can follow the various types of orbits that are consistent with the corresponding snapshots of the potential.

As an example, four different snapshots are shown in Figure 10a–d at times $t = 10$, $t = 100$, $t = 200$, $t = 280$, respectively, where all the orbits of the real particles of the system are plotted. In these figures the areas of box, HORT and ILAT orbits are moving upwards approaching the line of the SAT orbits (compare Figure 10a–c). The number of all the other types of orbits decreases while the number of SAT orbits increases.

At the snapshot of $t = 280$ (Figure 10d) the majority of the orbits appear along the line of SAT orbits. There is also a smaller number of orbits along the line of OLAT orbits.Part of these orbits are expected to be (partially) chaotic with small values of their LCN. In any case a component of angular momentum along the shortest axis for the SAT orbits and along the longest axis for the OLAT orbits is approximately conserved.

The fraction of mass in chaotic motion at this snapshot ($t = 280$) is found $\approx 22\%$ and remains constant at this level. The system has achieved an almost oblate spheroidal equilibrium configuration supported by SAT orbits.

The time scale needed for the CM model to reach the equilibrium configuration is comparable with the Hubble time. It is remarkable that cuspy triaxial models with central mass of the same relative size, i.e. $m = 0.01$ investigated by other authors, (e.g. Holley-Bockelmann et al., 2002) appear considerably more stable compared with the CM model here. In Kalapotharakos et al. (2004), the CM model (called Q100) is compared with another model (called C100) containing a central mass of the same size. The C100 model evolves about six times more slowly than the Q100 model. The reason for this slow evolution is the fact that the C100 model contains initially (before the central mass is inserted) considerably less mass in box orbits and more mass in tube-like orbits. Stability in such models has a sensitive

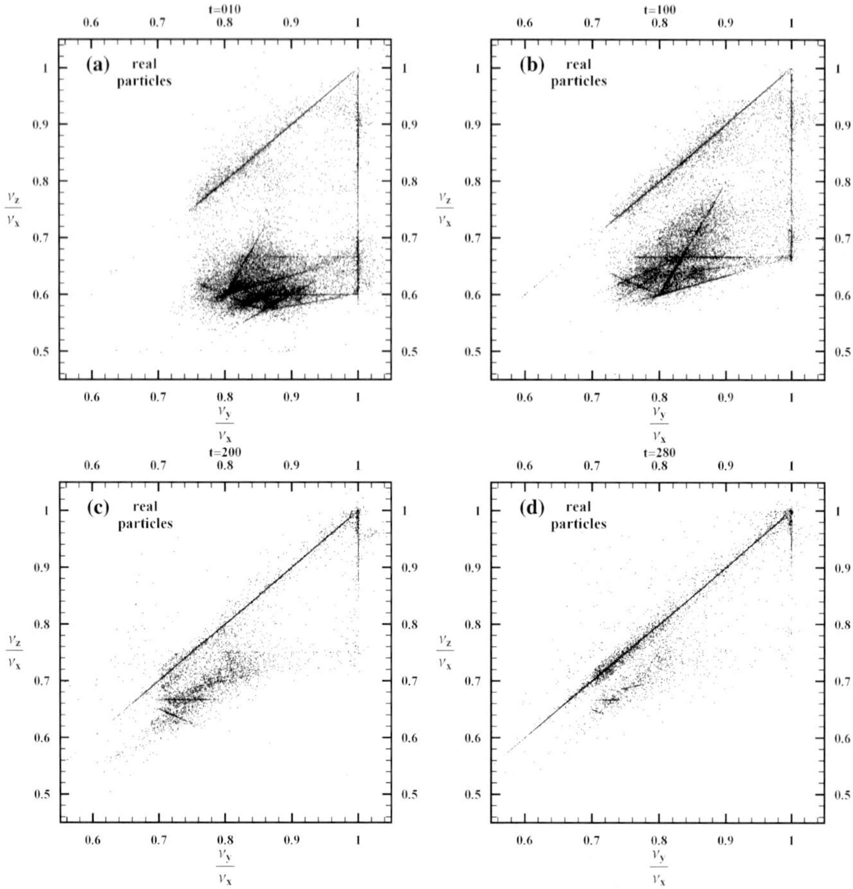

Figure 10. Four snapshots of the rotation number plane of the real particles of the CM model at times $t = 10$ (a), $t = 100$ (b), $t = 200$ (c), $t = 280$ (d), indicating the secular evolution of the system. Most of the chaotic orbits approach gradually the line of the SAT orbits and they are trapped on this line until an almost oblate spheroidal equilibrium configuration is achieved.

dependence on the number of box orbits supporting the triaxiality of the system. If triaxiality is mainly supported by combinations of tube orbits, rather than by box orbits, stable triaxial configurations are possible even with large central black holes.

The mechanism by which the chaotic orbits are converted to ordered orbits of the SAT type is discussed in detail in Kalapotharakos et al. (2004).

In brief, this mechanism is as follows: Due to the deflection of the box or box-like (ILAT) orbits as they pass near the center, the libration of their major axis around the z axis becomes chaotic. Provided that the central mass is large enough the chaotic libration of the major axis, after a transient period, turns to rotation.

This rotation takes place preferably close to the $y-z$ plane, i.e. around the shortest axis (x-axis) of the system, because this orientation is more stable than others. Since the major axis of the orbit is no more trapped near the z-axis of the system, it can be directed along all the azimuthal angles on the $y-z$ plane. If this happens for a good number of orbits the predominance of the velocity dispersion along z is lost and the two directions y and z tend to be equivalent as regards the dispersion of velocities. As a response the equidensity surfaces and the self-consistent equipotential surfaces approach an oblate spheroidal shape. Due to this change of the potential the chaotic orbits are gradually trapped and organized as SAT orbits. In their new SAT form these orbits do not approach the center any more.

The secular evolution ceases when the remaining chaotic orbits are isotropically distributed, or the rate of their passing near the central mass is too small to cause any considerable change on the system.

The secular evolution of the CM model can also be seen in terms of the triaxiality index T defined as

$$T = \frac{c^2 - b^2}{c^2 - a^2}, \tag{13}$$

where a,b,c are the lengths of the principal axes of an equidensity surface with major axis $c = 1$ in units of the half mass radius. The triaxiality index T is equal to 0 for an oblate spheroidal surface and it is equal to 1 for a prolate spheroidal surface, while it is 0.5 for a maximally triaxial surface.

In Figure 11 the evolution of this index is plotted (in our two models for comparison) for times much longer than the Hubble time.

The SC model preserves the initial value of the triaxiality index $T \approx 0.9$, indicating that this model preserves its initial triaxial shape (close to a prolate shape) during all this time.

On the contrary, the triaxiality index of the CM model, starting from the same initial value, decreases tending to a small constant value. This decrease is irreversible. The index T remains permanently close to zero, indicating that the system has achieved an almost oblate spheroidal equilibrium configuration. In this configuration the majority of orbits are of the SAT type as we have seen above (Figure 10c).

7. Summary and Conclusions

In terms of the fundamental frequency analysis we have investigated all the types of orbits that are described by the particles in the two examined N-body systems, i.e. a SC and a model with a large CM.

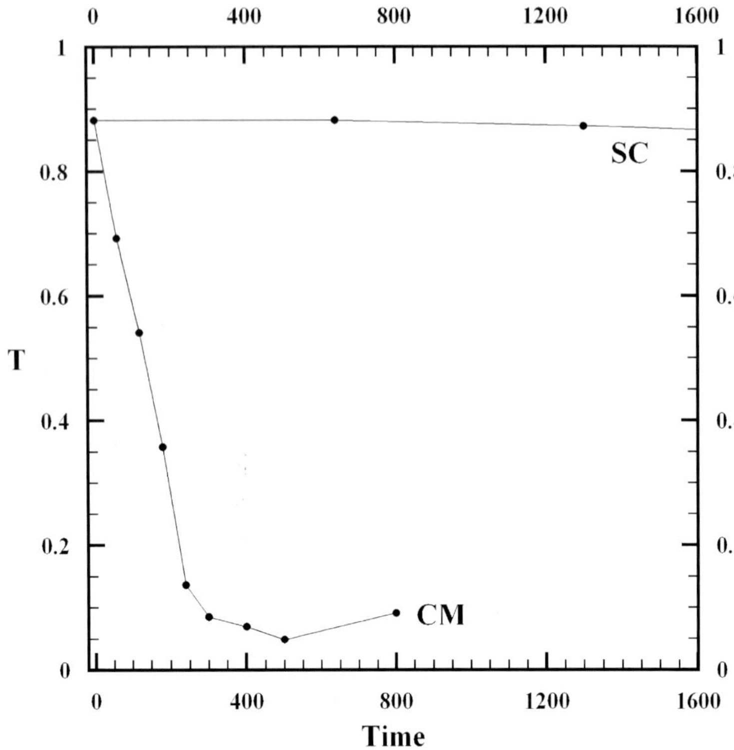

Figure 11. The secular evolution of the CM model in comparison with the stability of the SC model for times much longer than a Hubble time expressed in terms of the triaxiality index T. In the SC model T remains remarkably constant at the initial value $T \approx 0.9$ for long time. In the CM model T starts from this value and decreases approaching the zero value in a time scale of about one Hubble time when an almost oblate spheroidal equilibrium configuration is established.

The investigation combines the results of the fundamental frequency analysis with the results obtained by distinguishing chaotic from ordered orbits in terms of L_j, i.e. the Specific Finite Time Lyapunov Characteristic Number (S-FT-LCN) and the Alignment Index AI_j, a method introduced in a previous paper. The use of L_j has several pros and cons with respect to the use of L_{cuj}, i.e. the Finite Time Lyapunov Characteristic Number in common units (CU-FT-LCN), that are discussed in Section 3. We show that the alignment index can make a considerably sharper distinction than the distinction obtained by the variability of frequencies $\delta\omega_j$ or by the rate of logarithmic divergence L_j of the orbits (Figure 2).

We find that the SC model is stable. It contains a large fraction of ordered orbits ($\approx 68\%$ of the total mass). The main part of them are box orbits, but there are also many resonant tube orbits of various types, i.e. ILAT, OLAT,

SAT and HORT. On the rotation number plane these types of orbits occupy different loci, so they can be easily recognized, e.g. all the tube orbits are plotted along their characteristic resonant lines. At low energy levels only box and HORT orbits are present. As the energy increases the other types of tube orbits appear in the sequence ILAT, SAT, OLAT.

Among the chaotic orbits in this model, many of them are weakly or partially chaotic. Part of them resemble box orbits, but another part is trapped along resonant lines. More strongly chaotic orbits have small probability of being trapped along resonant lines.

In spite of the noise of the potential in a self-consistent run of the SC model, the number of particles occupying the various types of orbits is remarkably constant. This is obtained by a dynamical equilibrium established between the orbits that escape from a resonant line and the orbits that are trapped in this resonant line.

Comparing the distribution of the orbits of the real particles on the rotation number plane with the orbits of test particles, we conclude that extensive stable regions of phase space remain empty, i.e. they are not occupied by real particles of the system. This effect is due to the self-consistency of the system. Self-consistent equilibrium imposes serious limitations on the distribution of the real particles. These limitations are more severe than the limitations imposed by the stability of the orbits. Thus, finding stable orbits in a given galactic potential does not guarantee the existence of mass there and finding unstable regions does not guarantee the lack of mass there. This remark underlines the significance of the self-consistent models in studying galactic structures.

The empty stable regions are more pronounced in the low energy orbits, where chaos is weak and the chaotic orbits are rare. As the binding energies increase the orbits of real particles are in a better agreement with the orbits of test particles. In high binding energies real particles tend to occupy most the available phase space, either by ordered or by chaotic orbits.

The CM model is initially unstable. It presents a secular evolution due to the large amount of mass in chaotic motion produced by the central mass. This is initially about 80%. Most of this mass (about 55%) is characterized by Lyapunov times $L_{\text{cuj}}^{-1} < 100T_{\text{hmr}}$, i.e. less than one Hubble time, and it has initially an anisotropic spatial distribution forming the bar of the system. Due to the diffusion of these orbits, the initially bar-like equipotential surfaces are gradually deformed approaching oblate spheroidal shapes with their flatness along the x axis. As we have seen, by representing the orbits on the rotation number plane at various successive snapshots, during this deformation of the system, chaotic orbits are gradually converted to ordered orbits mainly of SAT type.

The secular evolution of the CM model ceases and equilibrium is established, when the remaining number of chaotic orbits is no more efficient to cause any considerable changes on the system, either because it is isotropically distributed, or because the rate of orbits that pass near the center is small.

The oblate spheroidal equilibrium configuration, established by the end of the secular evolution, is supported mainly by SAT orbits preserving the component of their angular momentum along the shortest axis. The orbital structure of this system is remarkably simpler than the orbital structure of a triaxial configuration.

Acknowledgements

We wish to thank Drs Allen, Palmer and Papaloizou as well as Sidlichovsky and Nesvorny for their codes. We also wish to thank the referee for his comments that helped to improve the presentation of the paper and prof. G. Contopoulos for useful discussions. This work is supported by a research program of the EMPEIRIKEION Foundation. C.K. wishes to thank the Greek State Scholarship Foundation (I.K.Y) for financial support.

References

Allen, A. J., Palmer, P. L. and Papaloizou, J.: 1990, 'A conservative numerical technique for collisionless dynamical systems – Comparison of the radial and circular orbit instabilities', *MNRAS* **242**, 576–594.

Contopoulos, G., Voglis, N., Kalapotharakos, C.: 2002, 'Order and chaos in self-consistent galactic models', *Celest. Mech. Dynam. Astron.* **83**, 191–204.

Cretton, N. and van den Bosch, F. C.: 1999, 'Evidence for a massive black hole in the S0 galaxy NGC 4342', *Astrophys. J.* **514**, 704–724.

de Zeeuw, T.: 1985, 'Elliptical galaxies with separable potentials', *MNRAS* **216**, 273–334.

Faber, S. M., Tremaine, S., Ajhar, E. A., Byun, Y., Dressler, A., Gebhardt, K., Grillmair, C., Kormendy, J., Lauer, T. R. and Richstone, D.: 1997, 'The centers of early-type galaxies with HST. IV. central parameter relations', *Astron. J.* **114**, 1771–1796.

Ferrarese, L., van den Bosch, F. C., Ford, H. C., Jaffe, W. and O'Connell, R. W.: 1994, 'Hubble space telescope photometry of the central regions of Virgo cluster elliptical galaxies. 3: Brightness profiles' *Astron. J.* **108**, 1598–1609.

Fridman, T. and Merritt, D.: 1997, 'Periodic orbits in triaxial galaxies with weak cusps' *Astron. J.* **114**, 1479–1487.

Gebhardt, K., Richstone, D., Ajhar, E. A., Lauer, T. R., Byun, Y., Kormendy, J., Dressler, A., Faber, S. M., Grillmair, C. and Tremaine, S.: 1996, 'The centers of early-type galaxies with HST. III. non-parametric recovery of stellar luminosity distribution', *Astron. J.* **112**, 105–113.

Gebhardt, K., Richstone, D., Kormendy, J., Lauer, T. R., Ajhar, E. A., Bender, R., Dressler, A., Faber, S. M., Grillmair, C., Magorrian, J. and Tremaine, S.: 2000, 'Axisymmetric, three-integral models of galaxies: A massive black hole in NGC 3379', *Astron. J.* **119**, 1157–1171.

Gerhard, O. E. and Binney, J.: 1985, 'Triaxial galaxies containing massive black holes or central density cusps', *MNRAS* **216**, 467–502.

Holley-Bockelmann, K., Mihos, J. C., Sigurdsson, S. and Hernquist, L.: 2001, 'Models of cuspy triaxial galaxies', *Astrophys. J.* **549**, 862–870.

Holley-Bockelmann, K., Mihos, J. C., Sigurdsson, S. Hernquist, L., Norman, C.: 2002, 'The evolution of cuspy triaxial galaxies harboring central black holes', *Astrophys. J.* **567**, 817–827.

Kalapotharakos, C., Voglis, N. and Contopoulos, G.: 2004, 'Chaos and secular evolution of triaxial N-body galactic models due to an imposed central mass', *Astron. & Astrophys.* **428**, 905–923.

Kandrup, H. E. and Sideris, I. V.: 2002, 'Chaos in cuspy triaxial galaxies with a supermassive black hole: A simple toy model', *Celest. Mech. Dynam. Astron.* **82**, 61–81.

Kandrup, H. E. and Siopis, Ch.: 2003, 'Chaos and chaotic phase mixing in cuspy triaxial potentials', *MNRAS* **345**, 727–742.

Kormendy, J. and Richstone, D.: 1995, 'Inward bound – the search for supermassive black holes in galactic nuclei', *Ann. Rev. Astron. Astrophys.* **33**, 581–624.

Kormendy, J., Bender, R., Magorrian, J., Tremaine, S., Gebhardt, K., Richstone, D., Dressler, A., Faber, S. M., Grillmair, C. and Lauer, T. R.: 1997, *Astrophys. J. Letters* **482**, 139–144.

Kormendy, J., Bender, R., Evans, A. S. and Richstone, D.: 1998, 'The mass distribution in the elliptical galaxy NGC 3377: Evidence for a $2 \times 10^8 M_\odot$ black hole', *Astron. J.* **115**, 1823–1839.

Laskar, J.: 1990, 'The chaotic motion of the solar system – A numerical estimate of the size of the chaotic zones', *Icarus* **88**, 266–291.

Laskar, J.: 1993a, 'Frequency analysis of a dynamical system', *Celest. Mech. Dynam Astron.* **56**, 191–196.

Laskar, J.: 1993b, 'Frequency analysis for multi-dimensional systems. Global dynamics and diffusion', *Physica D* **67**, 257–281.

Laskar, J., Froeschle, C. and Celletti, A.: 1992, 'The measure of chaos by the numerical analysis of the fundamental frequencies. Application to the standard mapping', *Physica D* **56**, 253–269.

Lauer, T. R., Ajhar, E. A., Byun, Y. I., Dressler, A., Faber, S. M., Grillmair, C., Kormendy, J., Richstone, D. and Tremaine, S.: 1995, 'The Centers of Early-Type Galaxies with HST.I.An Observational Survey', *Astron. J.* **110**, 2622–2654.

Lichtenberg, A. J. and Lieberman, M. A.: 1992, Regular and Chaotic Dynamics', Springer-Verlag, New York..

Magorrian, J., Tremaine, S., Richstone, D., Bender, R., Bower, G., Dressler, A., Faber, S. M., Gebhardt, K., Green, R., Grillmair, C., Kormendy, J. and Lauer, T.: 1998, 'The demography of massive dark objects in galaxy centers', *Astron. J.* **115**, 2285–2305.

Merritt, D. and Ferrarese, L.: 2001, 'Black hole demographics from the $M_\bullet - \sigma$ relation', *MNRAS* **320**, 30–34.

Merritt, D. and Fridman, T.: 1996, 'Triaxial galaxies with cusps', *Astrophys. J.* **460**, 136–162.

Merritt, D. and Quinlan D. G.: 1998, 'Dynamical evolution of elliptical galaxies with central singularities', *Astrophys. J.* **498**, 625–639.

Merritt, D. and Valluri, M.: 1996, 'Chaos and mixing in triaxial stellar systems', *Astrophys. J.* **471**, 82–105.

Muzzio, J. C. and Mosquera, D. E.: 2004, 'Spatial structure of regular and chaotic orbits in self-consistent models of galactic satellites' *Celest. Mech. Dynam. Astron.* **88**, 379–396.

Muzzio, J. C., Carpintero, D. D. and Wachlin, F. C.: 2005, 'Spatial structure of regular and chaotic orbits in a self-consistent triaxial stellar system' *Celest. Mech. Dynam. Astron.* **91**, 173–190.

Papaphilippou, Y. and Laskar, J.: 1996, 'Frequency map analysis and global dynamics in a galactic potential with two degrees of freedom', *Astron. & Astrophys.* **307**, 427–449.
Papaphilippou, Y. and Laskar, J.: 1998, 'Global dynamics of triaxial galactic models through frequency map analysis', *Astron. & Astrophys.* **329**, 451–481.
Poon, M. Y. and Merritt, D.: 2002, 'Triaxial black hole nuclei', *Astrophys. J.* **568**, 89–92.
Poon, M. Y. and Merritt, D.: 2004, 'A self-consistent study of triaxial black hole nuclei', *Astrophys. J.* **606**, 774–787.
Siopis, Ch.: 1999, PhD thesis, University of Florida.
Siopis, Ch. and Kandrup, H. E.: 2000, 'Phase-space transport in cuspy triaxial potentials: Can they be used to construct self-consistent equilibria?', *MNRAS* **319**, 43–62.
Sidlichovsky, M. and Nesvorny, D.: 1997, 'Frequency modified Fourier transform and its applications to asteroids', *Celest. Mech.* **65**, 137–148.
Skokos Ch.: 2001, 'Alignment indices: A new, simple method for determining the ordered or chaotic nature of orbits', *J. Phys. A* **34**, 10029–10043.
Statler, T. S.: 1987, 'Self-consistent models of perfect triaxial galaxies', *Astrophys. J.* **321**, 113-152.
Valluri, M. and Merritt, D.: 1998, 'Regular and chaotic dynamics of triaxial stellar systems', *Astrophys. J.* **506**, 686–711.
van der Marel, R. P., de Zeeuw, P. T. and Rix, H. W.: 1997, 'Improved evidence for a black hole in M32 from HST/FOS spectra. I. Observations', *Astrophys. J.* **488**, 119–135.
van der Marel, R. P., and van den Bosch F. C.: 1998, 'Evidence for a $3 \times 10^8 M_\odot$ black hole in NGC 7052 from hubble space telescope observations of the nuclear gas disk', *Astron. J.* **116**, 2220-2236.
Voglis, N., Kalapotharakos, C.: 2005, 'The role of ordered and chaotic motion in N-body models of elliptical galaxies', in A. M. Fridmann, M. Ya Marov and I. G. Kovalenko (eds.), *Progress in the Study of Astrophysical Disks: Collective and Stochastic Processes and Computation Tools*, ASSL series, 2005, Kluwer Academic Press (in press).
Voglis, N., Contopoulos, G. and Efthymiopoulos, C.: 1998, 'Method for distinguishing between ordered and chaotic orbits in four-dimensional maps', *Phy. Rev. E* **57**, 372–377.
Voglis, N., Contopoulos and G., Efthymiopoulos, C.: 1999, 'Detection of ordered and chaotic motion using the dynamical spectra', *Celest. Mech. Dynam. Astron.* **73**, 211–220.
Voglis, N., Kalapotharakos, C. and Stavropoulos, I.: 2002, 'Mass components in ordered and in chaotic motion in galactic N-body models', *MNRAS* **337**, 619–630.
Wachlin, F. C. and Ferraz-Mello, S.: 1998, 'Frequency map analysis of the orbital structure in elliptical galaxies', *MNRAS* **298**, 22–32.

Celestial Mechanics and Dynamical Astronomy (2005) 92:189–217
DOI 10.1007/s10569-004-6878-8

CHAOTIC MOTIONS IN THE FIELD OF TWO FIXED BLACK HOLES

G. CONTOPOULOS and M. HARSOULA
Academy of Athens, Center for Astronomy
Soranou Efessiou 4, GR-11527, Athens, Greece

(Received: 6 September 2004; accepted: 29 November 2004)

Abstract. We study the chaotic motions in the field of two fixed black holes M_1, M_2 by calculating (a) the asymptotic curves from the main unstable periodic orbits, (b) the asymptotic orbits, with particular emphasis on the homoclinic and heteroclinic orbits, and (c) the basins of attraction of the two black holes. The orbits falling on M_1 and M_2 form fractal sets. The asymptotic curves consist of many arcs, separated by gaps. Every gap contains orbits falling on a black hole. The sizes of various arcs decrease as the mass of M_1 increases. The basins of attraction of the black holes M_1, M_2 consist of large compact regions and of thin filaments. The relative area of the basin M_2 tends to 100% as $M_1 \to 0$, and it decreases as M_1 increases. The total area of the basins is found analytically, while the relative area of the basin M_2 is given by an empirical formula. Further empirical formulae give the exponential decrease of the number of asymptotic orbits that have not yet reached a black hole after n iterations.

Key words: Chaotic motion, two fixed black holes problem

1. Introduction

The problem of two fixed black holes refers to the motions of particles of infinitesimal mass (time-like geodesics) and photons (null geodesics) in the field of two fixed centres M_1 and M_2. It is known that the classical problem of two fixed centres is integrable (Charlier, 1902; Deprit, 1960), while the relativistic problem is chaotic (Contopoulos, 1990, 1991); namely the motions of photons are completely chaotic, while the motions of particles are chaotic to a large degree.

If we use a coordinate system x–z (Figure 1) where the black holes of masses M_1, M_2 are placed at the points $z = \pm 1$ the potential at the point P is

$$V = -\left[\frac{M_1}{r_1} + \frac{M_2}{r_2} + 1\right] \tag{1}$$

Then the Lagrangian in prolate spheroidal coordinate, where

$$x = \sinh\psi \sin\theta \cos\varphi, \quad y = \sinh\psi \sin\theta \sin\varphi, \quad z = \cosh\psi \cos\theta \tag{2}$$

is

$$2L = \frac{\dot{t}^2}{V^2} - V^2[Q(\dot{\psi}^2 + \dot{\theta}^2) + \sinh^2\psi \sin^2\theta\dot{\phi}^2] = \delta_1 \tag{3}$$

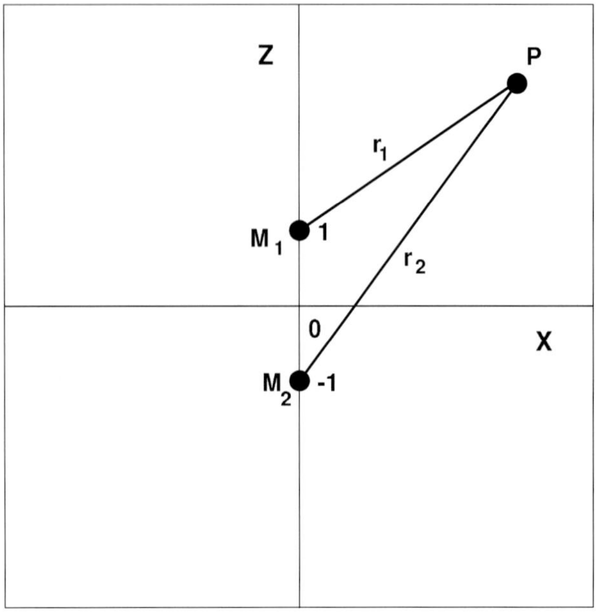

Figure 1. The position of a particle P in the coordinate system x–z.

(Chandrasekhar, 1989), where $\delta_1 = 0$ for photons and $\delta_1 = 1$ for particles, and

$$Q = \cosh^2\psi - \cos^2\theta > 0 \tag{4}$$

while the dots represent derivatives with respect to an affine parameter. From this Lagrangian we derive the energy E and the angular momentum L_z integrals.

From now on we consider orbits on a meridian plane and take $L_z = 0$.

In the case of particles the energy E is given by the equation

$$E^2 = Q(\dot{\psi}^2 + \dot{\theta}^2) + V^{-2}. \tag{5}$$

If the energy is of the elliptic type we have $0 < E < 1$ and the particles cannot escape to infinity. Then there are some simple periodic orbits around one or the other black hole. These orbits intersect perpendicularly the z-axis.

In Figure 2 we give the characteristics of these periodic orbits, i.e., the values of z for various values of M_1, assuming that $M_2 = 1$ and the energy E is fixed (in Figure 2, $E = \sqrt{0.5}$). We see that for $0 < M_1 < M_{1\max} \approx 1.3258$ there are two closed orbits of period 1 around M_1 that we call a (inner) and a' (outer) (in our previous papers we called both of them a'). The orbits a and a' join at a maximum M_1 and near this point the orbits a' are stable (Contopoulos, 1991).

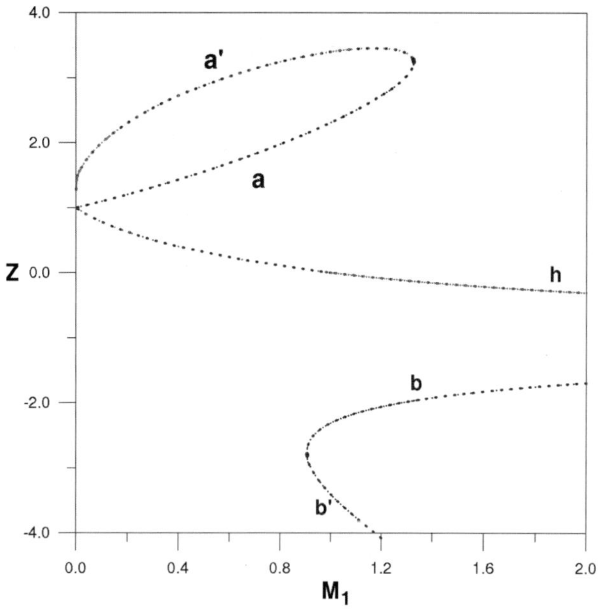

Figure 2. The characteristics of the periodic orbits a, a', h, b and b' for an energy $E = \surd 0.5$.

For $M_1 > 0.9061$ there are two orbits around M_2 that we call b (inner) and b' (outer). These orbits join at a minimum M_1 and near this point the orbits b' are stable. Orbits of types b and b' do not exist for values of M_1 close to zero. However for larger values of the energy E there are orbits of this type for smaller M_1 and when $E \to 1$ the orbits of types b, b' approach the limit $M_1 \to 0$.

The orbits a, a', b, b' can be described in the opposite direction and then they are called $\bar{a}$, $\bar{a}'$, $\bar{b}$, and $\bar{b}'$. There is also a simple orbit like an arc of hyperbola (orbit h) that is described in both directions.

The inner orbits a, $\bar{a}$ or b, $\bar{b}$ are very important, because any nonperiodic orbit crossing them inwards falls into the black holes M_1, M_2, respectively.

It is remarkable that in the classical case there are no simple periodic orbits around one black hole, i.e., satellite orbits closing after only one rotation around M_1, or M_2. This fact was established only relatively recently (Contopoulos, 1990).

One way to establish the existence of chaos in a dynamical system is by studying the homoclinic and heteroclinic intersections of the asymptotic curves from the unstable periodic orbits. On a surface of section $(z,\dot{z})$ for $x = 0$ such intersections represent doubly asymptotic orbits. Another phenomenon related to chaos is the fractal structure of the basins of attraction of the orbits falling into the black holes M_1 and M_2. In this case many orbits fall into the black holes and there are no Poincaré surfaces of section. Thus, the problem of two fixed black holes has certain properties of dissipative systems.

In the present paper we study the asymptotic curves of the periodic orbits a, b, a', with emphasis on the homoclinic and heteroclinic orbits.

Our study extends the results of a previous paper of ours on the same subject (Contopoulos and Harsoula, 2004).

In Section 2 we find the forms of the asymptotic curves and in Section 3 we study the forms of the asymptotic orbits. The asymptotic orbits falling into the black holes M_1 and M_2 form sets with a fractal structure. Then in Section 4 we find the basins of attraction of the black holes M_1 and M_2 for various values of M_1 (with $M_2 = 1$) and in Section 5 we formulate our conclusions.

2. Asymptotic Curves

The asymptotic curves $(z,\dot{z})$ (for $x = 0$, $\dot{x} > 0$) of the periodic orbit O (of type a) are shown in Figure 3 for $M_1 = M_2 = 1$. There are two unstable asymptotic curves U, UU and two stable asymptotic curves S, SS that intersect with the U, UU curves at the homoclinic points H_1, H_2, and further

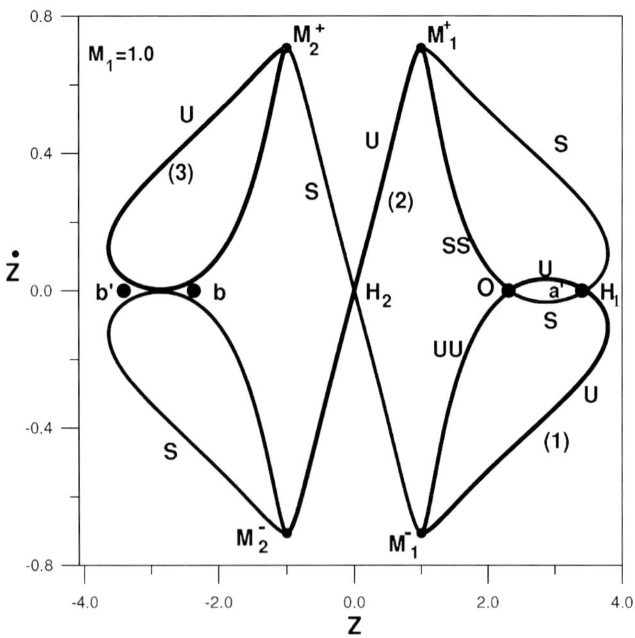

Figure 3. Parts of the asymptotic curves U, UU (unstable) and S, SS (stable) from the unstable periodic orbit O≡a for $M_1 = M_2 = 1$ and $E = \sqrt{0.5}$. These unstable asymptotic curves intersect the stable asymptotic curves at the homoclinic points H_1, H_2 and other points not marked in this figure. The periodic orbits a', b and b' are also given.

points not shown in Figure 3. The curves S, SS are symmetric to the curves U, UU with respect to the axis $\dot{z} = 0$.

There is no symmetry with respect to the centre (0,0). The arc (1) of U intersects its symmetric arc S above it, but the arc (3) of U is completely above the $\dot{z} = 0$ axis and does not intersect its symmetric arc S below it. However, if we continue the asymptotic curve U of Figure 3 we find, in the case $M_1 = M_2 = 1$, further arcs close to the arcs (1), (2), (3) that intersect the symmetric arcs S at further homoclinic points. In particular there are homoclinic points close to b and b', not marked in Figure 3.

We see that the asymptotic curves pass through the black holes M_1 and M_2 with velocities $\dot{z} = \pm 0.71$. In the Appendix of paper I we have shown that these values of $\dot{z}$ are equal to $\dot{z} = \pm E$.

In Figures 4a, 5 and 6 we give one asymptotic curve U from the point O for $M_1 = 1.1$ (Figure 4a), $M_1 = 1.2$ (Figure 5) and $M_1 = 1.24$ (Figure 6), while $M_2 = 1$ in all cases. In Figure 4b we give both curves U and S for $M_1 = 1.1$.

The asymptotic curve U of Figure 4a starts at the point O upwards and to the right. This curve consists of the 4th images of points starting close to O at distances $m \times 10^{-8}$ along the asymptotic direction from O with m between 0 and 20,000.

The asymptotic curve consists of several arcs reaching one or two black holes, i.e., the points $M_1^-, M_1^+, M_2^-, M_2^+$. The numbers in Figure 4a give the values of m at the 4th iterations of the points $m \times 10^{-8}$.

The first arc (1) starts at O and reaches the point M_1^- (point 207). Then the orbits with $m = 208$–279 do not have a 4th intersection with the surface of section $x = 0$ ($\dot{x} > 0$). Thus we have a gap in the initial conditions consisting of orbits that fall into a black hole before a 4th intersection with the surface of section. Then follows the arc (2) starting at M_1^+ with $m = 280$ and terminating at M_1^- with $m = 340$. Further on we have a gap between $m = 341$ and $m = 412$, an arc (3) from M_2^+ to the same point M_2^+ with $m = 413$ up to $m = 2054$, and so on.

The successive arcs start at a black hole, M_1 or M_2, and terminate at the same, or the other black hole. Between any two arcs there are gaps, i.e., intervals of values of m without a 4th intersection.

In particular we notice that the arc (5) is a little below the arc (1). While the arc (1) is described clockwise, the arc (5) is described counterclockwise. It starts at M_1^- and returns again to M_1^-. The point marked 2423 is on the arc (5) very close to the original point O.

The various iterations of an initial point $m \times 10^{-8}$ are at distances $\lambda m \times 10^{-8}$, $\lambda^2 m \times 10^{-8}, \ldots,$ where λ is the largest eigenvalue of the unstable orbit O.

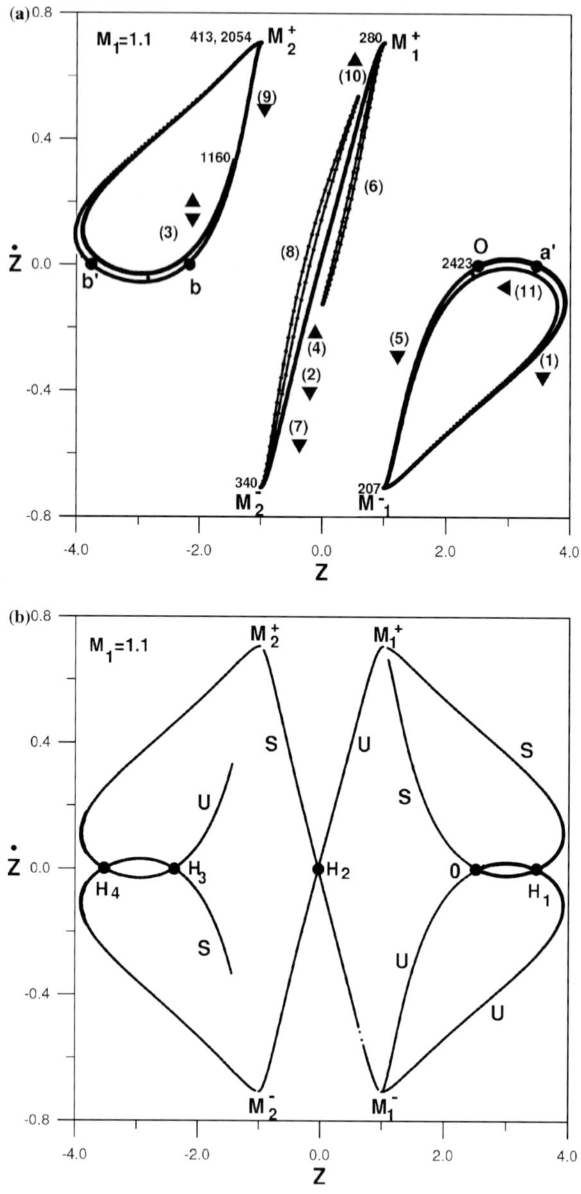

Figure 4. (a) Part of the asymptotic curve U from O(≡a), giving the 4th iterations of points along the unstable asymptotic curve U from O at distances $m \times 10^{-8}$, where m varies from $m = 0$ up to $m = 20{,}000$ for $M_1 = 1.1$, $M_2 = 1$, $E = \sqrt{0.5}$. This curve consists of 11 successive arcs. The orbits a', b and b' are close but not exactly on the asymptotic curves from O. (b) the arcs (1), (2), (3) and the corresponding arcs of the stable asymptotic curve S intersect at the homoclinic points H_1, H_2, H_3, H_4.

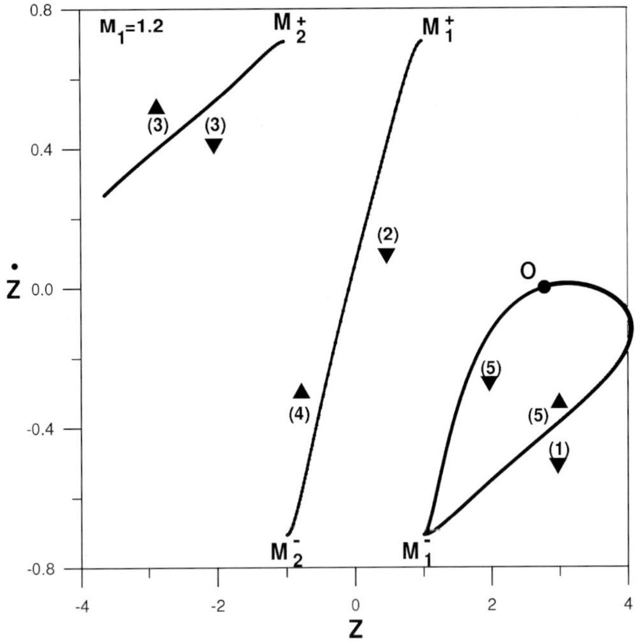

Figure 5. As in Figure 4 for $M_1 = 1.2$, $M_2 = 1$, $E = \sqrt{0.5}$.

In the case $M_1 = 1.1$ we have $\lambda = 32.33$. Thus a point marked m in Figure 4a (4th iteration) is marked $m_3 = \lambda m$ at its 3rd iteration, $m_5 = m/\lambda$ at its 5th iteration, etc. Therefore the figure giving the 3rd, 5th etc. iterations is the same as Figure 4a, and only the numbers denoting particular points are different.

If we increase m beyond $m = 20{,}000$ we find further arcs. But if the distances $m \times 10^{-8}$ become larger than 2×10^{-4} we have better accuracy if we start closer to O, at distances m/λ or m/λ^2, etc. and take the 5th, or 6th etc. iterations of the initial points.

In Figure 4b we see the first 3 arcs of the unstable asymptotic curve U from O and the corresponding 3 arcs of the stable curve S. The arc (3) of U intersects its symmetric arc S, in contrast with Figure 3 for $M_1 = 1$, and with similar figures with $M_1 < 1$.

As we increase M_1 the loops that we see in Figure 4a become shorter (Figures 5 and 6). For Example, the arc (3) in Figure 4a reaches the point $m = 1160$ and returns along a curve very close to the curve from $m = 413$ to $m = 1180$, reaching M_2^+ at $m = 2054$.

In the case $M_1 = 1.2$ (Figure 5) this arc is much shorter, and in the case $M_1 = 1.24$ the arc (3) does not exist at all. In this last case the arcs (2) and (4) of Figures 4a and 5 join and form a thin loop from M_1^+ to the same point M_1^+ without reaching the point M_2^-.

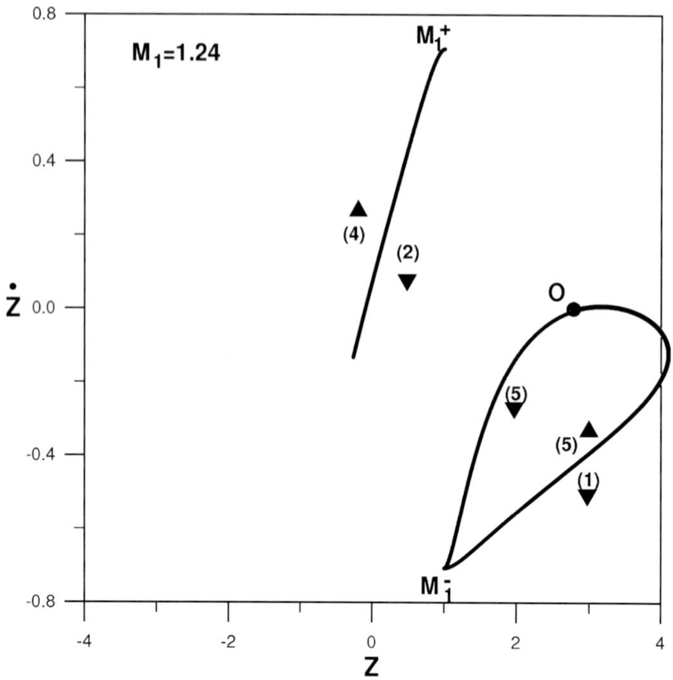

Figure 6. As in Figure 4 for $M_1 = 1.24$, $M_2 = 1$, $E = \sqrt{0.5}$. In this case the arc (3) does not exist and the arcs (2) and (4) are joined.

For a little larger value of M_1 even this loop (2)–(4) does not exist. For Example, in Figure 7 the arc (1) does not reach the point M_1^- from the right, but joins the arc (5) and forms a lobe from H_1 to H_1^-. The points H_1, H_1^-, H_1', are homoclinic points, i.e., intersections of the unstable (U) and stable (S) asymptotic curves from O. At the point H_1 the curve U goes outwards from S, while at H_1^- the U curve goes inwards. The point H_1' is the first image of H_1. Beyond the point H_1' the curve U goes to a minimum $\dot{z}$ on the left and below the boundaries of the figure. Then it returns close to O forming a number of oscillations.

When $M_1 = 1.325$ (Figure 8) the lobe from H_1 to H_1^- is much smaller. In this case between O and H_1 there is an island of stability around the stable periodic orbit a'. This island consists of invariant curves closing around the point a' and is surrounded by four smaller islands. Outside the islands there is chaos around a' and also inside the lobes. A further discussion of the islands of stability for various values of M_1 is given by Contopoulos (1991).

The periodic orbits a and a' join for a value of M_1 between $M_1 = 1.325$ and $M_1 = 1.326$ (Figure 2) and for $M_1 = 1.326$ these periodic orbits do not exist any more. Then the whole set of asymptotic orbits disappears.

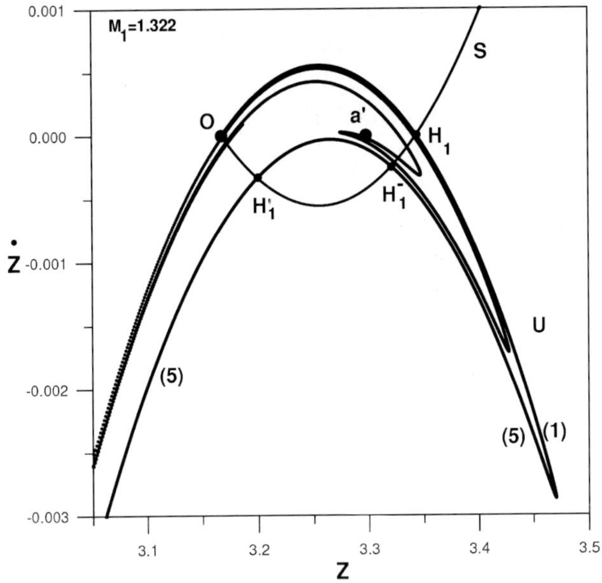

Figure 7. Parts of the asymptotic curves U and S from O for $M_1 = 1.322$, $M_2 = 1$, $E = \sqrt{0.5}$ close to O. The orbit a' is unstable.

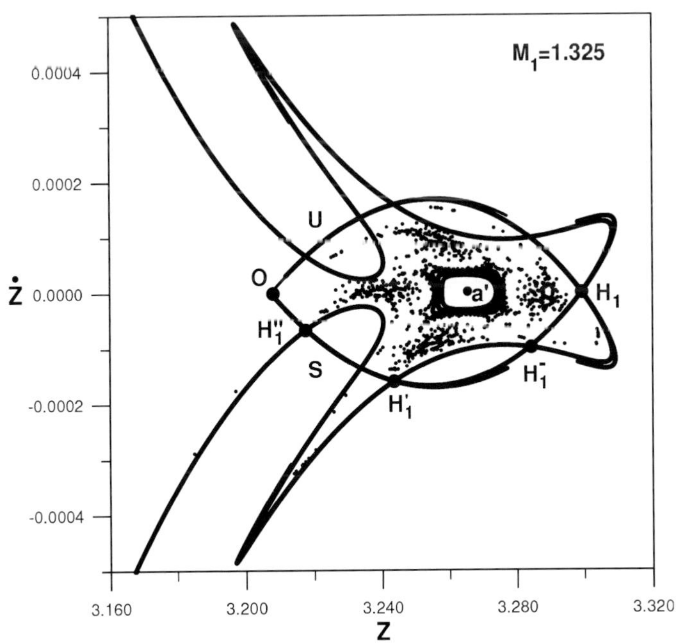

Figure 8. As in Figure 7 for $M_1 = 1.325$, $M_2 = 1$, $E = \sqrt{0.5}$. The periodic orbit a' is now stable, and it is surrounded by closed invariant curves. Further out there are four small islands. The scattered points belong to a single chaotic orbit outside the islands.

The orbits in the region of the gap $m = 208$–279 are shown in Figure 9. These orbits start close to the periodic orbit a, above M_1, with $x = 0$ and $\dot{x} > 0$. All of them make three complete rotations clockwise around M_1 and then they deviate from each other. The orbits with a little smaller m than $m = 208$ have a 4th intersection with the z-axis a little above M_1, and they have $\dot{z} < 0$. The orbit $m = 208$ does not have a 4th intersection. In fact all the orbits $m = 208$–279 reach the black hole M_1 from the left without intersecting the z-axis a 4th time. On the other hand the orbits with a little larger m than $m = 279$ have again 4th intersections with the z-axis, and $\dot{z} > 0$.

If we now take a surface of section $(z,\dot{z})$ with $x = -0.001$ $(\dot{x} > 0)$, all the orbits of the gap have 4th intersections with this surface of section at points intermediate between M_1^- and M_1^+ (Figure 10). Thus the surface of section $x = -0.001$ does not indicate a gap in this case. Nevertheless these orbits do not have a 5th intersection with the surface $(z,\dot{z})$ $(x = -0.001,\ \dot{x} > 0)$ therefore this surface is not a Poincaré surface of section. Furthermore some gaps are not eliminated by using a surface of section $x = -0.001$, because there are orbits reaching the black hole without a 4th intersection with any of the surfaces $x = 0$ or $x = -0.001$.

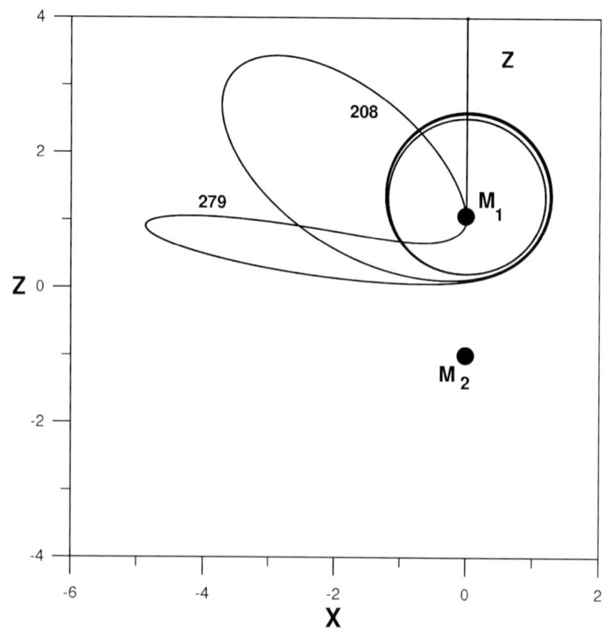

Figure 9. Asymptotic orbits in the plane (x,z). The orbits with m between $m = 208$ and $m = 279$ do not have a 4th intersection with the z-axis before falling on M_1. The orbits $m = 207$ and $m = 341$ have a 4th intersection close to M_1, the first above it and the second below it.

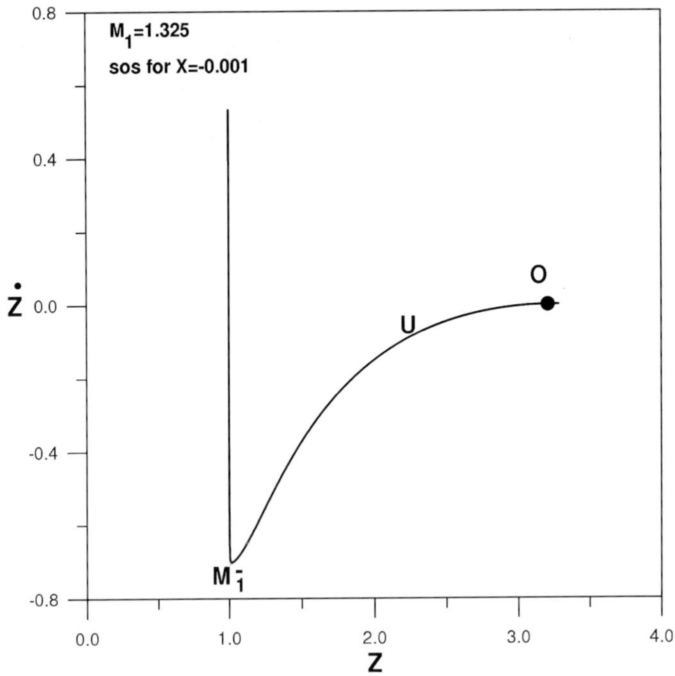

Figure 10. Part of the asymptotic curve U from O for $M_1 = 1.325$, on the surface of section $x = -0.001$. On the right of O there are details very similar to Figure 8, but they are not distinguished in this scale. The lowest point is very close to the point M_1^- of Figures 3–6 that appears for $x = 0$.

In the case $M_1 = 1.325$ all the oscillations of the asymptotic curve U on the right of O (Figure 8) are contained in the small extension of the curve U in Figure 10 beyond O. On the other hand the orbits on the almost vertical curve from M_1 do not have a 4th intersection with the $x = 0$ axis, but they do have such an intersection with the $x = -0.001$ axis. Nevertheless in the case $M_1 = 1.325$ there are no orbits coming close to M_1^+, but the set of orbits reaches a maximum $\dot{z}$, below M_1^+ ($z = 1$, $\dot{z} = 0.71$), and then it continues with smaller values of $\dot{z}$.

The asymptotic curves of the periodic orbit a' (Figure 11), when this is unstable, are very close to the asymptotic curves of the orbit a (Figure 4b). However, the orbit a' is inverse hyperbolic, i.e., it has $\lambda < 0$, and successive iterates of an initial point on an asymptotic curve alternate on both sides of a'.

The asymptotic curves of the orbit b have also a similar overall form with the asymptotic curves of the orbit a of Figure 4a, except that the various arcs are described with a different sequence. Thus when M_1 becomes equal to $M_1 = 1.324$ the asymptotic curve of the orbit b (Figure 12) has a very similar form, as the asymptotic curve of Figure 4a that starts from the orbit a. Only the sequence of arcs in Figure 11 is different from that of the asymptotic

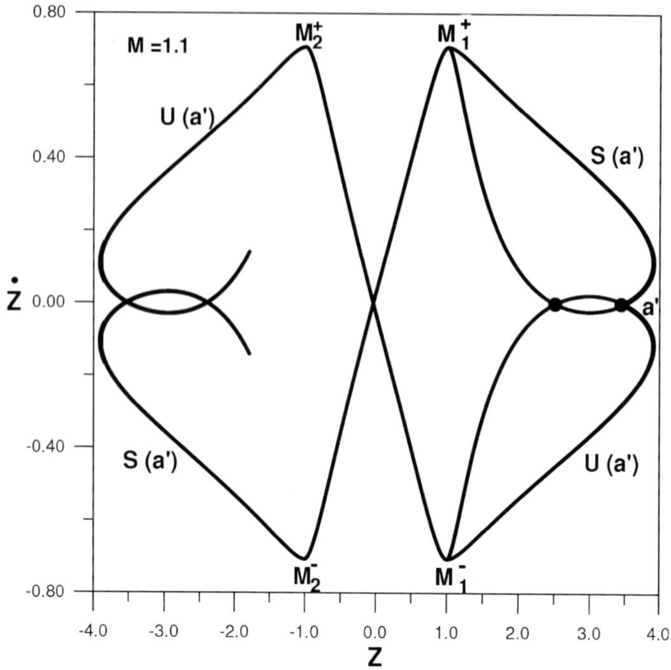

Figure 11. Parts of the unstable and stable asymptotic curves U and S from the periodic orbit a' for $M_1 = 1.1$.

curve from a. The asymptotic curves of the orbit b', when this is unstable, are also very similar.

3. Asymptotic, Homoclinic and Heteroclinic Orbits

The asymptotic curves start at the unstable periodic point O along two eigendirections, one unstable (U) and the other stable (S).

Further away these curves deviate from straight lines. An orbit starting at a point of an asymptotic curve is asymptotic to the periodic orbit O in the past, for $t \to -\infty$ (in the case U), or in the future, for $t \to +\infty$ (in the case S).

The successive iterates of an orbit in U are either all on the same side of O (regular hyperbolic point), or alternatively on both sides of O (inverse hyperbolic point). In the present case the orbits a and b are regular hyperbolic, while the orbits a' and b' are inverse hyperbolic.

The intersections of the asymptotic curves U and S are homoclinic points, like the points H_1, H_2, of Figure 3 and the points H_3, H_4 of Figure 4b. The corresponding orbits are asymptotic to the same orbit both in the past ($t \to -\infty$) and in the future ($t \to +\infty$).

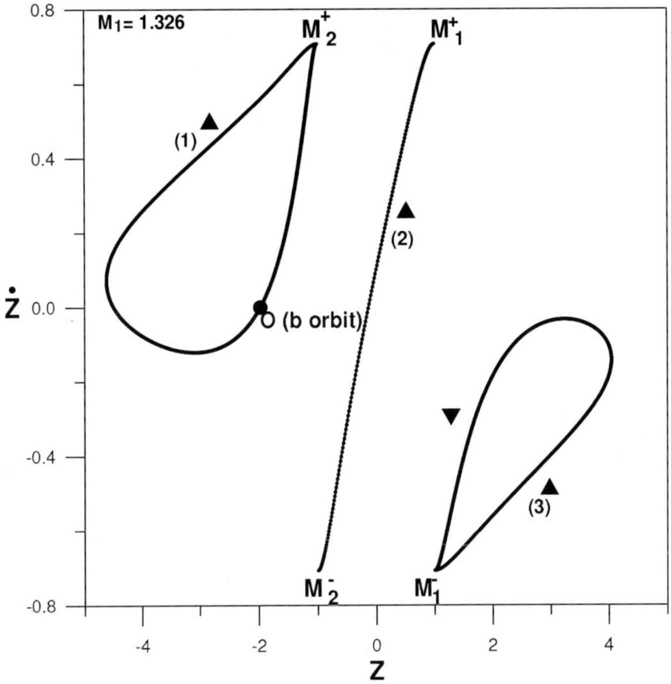

Figure 12. Parts of the asymptotic curve of the orbit b for $M_1 = 1.4$. If $M_1 \geqslant 1.326$ ($M_2 = 1$, $E\sqrt{0.5}$) there are no orbits a and a' and no corresponding asymptotic curves. However the asymptotic curves of b are similar to the asymptotic curves of a for smaller values of M_1 (Figures 3–6).

The intersection of the unstable asymptotic curve U with the stable asymptotic curve S' of a different periodic orbit are heteroclinic points and the corresponding orbits are heteroclinic orbits, approaching O for $t \to -\infty$, and a different periodic orbit for $t \to +\infty$. Similarly we have heteroclinic intersections between S and an unstable curve U' of a different orbit. On the other hand the curve U cannot intersect itself or another unstable curve U', and the same applies to S and S'.

The forms of the asymptotic orbits starting on the curve U are of two basic types, those falling on the black hole M_1 (type I orbits), and those falling on the black hole M_2 (type II orbits). Such orbits are shown in Figure 13. In this figure we give some asymptotic orbits with m corresponding to points on the arc (2) of Figure 4a. These orbits are given for initial distances $m \times 10^{-8}$ with $\Delta m = 10$ from $m = 280$ to $m = 340$. The first four orbits fall into M_1 and the last three into M_2. However close to $m = 310$ there are many (infinite) transitions of type I and type II orbits. The details of these transitions are discussed by Contopoulos and Harsoula (2004). Here we give the main conclusions of this study.

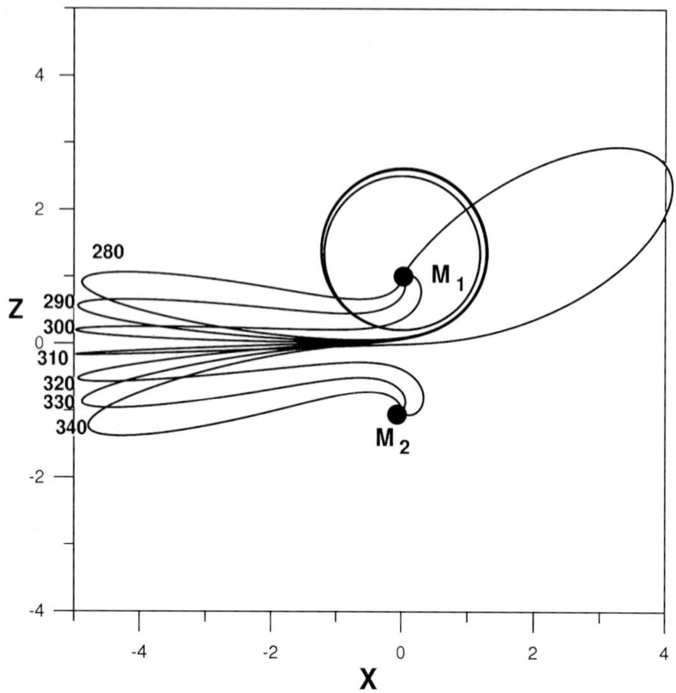

Figure 13. Some asymptotic orbits corresponding to points m on the arc (2) of Figure 4. These orbits fall either on M_1 (if $m \leqslant 300$), or on M_2 (if $m \geqslant 320$). But close to $m = 310$ there are infinite sets of orbits falling into M_1 and M_2.

1. The sets of types I and II are fractal consisting of infinite subsets. Every subset is limited by homoclinic or heteroclinic orbits.
2. Near every homoclinic and heteroclinic orbit there are infinite further homoclinic and heteroclinic orbits and infinite subsets of type I and II orbits.
3. All the orbits of the arc (2) with $m < m^* \approx 307.643390$ are of type I (they fall into M_1) and all the orbits of this arc with $m > m^{**} \approx 314.384627$ are of type II (they fall into M_2).

In particular the orbit with $m = m^*$ (Figure 14) is a heteroclinic orbit joining the orbit a (for $t \to -\infty$) to the orbit $\bar{a}$ (for $t \to +\infty$), i.e., the same orbit described in the opposite direction (for this reason the orbit $\bar{a}$ is considered as different form a). The orbit $m = m^*$ after infinite rotations clockwise reaches a minimum x at a point with $\dot{z} = 0$, therefore it retraces its path in the opposite direction, reaching the orbit $\bar{a}$ after infinite rotations counterclockwise.

If we decrease m all the orbits fall into M_1, as m decreases until $m = 75$. The interval $75 < m < m^*$ contains not only orbits of the upper part of the arc (2) of Figure 4a but also all the orbits of the gap between the arcs (1) and

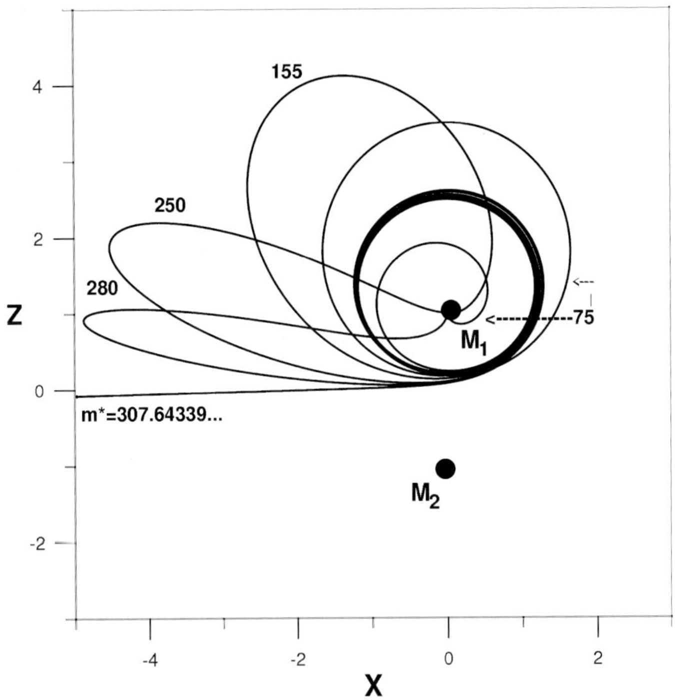

Figure 14. All the asymptotic orbits from $m = 75$ to $m = m^* = 307.643390$ fall into M_1.

(2), and many orbits of the arc (1). On the other hand orbits with $m < 75$ are both of type I and II.

In an analogous way the orbit with $m = 314.384628$ (Figure 15) is very close to the orbit $m = m^{**}$ which is heteroclinic between the orbit a (for $t \to -\infty$) and the orbit $\bar{b}$ (for $t \to +\infty$) The orbits with $m > m^{**}$, up to $m = 559.1297$ fall into the black hole M_2. These orbits fall not only in the lower part of the arc (2) of Figure 4a, but also in the gap between the arcs (2) and (3) and in a part of the arc (3).

The orbit $m = 559.1297$ is close to a heteroclinic orbit m^{***} between the orbits a and b. But beyond m^{***} there are orbits of both types I and II.

Besides the above heteroclinic orbits there are infinite more heteroclinic and homoclinic orbits. An orbit very close to a homoclinic orbit, from a to a, is shown in Figure 16. This orbit starts with infinite clockwise rotations around M_1, reaches a minimum x with $\dot{z} < 0$, then a maximum x with $\dot{z} > 0$, and then makes infinite clockwise rotations around M_1 reaching finally again asymptotically the orbit a. A heteroclinic orbit from a to $\bar{a}'$ is shown in Figure 17. Similar heteroclinic orbits exist also between a and $\bar{a}'$, $\bar{b}$, b' and $\bar{b}'$.

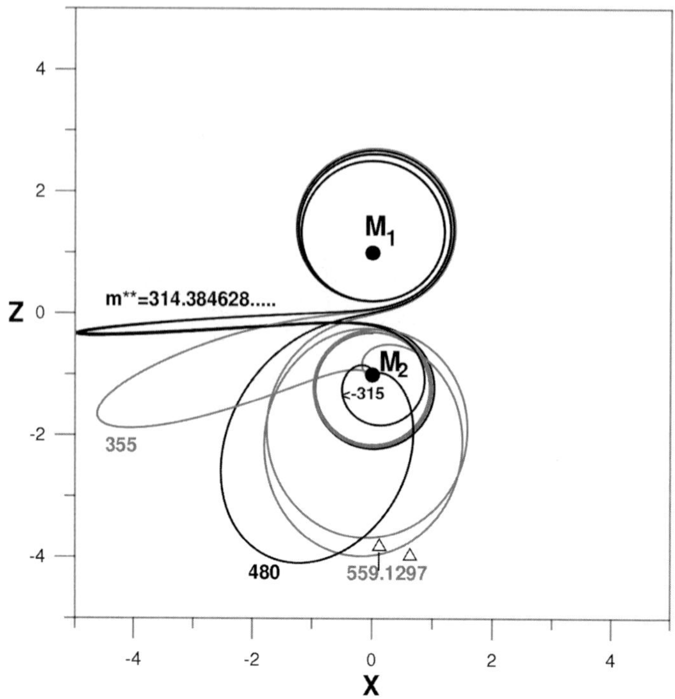

Figure 15. All the asymptotic orbits from $m = m^{**} = 314.384627\ldots$ to $m^{***} = 559.1297\ldots$ fall into M_2.

Further details about the homoclinic and heteroclinic orbits and the fractal structure of the sets of orbits I and II are given by Contopoulos and Harsoula (2004).

4. Basins of Attraction

The basins of attraction of two and three black holes and their fractal structure were considered by Dettmann et al. (1994, 1995). In the present paper we consider the basins of attraction for various values of the mass M_1 and their relations with the asymptotic manifolds discussed in the previous sections. If we calculate all orbits with $M_1 = 1.1$, $M_2 = 1$ and $E = \sqrt{0.5}$ and initial conditions on the surface of section $(z,\dot{z})$ (with $x = 0$ and $\dot{x} > 0$)we find (Figure 18) large sets falling directly into M_1 and M_2 without any intersection with the surface of section (black and gray respectively) and smaller sets that fall into M_1, or M_2, after 1, or 2, or more intersections. These sets consist of large compact regions and of thin filaments. A few filaments can be seen in Figure 18.

These filaments form fractal sets. They correspond to the sets Δm of values of m along the unstable asymptotic curve from the orbit a that are terminated

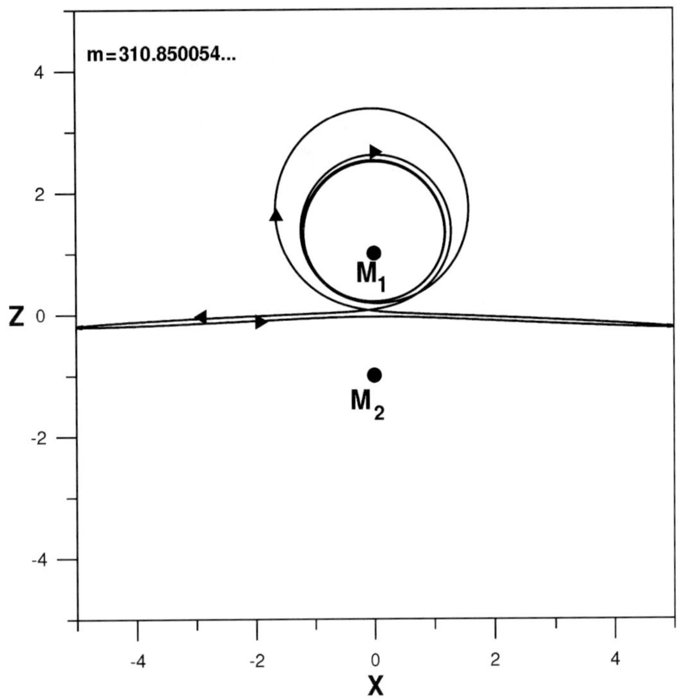

Figure 16. A homoclinic orbit starting asymptotically close to the periodic orbit a (as $t \to -\infty$) and reaching again asymptotically the same orbit a (as $t \to \infty$).

by homoclinic or heteroclinic orbits and fall into M_1, or M_2. Figures 18–21 are calculated by taking initial conditions on the surface of section at a grid of points with sizes $\Delta z = 10^{-2}$ and $\delta\dot{z} = 0.8 \times 10^{-2}$. If we compare Figure 18 with Figure 4a we see two filaments passing through the central part of the arc (2), namely in the region between the first heteroclinic point $m = m^*$ and the last heteroclinic point $m = m^{**}$ in this region. More detailed figures of the central part of Figure 18 show many more filaments that are limited by the homoclinic and heteroclinic points along the arc (2).

The upper part of the arc (2) of Figure 4a, i.e., with $m < m^*$, does not contain any homoclinic or heteroclinic orbits. These orbits are in the black region of Figure 18, therefore they escape immediately into the black hole M_1. In fact the points on the arc (2) are the 4th iterates of points close to O along the asymptotic curve U at distances $m \times 10^{-8}$, and the starting orbits in this part of the arc (2) do not have a 5th intersection with the surface of section, but fall into M_1 after the 4th intersection. In the same way at the points of the lower part of the arc (2), with $m > m^{**}$, start orbits that fall directly into the black hole M_2 without a 5th intersection with the surface of section.

Two more regions with many (infinite) filaments of orbits escaping to M_1 and M_2 form loops in the upper right part and in the lower left part

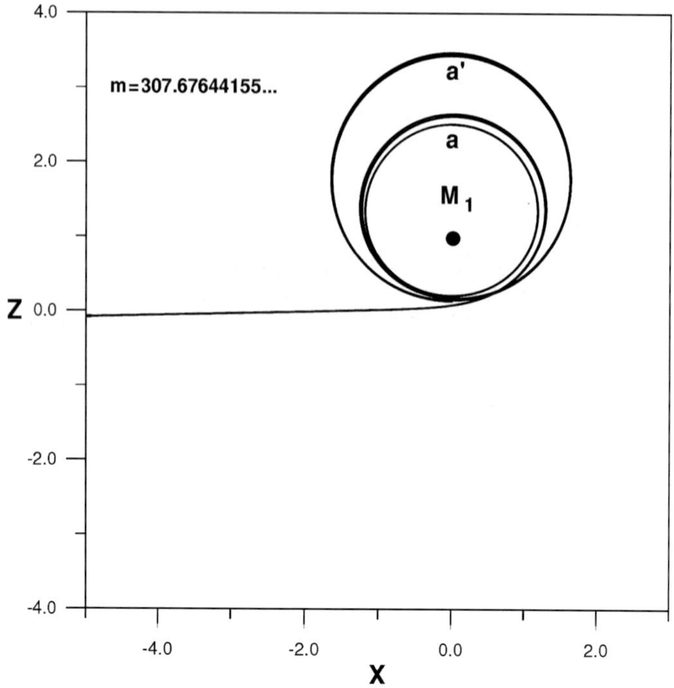

Figure 17. A heteroclinic orbit between a and a'. This orbit starts asymptotically from the orbit a for $t \to -\infty$ and terminates asymptotically on the orbit a' for $t \to \infty$.

of Figure 18. These loops surround a compact gray region on the upper right (type II orbits) and a compact black region on the lower left (type I orbits). The loops contain filaments with orbits falling into either M_1 or M_2, after one or more iterations. The limits of the filaments falling into M_1 and M_2 pass through the homoclinic and heteroclinic points along the asymptotic curves U and S of Figure 4b.

As M_1 increases the area covered by orbits on the surface of section increases (see Appendix A), while the points M_1^-, M_1^+, M_2^-, M_2^+, remain the same. Examples are shown in Figures 19 and 20. We see that, as M_1 increases the black regions increase, therefore more orbits fall into M_1 than into M_2. In particular the gray upper right region becomes smaller and for $M_1 \geqslant 1.326$ it does not reach the axis $\dot{z} = 0$. This corresponds to the fact that for $M_1 \geqslant 1.326$ the periodic orbits a, a' and their asymptotic curves with the corresponding homoclinic and heteroclinic intersections disappear.

On the other hand the black region on the lower left side becomes larger as M_1 increases. This region intersects the $\dot{z} = 0$ axis and just outside it are the periodic points b and b', with their asymptotic curves and their homoclinic and heteroclinic intersections.

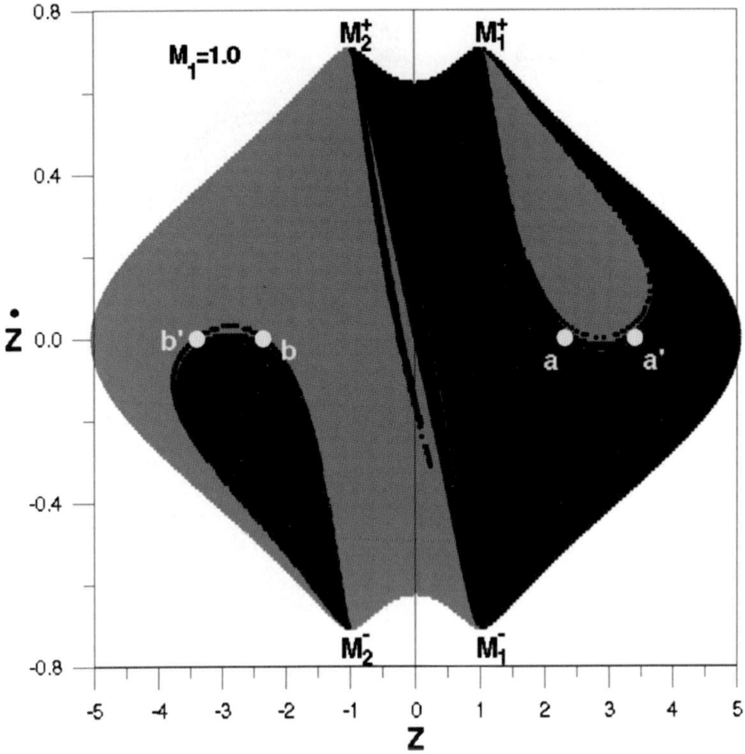

Figure 18. Basins of attraction of the black holes M_1 (black) and M_2 (gray) for $M_1 = M_2 = 1$, $E = \sqrt{0.5}$.

As M_1 decreases below $M_1 = 1$ the total area on the surface of section decreases and the black regions (orbits falling into M_1) decrease (Figures 21 and 22). When $M_1 \to 0$ there are no points falling on M_1 and the black region vanishes.

The total area covered by the orbits (permissible area) on the surface of section $(z,\dot{z})$ (with $x = 0$, $E = \sqrt{0.5}$ and $\dot{x} > 0$) is given analytically in Appendix A and in Figure 23. This area is equal to about 4.2 units for $M_1 = 0$ and increases with M_1. For relatively large M_1 the area increases almost linearly with M_1. The boundary of this permissible area is also calculated in Appendix A. For small M_1 there is a broad peak around $z = -1$ and a narrow peak around $z = 1$. For larger M_1 the peak around $z = 1$ is broader and for $M_1 = 1$ it is equal to the peak around $z = -1$. For $M_1 > M_2$ the peak around $z = 1$ becomes even broader and the peak around $z = -1$ becomes narrower (Figures 19–20).

The proportion of the orbits falling on M_2 is the gray area of figures like Figures 18–22, divided by the total area on the surface of section (Figure 24).

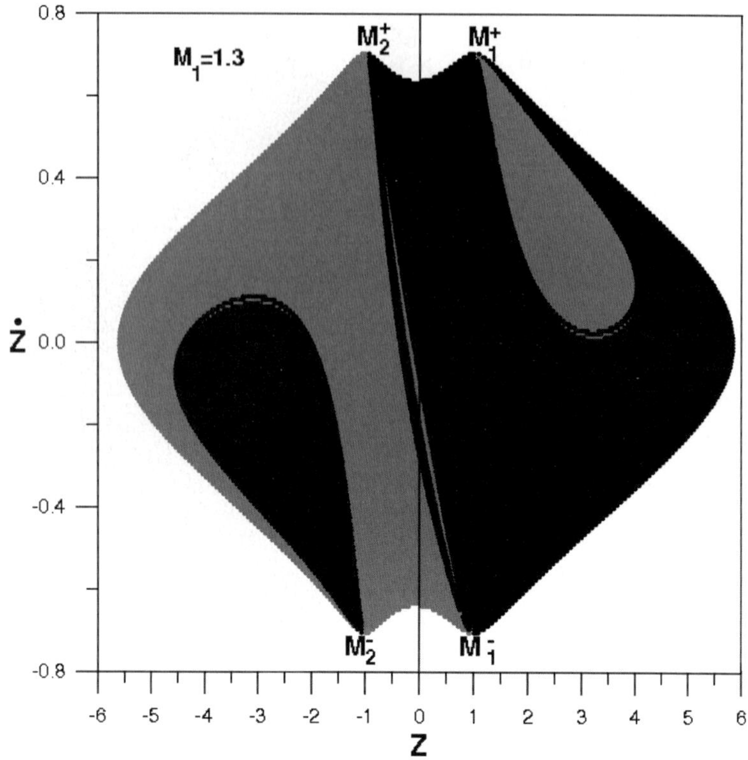

Figure 19. As in Figure 18 for $M_1 = 1.3$, $M_2 = 1$, $E = \sqrt{0.5}$.

If $M_1 \to 0$ this proportion goes to 1. However if M_1 is exactly equal to zero only a set of orbits of measure zero fall into M_2.

For small M_1 most of the permissible area is filled with orbits falling into M_2 (Figures 21 and 22). In particular the black region close to M_2^- (that contains orbits falling into M_1) is very small, and the black region on the right of the line $M_1^- M_2^+$ is thinner than in the case $M_1 = 1$ (compare Figures 21 and 18). As M_1 tends to zero both these black regions tend to disappear (Figure 22).

On the other hand as M_1 increases the black regions increase (Figures 18–20), while the gray regions become relatively smaller.

The total per cent proportion P of the gray regions as a function of M_1 can be given approximately by the formula (Figure 24)

$$\log P = 1.9656 - 0.2546 M_1 \tag{6}$$

Most orbits of the black and gray regions fall immediately into the black holes M_1 and M_2 respectively, i.e., without any intersection with the surface of section $(z,\dot{z})$ $(x = 0,\ \dot{x} > 0)$. However a small proportion of

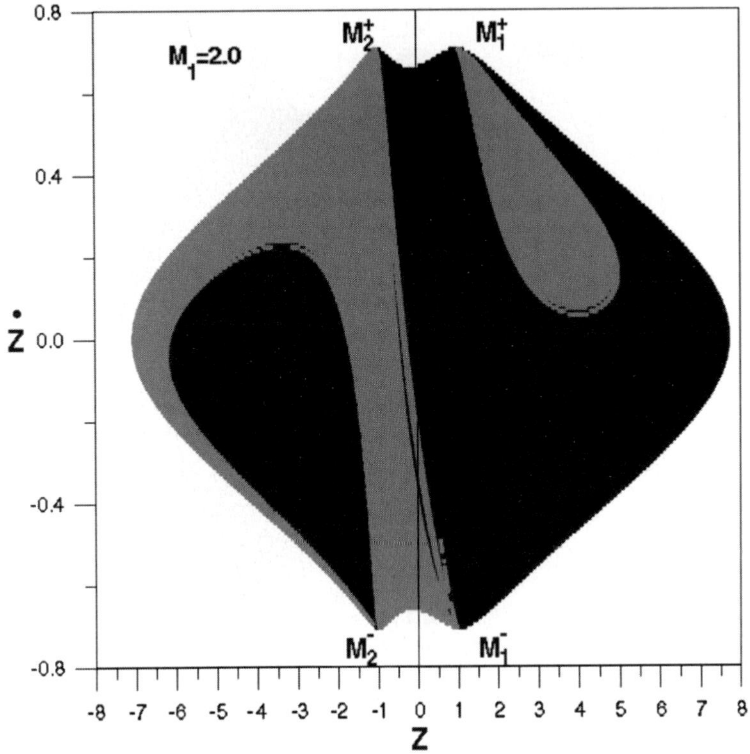

Figure 20 As in Figure 18 for $M_1 = 2$, $M_2 = 1$, $E = \sqrt{0.5}$. In this case the orbits a, a' do not exist. Nevertheless, the basins of attraction of M_1, M_2 are similar to those of Figures 18 and 19.

orbits have one, two, or more intersections with the surface of section $(z,\dot{z})$ before falling into M_1 or M_2. Finally a very small proportion of orbits are trapped around stable periodic orbits (when such orbits exist) and never fall into M_1 or M_2. An example of such trapped orbits, that form islands of stability and never fall into M_1 or M_2, is shown in Figure 8. Such orbits have been discussed by Contopoulos (1991). These orbits may define closed invariant curves around stable periodic orbits, or be chaotic, but without ever reaching either black hole.

In order to study further how various orbits fall into the black holes we have calculated the proportion of orbits, along the asymptotic curve from O(≡a) that have n intersections with the surface of section before falling into one of the black holes, where $n = 1,2,3$ etc. In Figure 25 we give the logarithm of the proportion of orbits remaining after n intersections as a function of n, for various values of M_1. We find approximately

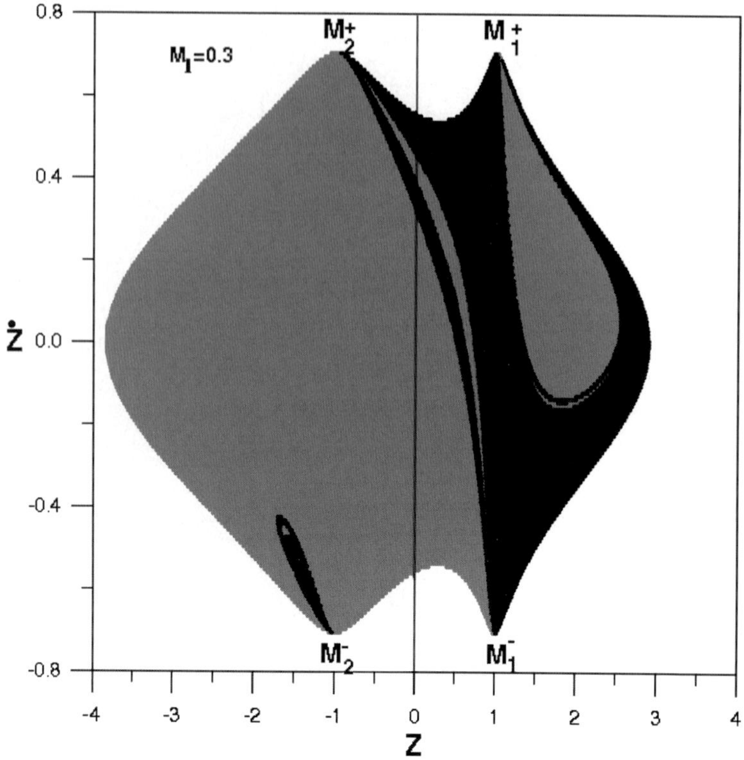

Figure 21. As in Figure 18 for $M_1 = 0.3$, $M_2 = 1$, $E = \sqrt{0.5}$.

$$\ln p = \mathrm{a} - \mathrm{b}n \tag{7}$$

i.e., the proportion p is decreasing exponentially with n for constant values of M_1, M_2, E. The slope of the rate of decrease b is near $b = 2$ for $M_1 = 1$ but increases for smaller M_1 and decreases for larger M_1 (see Table I and Figure 25). The last values of M_1 are close to the limiting value $M_{1\max}$, beyond which the periodic orbit a ceases to exist. In these cases the remaining orbits along the asymptotic curve decrease rather slowly as n increases, and more slowly as M_1 approaches the limiting value $M_{1\max}$. It seems that b tends to zero, as M_1 tends to $M_{1\max}$.

For M_1 close to $M_{1\max}$ the eigenvalue of the periodic orbit λ decreases abruptly and tends to 1 as $M_1 \to M_{1\max}$. The eigenvalue λ of the orbit a as a function of M_1 is given in Figure 26 and in Table I. If $\lambda - 1$ is small positive the orbits starting close to O along the unstable asymptotic curve from O deviate only slowly from the periodic orbit O, i.e., the orbits require a long time before they fall on the closest black hole (which is M_1). This explains why the value of b is smaller in this case.

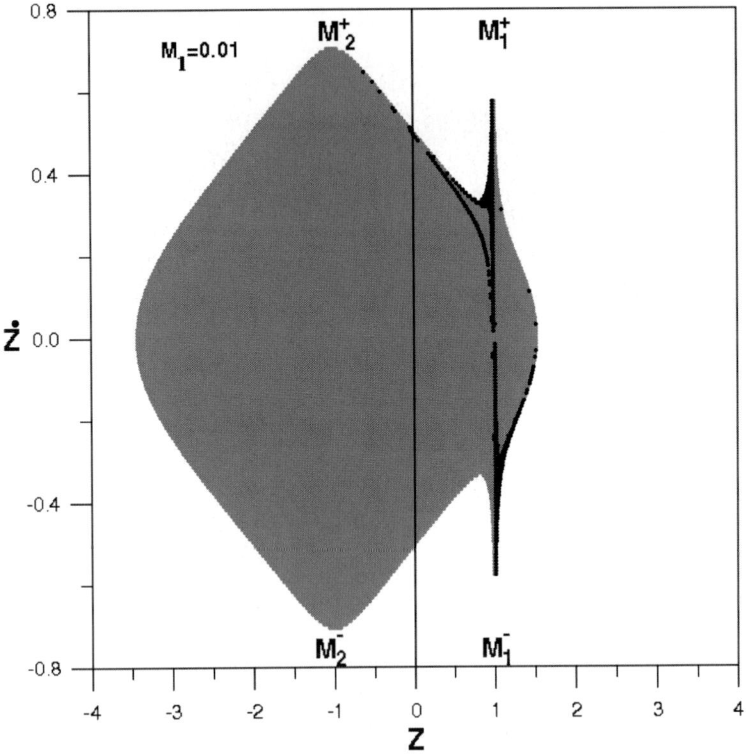

Figure 22. As in Figure 18 for $M_1 = 0.01$, $M_2 = 1$, $E = \sqrt{0.5}$.

5. Conclusions

1. In the case of two fixed black holes most of the chaotic orbits (but not all) fall into the black holes M_1 (orbits of type I), or M_2 (orbits of type II). These two sets of orbits are fractal, consisting of infinite subsets limited by homoclinic and heteroclinic orbits.
2. The homoclinic and heteroclinic points on a surface of section $(z,\dot{z})$ (with $x = 0$, $\dot{x} > 0$) are intersections of the asymptotic curves of the main unstable periodic orbits a, a', $\bar{a}$, $\bar{a}'$ (around M_1), h (hyperbolic-like) and b, b', $\bar{b}$, $\bar{b}'$ (around M_2). These asymptotic curves pass through the black holes M_1 ($z = 1$) and M_2 ($z = -1$) with velocities $\dot{z} = -E$ (points M_1^-, M_2^-) or $\dot{z} = +E$ (points M_1^+, M_2^+).
3. The asymptotic curves consist of several separate arcs, like (1), (2), (3), ... that terminate at one or two black holes. Between two successive arcs there are gaps containing orbits that fall into the black holes M_1 or M_2.
4. If we increase M_1, keeping $M_2 = 1$ and $E = \sqrt{0.5}$, we find that various arcs become shorter, and may not reach one, or the other black hole.

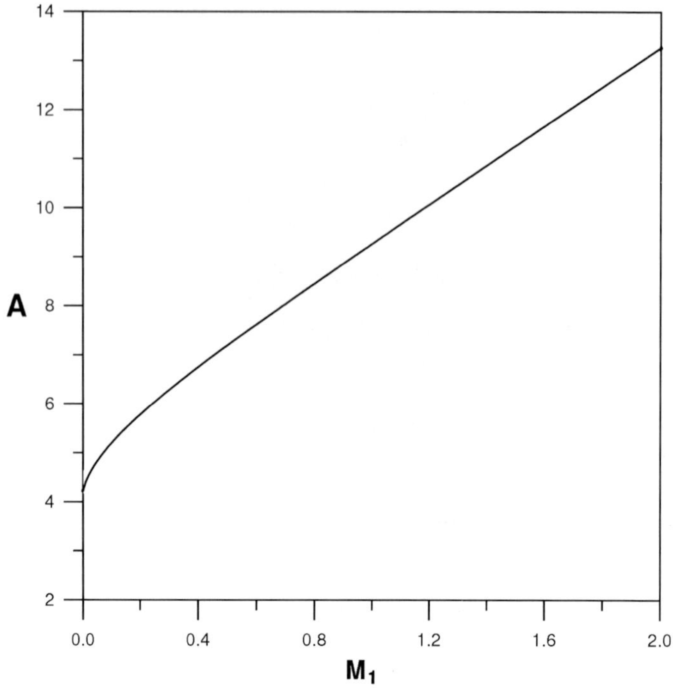

Figure 23. The total area on the surface of section $(z,\dot{z})$ (with $x = 0$, $E = \sqrt{0.5}$ and $\dot{x} > 0$) as a function of M_1.

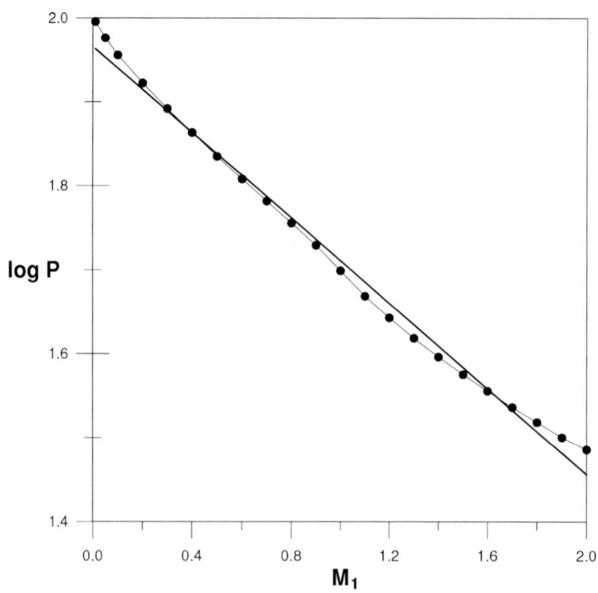

Figure 24. The logarithm of the percent proportion of the area of orbits falling on the black hole M_2.

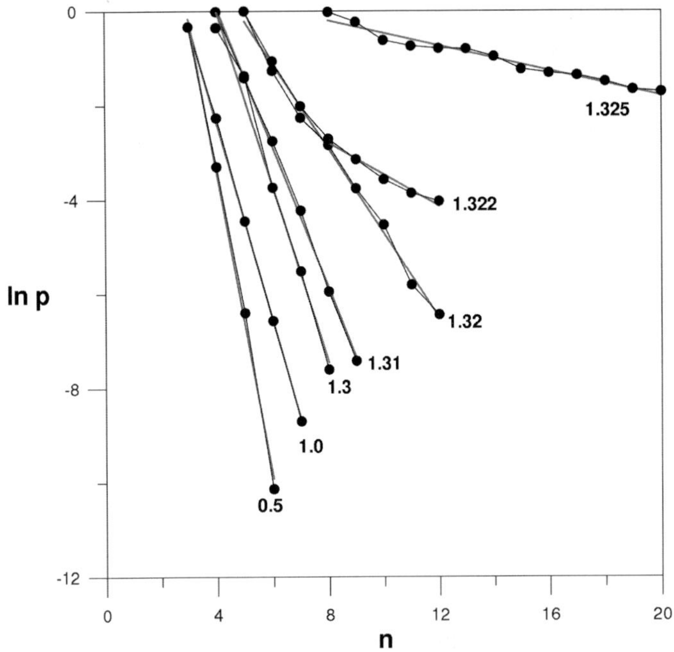

Figure 25. The logarithm of the proportion p of orbits starting along the asymptotic curve from O that remain after n intersections with the surface of section (before falling into a black hole). The curves $\ln p$ versus n for various values of M_1 are approximately straight lines with slopes b marked (Table I).

5. If we use a slightly different surface of section ($x = -0.001$) most (but not all) of the gaps disappear. Nevertheless, the orbits falling into the black holes do not have higher order intersections with the surfaces of section, therefore these surfaces are not Poincaré type surfaces of section.
6. As M_1 reaches a maximum value $M_{1\max}$ (≈ 1.3255 in the present case) the orbits a and a' join and for larger M_1 they disappear. For M_1 slightly smaller then $M_{1\max}$ the orbit a' is stable, and is surrounded by islands of stability.
7. The asymptotic curves of the orbits a', b, b' are very similar to those of orbit a.
8. The basins of attraction of the two black holes consist of large compact regions and of thin filaments forming fractal sets. The set of permissible orbits can be calculated analytically. It increases with M_1, almost linearly for large values of M_1.
9. The relative sizes of the set of orbits falling into M_1 increases, and the set of orbits falling into M_2 decreases as M_1 increases. If M_1 decreases and tends to zero, this set also decreases and tends to zero. The relative size of the set M_2 as a function of M_1 decreases almost exponentially.

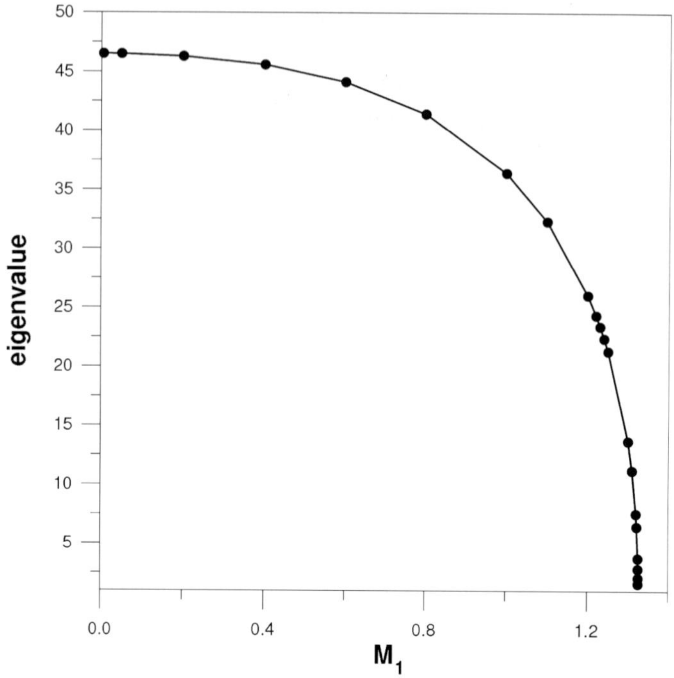

Figure 26. The eigenvalue λ of the orbits O($\equiv a$) as a function of M_1.

10. Finally we calculate the proportion p of orbits along an asymptotic curve that remains after $n = 1,2,3,\ldots$ intersections with the surface of section, before the orbits fall on a black hole. For fixed M_1, M_2 and E this proportion decreases exponentially with n ($\ln p = \mathrm{a} - \mathrm{b}n$). The slope b decreases as M_1 increases, and tends to zero as M_1 tends to $M_{1\max}$. This is explained by the fact that the eigenvalue λ of the orbit a decreases and tends to 1 as $M_1 \to M_{1\max}$.

Acknowledgements

This Research was supported by The Research Committee of the Academy of Athens (Program 200/557). We thank Dr. Efthymiopoulos for helping us in calculating Figure 23.

TABLE I

M_1	0.5	1.0	1.3	1.31	1.32	1.322	1.325
b	3.25	2.08	1.86	1.49	0.90	0.34	0.13
λ	44.98	36.43	13.71	11.23	7.58	6.48	3.79

APPENDIX A. Limiting Curves on the Surface of Section

The intersections of the orbits by the surface of section $x = 0$ are inside a limiting curve that is found as follows:

From Equation (2) we find that when $x = 0$ we have either $\theta = 0$, or $\psi = 0$.

In the first case we find from Equations (1)–(5):

$$V = -\left[\frac{M_1}{|z-1|} + \frac{M_2}{|z+1|} + 1\right], \tag{A.1}$$

$$Q\dot{\psi}^2 = \sinh^2\psi \cdot \dot{\psi}^2 = \dot{z}^2, \tag{A.2}$$

and

$$E^2 \leqslant \dot{z}^2 + V^{-2}. \tag{A.3}$$

Thus the limiting curve is

$$\dot{z}^2 = E^2 - V^{-2}. \tag{A.4}$$

Such curves are the boundaries of Figures 18–22.

In the second case

$$Q\dot{\theta}^2 = \sin^2\theta \cdot \dot{\theta}^2 = \dot{z}^2 \tag{A.5}$$

and we derive the same equation (A.4) for the limiting curve.

This curve goes through the points $M_1^-, M_1^+, M_2^-, M_2^+$, where $V = -\infty$ and $\dot{z} = \pm E$. When $z = 0$ we find

$$0 < \dot{z}^2 = E^2 - \frac{1}{(1 + M_1 + M_2)^2} < E^2. \tag{A.6}$$

The minimum value of $|\dot{z}|$ between $z = -1$ and $z = 1$ occurs when $|V|$ is minimum. This occurs for

$$z = \frac{(\sqrt{M_2} - \sqrt{M_1})}{\sqrt{M_2} + \sqrt{M_1}} \tag{A.7}$$

i.e., at a negative z if $M_1 > M_2$ and at a positive z if $M_1 < M_2$. This is seen in Figures 19–22.

When $\dot{z} = 0$ we find $z = z_{\max} > 1$ and $z = z_{\min} < -1$.

In the first case

$$\frac{M_1}{z_{\max} - 1} + \frac{M_2}{z_{\max} + 1} = \frac{1}{E} - 1 = \varepsilon. \tag{A.8}$$

Solving this equation we find

$$z_{\max} = \mu + \left[(\mu + 1)^2 - \frac{2M_2}{\varepsilon}\right]^{1/2}, \tag{A.9}$$

where

$$\mu = \frac{M_1 + M_2}{2\varepsilon}. \tag{A.10}$$

In the second case

$$\frac{M_1}{1 - z_{\min}} - \frac{M_2}{1 + z_{\min}} = \varepsilon. \tag{A.11}$$

Solving this equation we find

$$z_{\min} = -\mu - \left[(\mu - 1)^2 + \frac{2M_2}{\varepsilon}\right]^{1/2}. \tag{A.12}$$

If $M_1 > M_2$ we have $z_{\max} > |z_{\min}|$ and if $M_1 < M_2$ we have $|z_{\min}| > z_{\max}$.

In the particular case of Figure 16 where $M_1 = M_2 = 1$, $E = \sqrt{0.5} = 1/\sqrt{2}$ we have $\varepsilon = \sqrt{2}-1 = 0.414$, $\mu = 1/\varepsilon = \sqrt{2}+1 = 2.414$ and $z_{\max} = |z_{\min}| = \mu + (\mu^2 + 1)^{1/2} = 5.027$.

If $M_1 \to 0$ we have the same ε, but $\mu = 1/2\varepsilon = E/2(1 - E) = 1.207$ and $z_{\max} = 2\mu - 1 = 2E - 1/1 - E = 1.414$, while $z_{\min} = -2\mu - 1 = -1/1 - E = -3.414$. On the other hand if M_1 is large we have $\mu \approx M_1/2\varepsilon$ and $z_{\max} \approx 2\mu \approx M_1/\varepsilon$ while $z_{\min} \approx -2\mu \approx -z_{\max}$.

The area of initial conditions on the surface of section $(z, \dot{z})$ is

$$A = 2\int_{z_{\max}}^{z_{\min}} \dot{z}\, \mathrm{d}z, \tag{A.13}$$

where $\dot{z}$ is given from Equation (A.4), while V has the following expressions

$$\text{If } z > 1, \quad V = V_1 = -\left[\frac{M_1}{z - 1} + \frac{M_2}{z + 1} + 1\right], \tag{A.14}$$

$$\text{If } 1 > z > -1, \quad V = V_2 = -\left[\frac{M_1}{1 - z} + \frac{M_2}{1 + z} + 1\right], \tag{A.15}$$

$$\text{and if } -1 > z, \quad V = V_3 = -\left[\frac{M_1}{1 - z} - \frac{M_2}{1 + z} + 1\right]. \tag{A.16}$$

Then using the package of Mathematica we find the area as a function of M_1, M_2 and ε. The values of A as a function of M_1 for $M_2 = 1$ and $\varepsilon = \sqrt{2}-1$ are given in Figure 23.

The boundary of this area can also be calculated from Equation (A.4). For $M_1 = 0$ this boundary is symmetric around an axis $z = -1$. When M_1 is small there is a narrow peak around $z = 1$ reaching $\dot{z} = E = \sqrt{0.5}$ (Figure 22), while for larger M_1 this peak is broader (Figures 18–21).

References

Chandrasekhar, S.: 1989, *Proc. Roy. Soc. London A* **421**, 227.
Charlier, C. L.:1902, "Die Mechanik des Himmels", *von Veit,* Leipsig.
Contopoulos, G.: 1990, *Proc. Roy. Soc. London A* **431**, 183.
Contopoulos, G.: 1991, *Proc. Roy. Soc. London A* **435**, 551.
Contopoulos, G. and Harsoula, M.: 2004, *J. Math. Phys.* **45**, 4932.
Deprit, A.: 1960, Mathématiques du XXème sciècle, *Univ. Louvain* **1**, 45.
Dettman, C. P., Frankel, N. E. and Cornish, N. J.: 1994, *Phys. Rev. D* **50**, R618.
Dettman, C. P., Frankel, N. E. and Cornish, N. J.: 1995, *Fractals* **3**, 161.

Celestial Mechanics and Dynamical Astronomy (2005) 92:219–241
DOI 10.1007/s10569-005-2578-2

DISCRETE VIRIAL THEOREM

J. E. HOWARD
Laboratory for Atmospheric and Space Physics, University of Colorado, Boulder, CO 80309-0392 USA, e-mail: jhoward@colorado.edu

(Received: 5 May 2004; revised: 8 February 2005; accepted: 21 February 2005)

Abstract. We reexamine the classical virial theorem for bounded orbits of arbitrary autonomous Hamiltonian systems possessing both regular and chaotic orbits. New and useful forms of the virial theorem are obtained for natural Hamiltonian flows of arbitrary dimension. A discrete virial theorem is derived for invariant circles and periodic orbits of natural symplectic maps. A weak and a strong form of the virial theorem are proven for both flows and maps. While the Birkhoff Ergodic Theorem guarantees the existence of the relevant time averages for both regular and chaotic orbits, the convergence is very rapid for the former and extremely slow for the latter. This circumstance leads to a simple and efficient measure of chaoticity. The results are applied to several problems of current physical interest, including the Hénon–Heiles system, weak chaos in the standard map, and a 4D Froeschlé map.

Key words: chaos, Hamiltonian systems, symplectic maps, virial theorem

1. Introduction

Virial theorems (Clausius, 1850) are useful throughout the physical sciences, with important applications in statistical mechanics (Pathria, 1996), astrophysics, magnetohydrodynamics, quantum mechanics (Atkins, 1983), and many other fields. For example, in fluid models of galactic dynamics (Binney and Tremaine, 1987) a tensor virial theorem (Parker, 1954) has been derived and shown to yield important connections between the observed kinetic energy of a stellar system and its median radius. Shafranov (1966) has employed a magnetic virial theorem to prove that a plasma cannot be stably confined without the action of external forces. We present here a new kind of virial appropriate to symplectic maps modeled after "natural" Hamiltonian flows (Arnold, 1989). Symplectic maps are widely used to describe physical systems such as particle accelerators (Dragt, 1979), plasma wave heating (Howard et al., 1986), microwave ionization of Rydberg atoms (Blümel and Reinhardt, 1997), and many problems in celestial mechanics (Murray and Dermott, 1999).

We begin by revisiting the classical virial theorem for bounded orbits of natural Hamiltonian flows. A weak and a strong form of the virial theorem are proven and an asymptotic invariant function of the coordinates is obtained. For periodic orbits the relevant integrals are finite and offer a means of numerically determining the action variables. The existence of the time averages for both regular and chaotic orbits is assured by the Birkhoff Ergodic Theorem (Birkhoff, 1931; Robinson, 1998). Even though the virial integrals are extremely slow to converge for chaotic orbits, the running time-averages agree and, with the help of energy conservation, lead to an asymptotically invariant function of the coordinates alone. The results are applied to regular and chaotic orbits in the Hénon–Heiles system, for which an asymptotically invariant polynomial is found.

Next we define a discrete virial for natural symplectic maps and derive a discrete virial theorem which applies to all invariant circles, periodic orbits, and bounded chaotic orbits. The results are applied to a class of polynomial maps (Dullin and Meiss, 2000), the standard map (Lichtenberg and Lieberman, 1992), and a 4D Froeschlé Map (Froeschlé, 1984). In all cases dramatic differences in the convergence of the time averages of the kinetic energy are found between regular and chaotic orbits. For example, 10^{10} iterations of chaotic orbits are typically required to demonstrate convergence! This suggests that intermediate-time variations in momentum might offer an effective means of detecting chaotic motion. A new and simple measure of chaoticity is proposed and applied to the Hénon–Heiles flow and several 2D symplectic maps. Finally we suggest some generalizations to nonnatural systems. We expect that the theorem will find many such applications.

2. Natural Flows

We begin by briefly reviewing the classical virial theorem (Goldstein et al., 1989) in a form appropriate to a Hamiltonian flow, described by

$$H = T(\mathbf{q}, \mathbf{p}) + U(\mathbf{q}), \tag{1}$$

where $\mathbf{q}$ and $\mathbf{p}$ are canonical coordinates in $\mathbb{R}^n$. If the kinetic energy T is a positive definite quadratic form, i.e. $T = 1/2\mathbf{p}^T B\mathbf{p}$, where $B(\mathbf{q})$ is a positive definite, nondegenerate mass matrix, the flow is called "natural" (Arnold, 1989). The potential energy U is assumed to be smooth and to possess at least one local minimum, ensuring the existence of bounded orbits. Define

$$G = \mathbf{p}^T\mathbf{q} \tag{2}$$

so that

$$\dot{G} = \sum_1^n (p_i\partial_{p_i}H - q_i\partial_{q_i}H) = 2T - \mathbf{q}^T\nabla H. \tag{3}$$

Now integrate both sides from $t = 0$ to $t = \tau$ and divide by τ:

$$\frac{2}{\tau}\int_0^\tau T\,\mathrm{d}t - \frac{1}{\tau}\int_0^\tau \mathbf{q}^T\nabla H\,\mathrm{d}t = \frac{G(\tau)-G(0)}{\tau}. \tag{4}$$

If the motion is periodic of period τ, the RHS vanishes. Otherwise, since the motion is assumed bounded, the RHS tends to zero as $\tau \to \infty$, giving the classical virial theorem,

$$2\langle T\rangle = \left\langle \sum_i q_i \partial_{q_i} H\right\rangle = \langle \mathbf{q}^T \nabla H\rangle, \tag{5}$$

where $\langle f\rangle$ denotes the orbital time average of f and ∇ the spatial gradient. The expression $\langle \mathbf{q}^T\nabla H\rangle$ is the classical *virial of Clausius* (Clausius, 1850). If T is independent of $\mathbf{q}$, then energy conservation, $\langle T\rangle + \langle U\rangle = E$, may be used to derive a condition involving only coordinates, which takes on an especially simple form in $\mathbb{R}^2$:

$$\langle \nabla\cdot(\mathbf{q}U)\rangle = 2E. \tag{6}$$

As we shall see, this relation converges rapidly for both regular and chaotic orbits. We call such quantities "asymptotic invariants." In the important case of central forces (5) reduces to $2\langle T\rangle - \langle r\partial_r U\rangle$. For power law potentials, $U = U_0 r^m$, widely used as a paradigm for the classical virial theorem, (5) reduces to $2\langle T\rangle = m\langle U\rangle = 2m/(m+2)\ E$. These results readily generalize to a collection of particles in $\mathbb{R}^n$, a starting point for the kinetic theory of nonideal gases (Pathria, 1996).

A stronger form of the virial theorem may be derived for each component via (no sum on k)

$$G_k = p_k q_k. \tag{7}$$

Proceeding as above gives

$$2\langle T_k\rangle = \langle q_k \partial_{q_k} H\rangle, \tag{8}$$

which may be summed to obtain the weak form (5).

As an example consider the Hénon–Heiles system (Hénon and Heiles, 1964)

$$H = \frac{1}{2}(p_x^2 + p_y^2) + U(x,y) \tag{9}$$

with $U = 1/2(x^2+y^2) + x^2y - 1/3\ y^3$. The strong form of the virial theorem then implies

$$\langle p_x^2\rangle = \langle xU_x\rangle = \langle x^2 + 2x^2y\rangle, \quad \langle p_y^2\rangle = \langle yU_y\rangle = \langle y^2 + x^2y - y^3\rangle, \tag{10}$$

while the coordinate form (6) becomes

$$\langle P(x,y)\rangle = \langle 2x^2 + 2y^2 + 5x^2y - \frac{5}{3}y^3\rangle = 2E. \tag{11}$$

In order to apply these results let us examine a typical Poincaré section (p_y, y) with $x = 0$ and $p_x > 0$, with energy $E = 0.125$, as depicted in Figure 1.

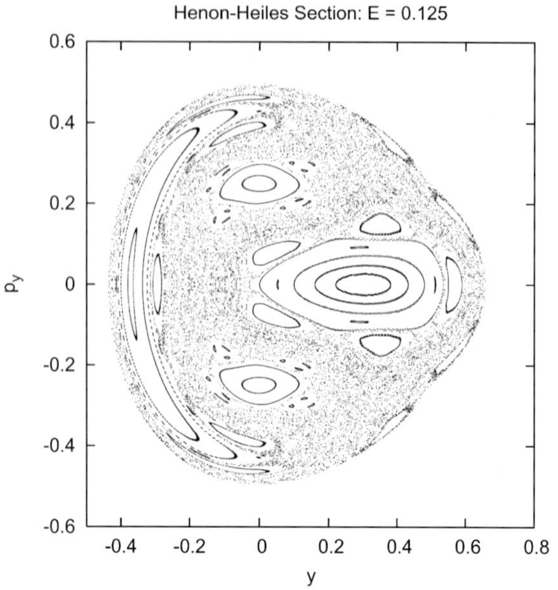

Figure 1. Poincaré section for Hénon–Heiles system for $E = 0.125$.

Choose one orbit at $y_0 = 0.4$ in the regular region and a second at $y_0 = -0.17$ in the large chaotic zone. Figure 2 compares the time evolution of the partial sums $\langle p_x^2 \rangle$ and $\langle xU_x \rangle$, while Figure 3 shows the polynomial (6) for each orbit. For the regular orbit, the virial sums rapidly converge to the same value, whereas for the chaotic orbit the same sums vary wildly over the time range considered. The asymptotic invariant polynomial $P(x, y)$, however, converges rapidly to $2E$ in both cases; the only apparent difference between the regular and chaotic orbits is the somewhat larger oscillations in the latter case. The convergence of the virial integrals for chaotic orbits is complicated by "sticky" regions, where an orbit can spend long times exploring "islands around islands" (Meiss, 1994). It is difficult to demonstrate convergence numerically for flows, owing to the extremely long integration times required. This regime is better explored using discrete mappings, which are discussed in the following sections.

For periodic orbits (4) is exact and yields a formula for the action integral $J_x = 1/2\pi \oint p_x \, dx$, so that one can obtain a good numerical value by integrating over a sufficiently long time. In general, when the motion is not periodic, there is still the possibility of regular quasiperiodic motion (Arnold, 1989). The existence of the time averages for almost all initial conditions is guaranteed by the Birkhoff Ergodic Theorem (Birkhoff, 1931; Robinson, 1998). However, the convergence rates differ greatly, very rapidly for regular orbits but extremely slowly for chaotic orbits.

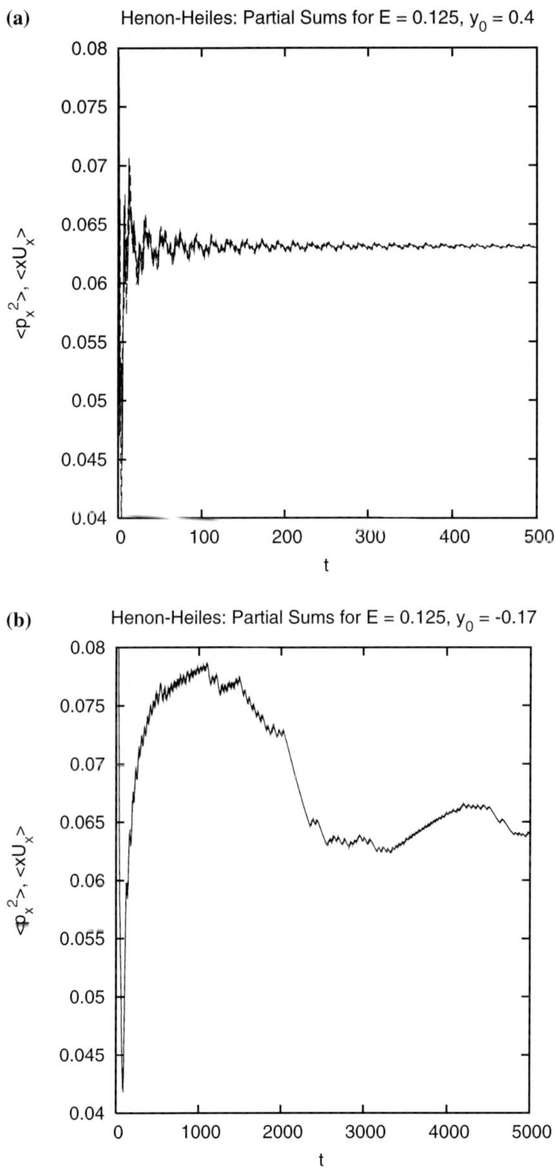

Figure 2. Time-dependent running averages of $\langle p_x^2 \rangle$ and $\langle xU_x \rangle$ for regular orbit (a) and chaotic orbit (b) for Hénon–Heiles system with $E = 0.125$. The individual sums are virtually identical in each case.

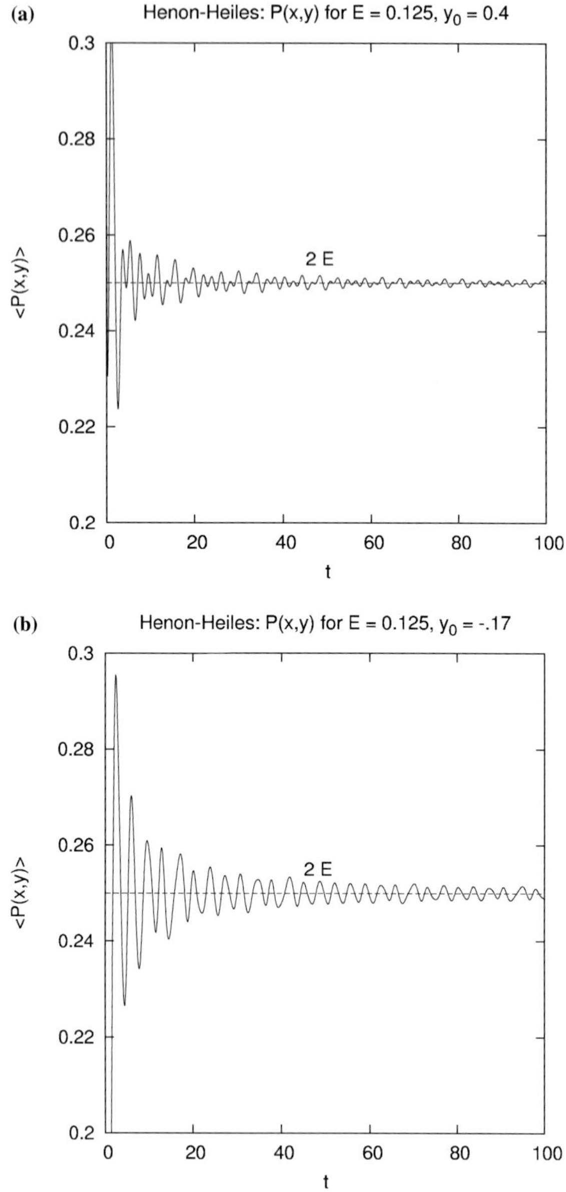

Figure 3. Averaged polynomial $\langle P(x,y)\rangle(t) = \langle 2(x^2+y^2) + 5(x^2y - y^3/3)\rangle$ for regular orbit (a) and chaotic orbit (b) for Hénon–Heiles system for $E = 0.125$.

For example, regular motion in the Hénon–Heiles system is two-frequency quasiperiodic, so that one may Fourier expand

$$p_x^2 = \sum_{j,k \in Z} a_{jk}(\mathbf{J}) e^{i(j\omega_1 t + k\omega_2 t)}, \tag{12}$$

where $\omega_1(\mathbf{J})$ and $\omega_2(\mathbf{J})$ are constant frequencies and $\mathbf{J} = (J_1, J_2)$ are the conserved (albeit unknown) actions. It follows that, in the absence of resonances:

$$\langle p_x^2 \rangle = a_{00}(\mathbf{J}). \tag{13}$$

Hence, for regular orbits, for which the actions $\mathbf{J}$ exist, the time average of p_x^2 tends to a constant. For chaotic orbits, on the other hand, such an expansion does not exist and $\langle p_x^2 \rangle$ wanders erratically for moderate times, as seen in Figure 2(b). Of course, convergence eventually obtains, although this is difficult to demonstrate numerically.

The dramatic difference in the time variations of $\langle p_x^2 \rangle$ suggests that they might serve as a useful indicator of chaotic motion. We shall return to this possibility in Section 5, after we have derived a virial theorem for discrete maps.

3. Natural Maps

The class of symplectic maps analogous to natural Hamiltonian flows are called natural maps (Howard and Dullin, 2001) and may be generated by the Hamiltonian generating function

$$F(\mathbf{q}_k, \mathbf{p}_{k+1}) = \mathbf{q}_k^T \mathbf{p}_{k+1} + H(\mathbf{p}_{k+1}, \mathbf{q}_k), \tag{14}$$

where $\mathbf{q}, \mathbf{p} \in \mathbb{R}^n$ and $H(\mathbf{q}, \mathbf{p})$ is given by (1). It follows that $\mathbf{p}_k = \partial F / \partial \mathbf{q}_k$ and $\mathbf{q}_{k+1} = \partial F / \partial \mathbf{p}_{k+1}$ which lead to the radial twist map (Lichtenberg and Lieberman, 1992),

$$\begin{aligned} \mathbf{p}_{k+1} &= \mathbf{p}_k - \nabla U(\mathbf{q}_k), \\ \mathbf{q}_{k+1} &= \mathbf{q}_k + B^{-1} \mathbf{p}_{k+1}. \end{aligned} \tag{15}$$

Physically, this map corresponds to a "kicked nonlinear oscillator," wherein a particle is subject to an infinite periodic string of impulsive kicks of strength given by ∇U (Lichtenberg and Lieberman, 1992). Even though the total energy is not conserved, we may nevertheless define local and averaged kinetic and potential energies and discover useful relationships among them.

Thus, analogous to (2) let us define (no sum on k)

$$G_k = \mathbf{p}_k^T \mathbf{q}_k \tag{16}$$

and analogous to $\dot{G}$,

$$G_{k+1} - G_k = \mathbf{p}_{k+1}^T (\mathbf{q}_{k+1} - \mathbf{q}_k) + (\mathbf{p}_{k+1} - \mathbf{p}_k)^T \mathbf{q}_k. \tag{17}$$

Time integration becomes a sum over N points of an orbit, so that, using the mapping (15),

$$\frac{1}{N}\sum_{1}^{N}[\mathbf{p}_{k+1}^{T}B^{-1}\mathbf{p}_{k+1}-\mathbf{q}_{k}^{T}\nabla H(\mathbf{q}_{k})]=\frac{1}{N}[G_{N+1}-G_{1}]. \tag{18}$$

By analogy with the continuous case we shall call the sum $\sum \mathbf{q}_k^T\nabla H$ the *discrete virial.* For a periodic orbit of length $N, G_{N+1}=G_1$ and the RHS vanishes. Otherwise, assuming that G_k remains bounded for all k we can make the RHS as small as we please for large enough N. Defining the 'discrete kinetic energy,' $T_k = 1/2\ \mathbf{p}_k^T B^{-1}\mathbf{p}_k$, we obtain the *discrete virial theorem* (weak form)

$$2\langle T\rangle = \langle \mathbf{q}^T\nabla U\rangle, \tag{19}$$

where we have assumed constant mass matrix B. As in the continuous case a strong form of the discrete may be derived by using $G_k^i = p_k^i q_k^i$ to obtain

$$2\langle T_i\rangle = \langle q^i\partial_i U\rangle. \tag{20}$$

Just as for the continuous case, these simple statements express a statistical balance between kinetic and potential energies, which becomes especially simple for power law potentials. The essential fact is that the LHS depends solely on $\mathbf{p}$ and the RHS solely on $\mathbf{q}$. For rotational orbits, unbounded in $\mathbf{q}$, G diverges but its average remains finite, so that the theorem generalizes to

$$\langle G\rangle = 2\langle T\rangle - \langle \mathbf{q}^T\nabla U\rangle. \tag{21}$$

As for continuous flows, convergence of the virial sums is assured almost everywhere by the Birkhoff Ergodic Theorem, with vast differences in convergence rates for regular and chaotic orbits. Note that in contrast to natural flows, there is no conserved energy for a natural map, only a locally conserved two-form (Howard and Dullin, 2001). Thus we do not obtain an asymptotic invariant analogous to (6). Note also that (21) holds for chaotic as well as for regular orbits, provided only that they are bounded in both $\mathbf{p}$ and $\mathbf{q}$. The most immediate application is to periodic orbits, giving exact new relations among the intermediate points, as we now illustrate by means of a few examples.

4. Examples

Power Law Map. Consider the 2D map defined by the power law potential, $U = U_0 q^m$. It easily follows that $2\langle T\rangle = m\langle U\rangle$, or

$$\lim_{N\to\infty}\frac{1}{N}\sum_{1}^{N}p_k^2 = mU_0\ \lim_{N\to\infty}\frac{1}{N}\sum_{i}^{N}q_k^m. \tag{22}$$

For even $m \neq 2$ this elementary polynomial map (Dullin and Meiss, 2000) has a fixed point at the origin, which is marginally stable (parabolic) for all U_0. In fact, it is easily shown that all power maps of the same degree $m \neq 2$ are scale equivalent: if $U_0 \mapsto \tilde{U}_0$ just scale $q = \alpha\tilde{q}, p = \alpha\tilde{p}$, with $\alpha = (\tilde{U}_0/U_0)^{1/(m-2)}$. Thus without loss of generality we can take $mU_0 = 1$, $M = m - 1$ and write

$$\begin{aligned} p' &= p - q^M, \\ q' &= q + p'. \end{aligned} \tag{23}$$

For $m = 2$ one obtains a one-parameter linear map. Figure 4(a) depicts the power law map for $m = 4$. The unbounded chaotic region surrounding the

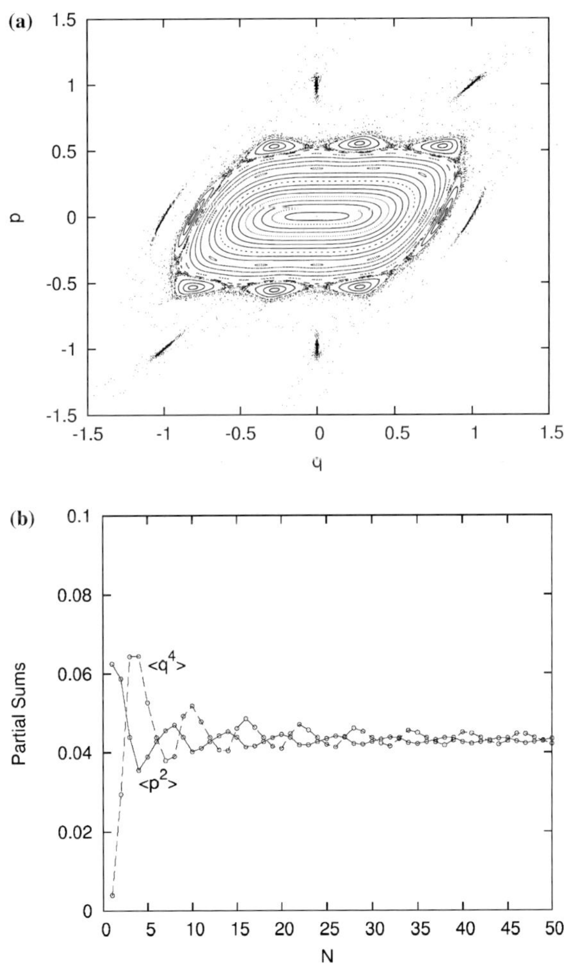

Figure 4. (a) Power-law map and (b) convergence of partial sums for $m = 4$, for initial conditions $q_i = 0, p_i = 0.25$. The solid curve is $\langle p^2 \rangle$ and the dashed curve $\langle q^4 \rangle$.

central island is generated by overlap of the 6- and 8-fold Birkhoff chains. There is also a very thin separatrix layer surrounding the 10-fold Birkhoff chain. Figure 4(b) illustrates the rapid convergence of the partial sums for a typical librational invariant circle. We have also verified (22) for a pair of coupled power maps based on the potential $U(x, y) = \frac{1}{4}(x^4 + y^4) - \frac{1}{2}hx^2y^2$.

Modified Power Law Map. We can create a larger bounded chaotic region by adding a term of lower degree to the power law map

$$\begin{aligned} p' &= p - q^M + \epsilon q^{M-2}, \\ q' &= q + p'. \end{aligned} \tag{24}$$

Figure 5 depicts this map for $M = 3, \epsilon = 0.31$, showing a bounded chaotic separatrix layer emanating from the hyperbolic point at the origin. The partial sums (22) become $\langle p^2 \rangle = \langle q^4 \rangle - \epsilon \langle q^2 \rangle$. Figure 6(a) illustrates the convergence of the partial sums for an invariant circle near the right-hand elliptic fixed point. The behavior is more complicated in the chaotic zone. As Figure 6(b) shows, the partial sums agree after a few hundred steps, but do not seem to converge to a constant, even after 5000 steps (Figure 6(c)). Nevertheless, convergence is assured by the Birkhoff Ergodic Theorem, but only after very long times. To demonstrate convergence, we have integrated the same chaotic orbit for 10^{11} steps. The results are shown logarithmically in Figure 6(d). The reason for this delayed convergence is thought to be the sticky regions near the boundary of the available space, involving "islands around islands" (Meiss, 1994).

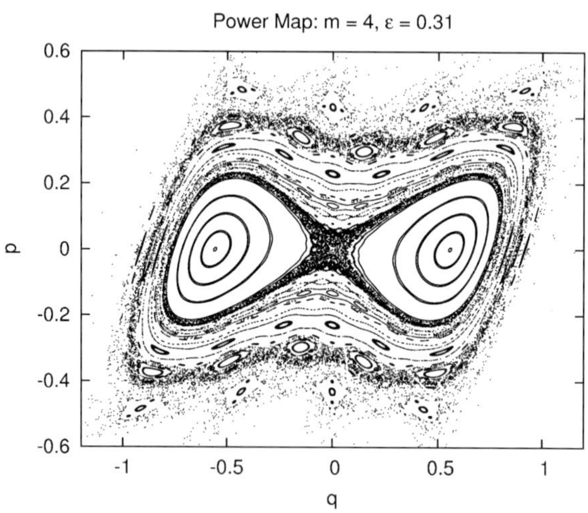

Figure 5. Modified power map for $m = 4$, $\varepsilon = 0.31$.

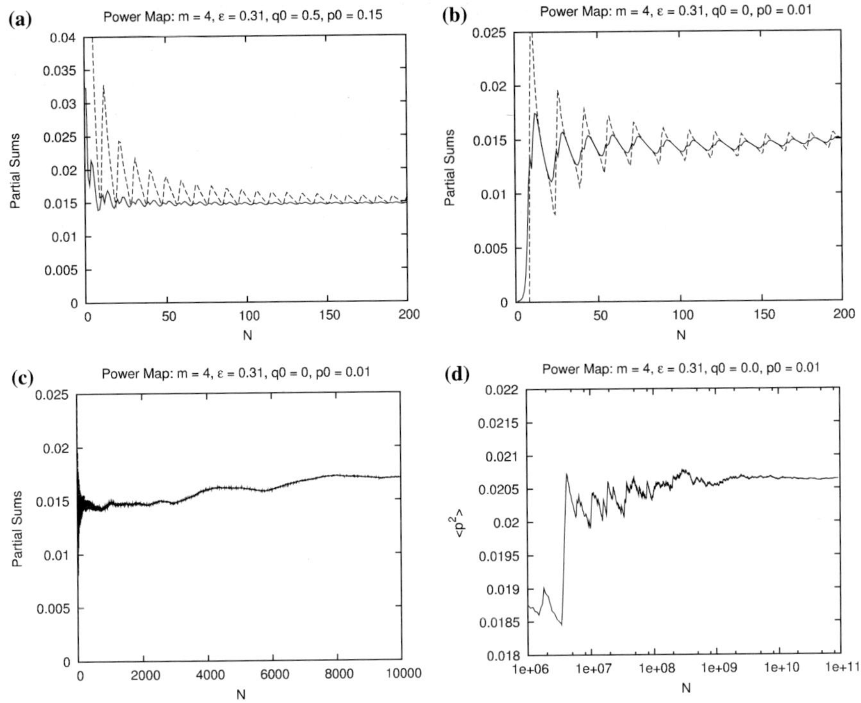

Figure 6. Convergence of partial sums for modified power map (a) regular orbit, (b) chaotic orbit (short time), (c) chaotic orbit (intermediate time), (d) chaotic orbit (long time log plot).

Standard Map. Next consider the standard map (Lichtenberg and Lieberman, 1992), with $U = -K\cos q$, shown in Figure 7(a) for $K = 0.95$. The virial theorem yields

$$\lim_{N\to\infty} \frac{1}{N}\sum_{1}^{N} p_k^2 = K \lim_{N\to\infty} \frac{1}{N}\sum_{1}^{N} q_k \sin q_k. \tag{25}$$

We have verified this result for a librational invariant circle and the eightfold Birkhoff chain in Figure 7(a). Figure 7(b) illustrates the rapid convergence of the two partial sums in (25) for the invariant circle; the finite-sum version was found to hold exactly for the island chain. As another check consider the period-two orbits which appear for $K > 4$ (see Figure 4.3 of Lichtenberg and Lieberman, 1992), which satisfy $q_1 = -q_2, p_1 = -p_2$, so that (25) reduces to $p_1^2 = Kq_1 \sin q_1$. But $p_1 = 2q_1$, which gives the known result $4q_1 = K \sin q_1$, thus confirming (25).

Froeschlé Map. As an example of a 4D natural map, consider the Symmetric Froeschlé Map (Froeschlé, 1984; Howard and Dullin, 2001) for which $U(x, y) = -K_1 \cos x - K_2 \cos y - h \cos(x + y)$, so that

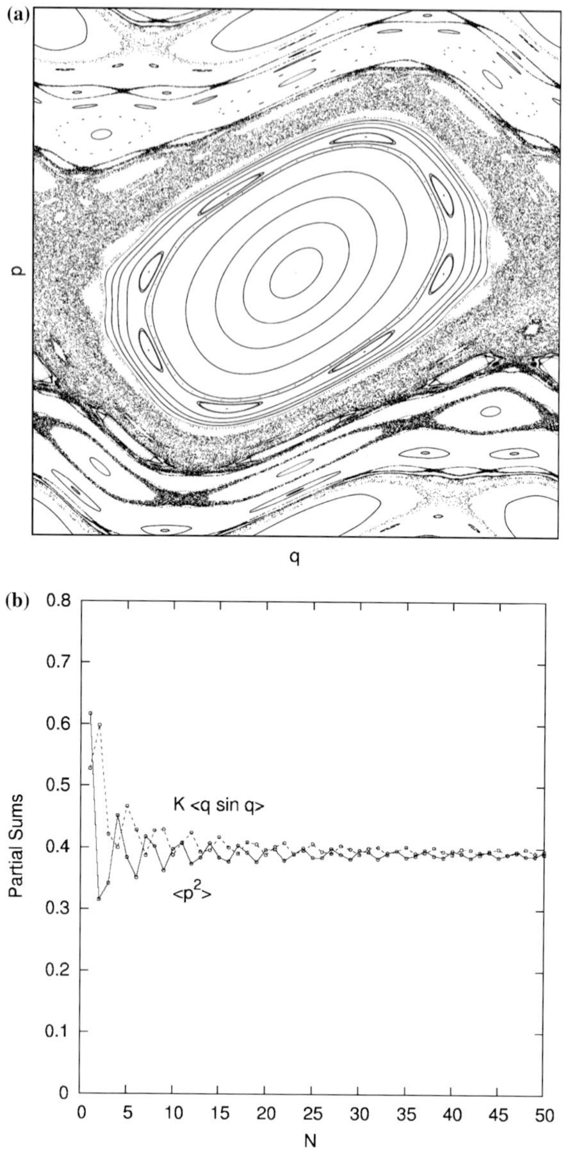

Figure 7. (a) Standard map for $K = 0.95$ and (b) partial sums for invariant librational circle.

$$\left.\begin{aligned} p_{x_{k+1}} &= p_{x_k} - K_1 \sin x_k - h \sin(x_k + y_k) \\ p_{y_{k+1}} &= p_{y_k} - K_2 \sin y_k - h \sin(x_k + y_k) \\ x_{k+1} &= x_k + p_{x_{k+1}} \\ y_{k+1} &= y_k + p_{y_{k+1}} \end{aligned}\right\} \bmod 2\pi. \tag{26}$$

The strong form of the virial theorem then reads

$$\lim_{N\to\infty} \frac{1}{N}\sum_{1}^{N} p_{x_k}^2 = \lim_{N\to\infty} \frac{1}{N}\sum_{1}^{N}[K_1 x_k \ \sin x_k + h x_k \ \sin(x_k + y_k)], \tag{27}$$

$$\lim_{N\to\infty} \frac{1}{N}\sum_{1}^{N} p_{y_k}^2 = \lim_{N\to\infty} \frac{1}{N}\sum_{1}^{N}[K_2 y_k \ \sin y_k + h y_k \ \sin(x_k + y_k)]. \tag{28}$$

Figure 8 depicts a regular orbit with $K_1 = K_2 = 2$ and $h = 0.0825$ and initial conditions $x_0 = 0$, $y_0 = 0$, $p_{x0} = \pi/4$, $p_{y0} = 0$. This invariant set maps into three disjoint curves on an invariant 2-torus. As in the 2D examples, the virial sums converge rapidly to the same value, as seen in Figure 9(a) for the x-component. For $K = 1$ and $h = 1.205$ we obtain the chaotic orbit depicted in Figure 10, whose virial sums also agree but converge much more slowly, as seen in Figure 9(b). In addition, plotting $\langle p_x^2 \rangle$ and $\langle p_y^2 \rangle$ together provides a vivid description of orbital trapping into resonances. For example, Figure 9(c) shows that these averages converge to the same value for a regular orbit trapped in the 1 : 1 resonance after about 1000 steps.

5. A New Chaos Indicator?

There are several competing methods for an efficient detector of chaoticity. The traditional method is to calculate the maximum Lyapunov exponent

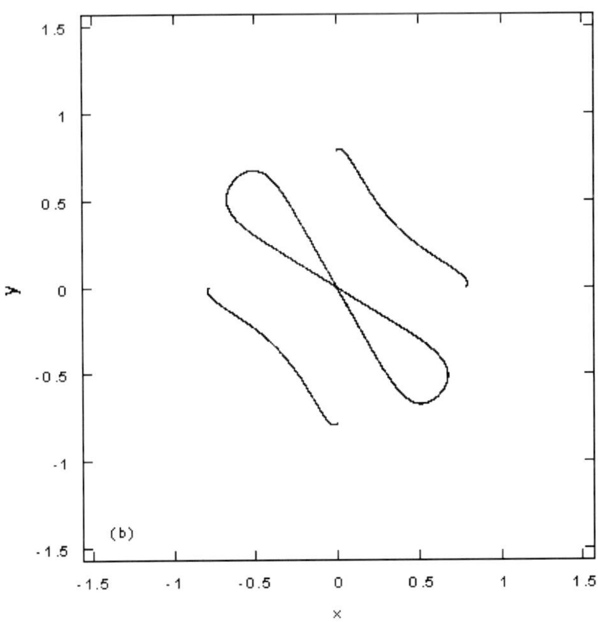

Figure 8. The x–y projection of regular orbit for Froeschlé map for $K = 2, h = 0.0825$.

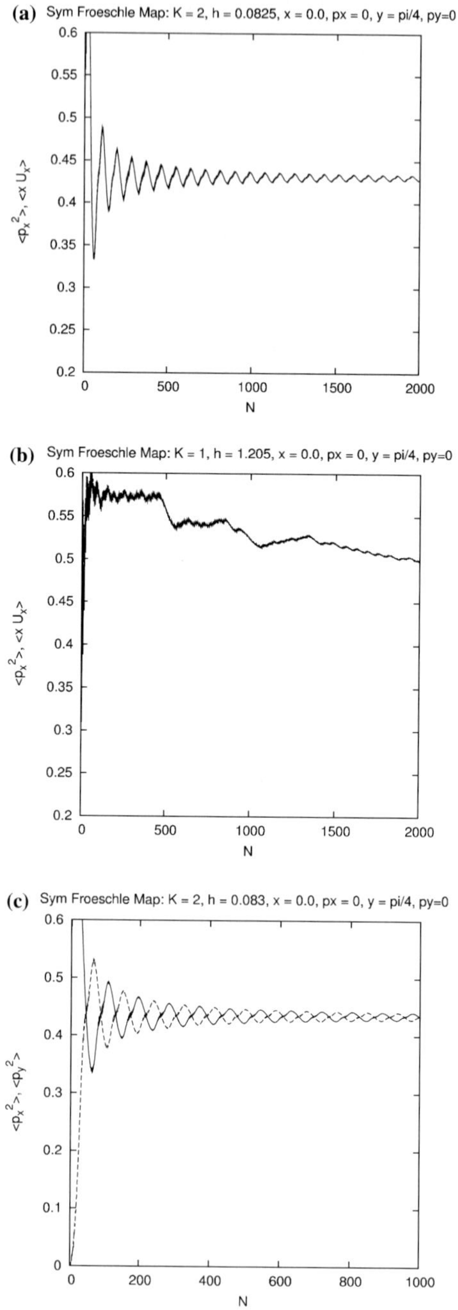

Figure 9. Virial sums (27) for Froeschlé map for (a) $K = 2,\ h = 0.0825$, (b) $K = 1,\ h = 1.205$; (c) convergence of $\langle p_x^2 \rangle$ and $\langle p_y^2 \rangle$ for regular orbit in 1 : 1 resonance.

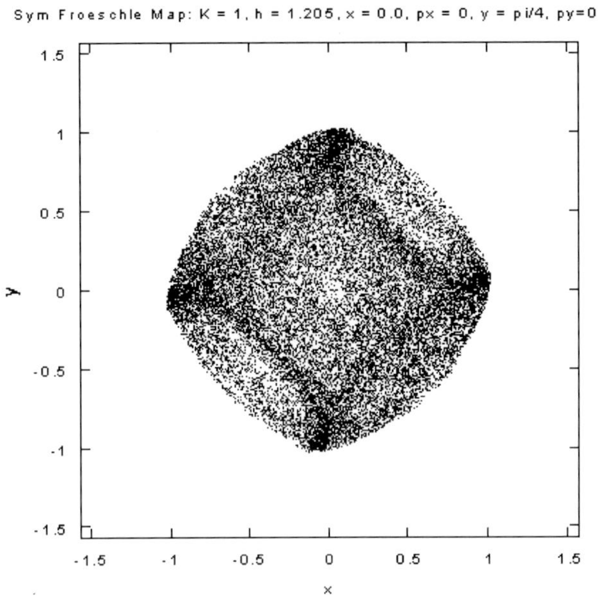

Figure 10. The x–y projection of chaotic orbit for Froeschlé map for $K = 1,\ h = 1.205$.

(Wolf et al., 1985), now largely superseded by Laskar's frequency map method (Laskar et al., 1992; Laskar, 1993), fast Lyapunov indicators (Lega and Froeschlé, 1997), and most recently, relative Lyapunov indicators (Sándor et al., 2000). In addition there is the twist method (Contopoulos and Voglis, 1997), the SALI method of Skokos (2001), and the "0–1" method of Gottwald and Melbourne (2004).

We have seen that the behavior of the time average of p_i^2 is radically different for regular and chaotic motion, for both natural flows and natural maps. Let us now consider how these differences might be quantified into a useful tool for distinguishing between regular and chaotic orbits. In order to be competitive with other methods, one must be able to decide within a few thousand steps for discrete maps, for which most of the published results have been carried out. In particular, much attention has been given to instances of so-called weak chaos (Zaslavskii et al., 1992; Laskar et al., 1992), where the Lyapunov exponents are small and difficult to distinguish from those for regular orbits, which rigorously tend to zero. We shall focus on two well-known cases of weak chaos, both for separatrix layers of the standard map.

First consider the standard map for $K = -0.3$, which places the main hyperbolic point at the origin. Figure 11(a) shows the very thin separatrix layer, where the Lyapunov exponents are known to be small. Enlarging this region reveals the presence of considerable complexity, with myriad small islands embedded in a chaotic sea, as seen in Figure 11(b). Plotting time

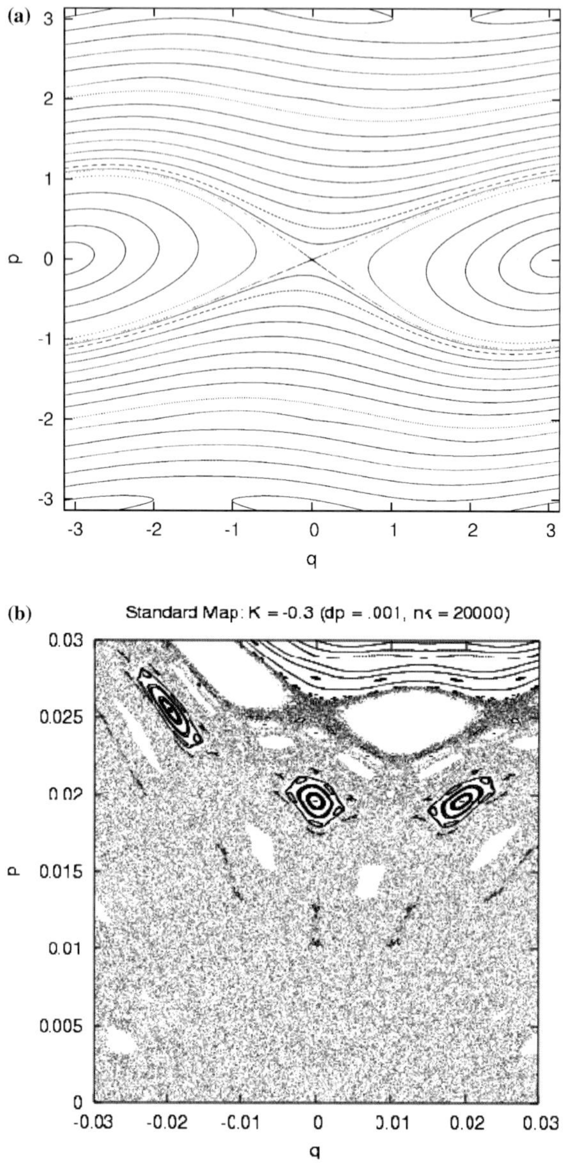

Figure 11. (a) Standard map for $K = -0.3$: (a) full-size and (b) enlarged separatrix layer.

histories of $\langle p^2 \rangle$ for several initial conditions along the $q = 0$ axis gives the curves plotted in Figure 12(a). Note that all curves suffer an initial transient phase (due to the $\Delta G/t$ term in (4)). After this, the curves corresponding to regular orbits become extremely flat after $\approx$1000 iterations, while the curves corresponding to chaotic orbits have not yet converged. We propose to utilize these variations as a measure of chaoticity via a quantity we call "meander,"

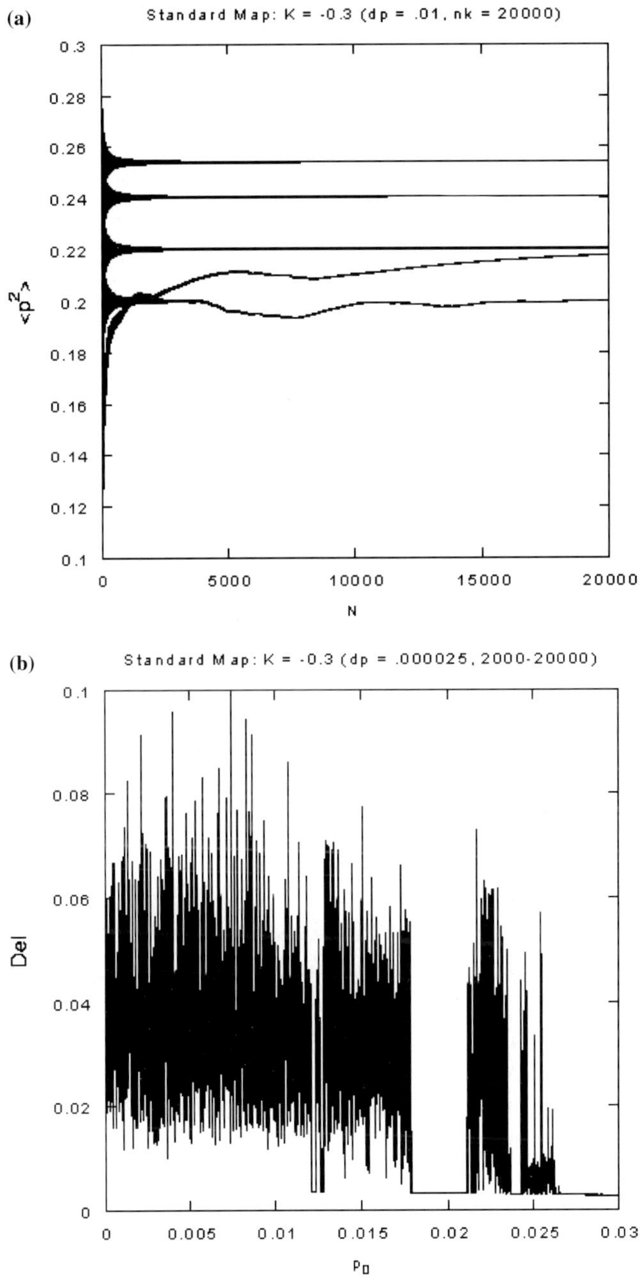

Figure 12. (a) Time histories with $K = -0.3$ of $\langle p^2 \rangle$ for several orbits in separatrix layer near the origin, (b) Meander scan along $q = 0$ line.

$$\Delta(N) = \frac{\langle p_{\max}^2 \rangle - \langle p_{\min}^2 \rangle}{\langle p_{\max}^2 \rangle + \langle p_{\min}^2 \rangle}, \tag{29}$$

where $p_{\max,\min}$ are the maximum and minimum values of p over a specified interval, typically the first 10000 steps, excluding the first few hundred steps to avoid the initial transient. No windowing is currently used in this calculation. Figure 12(b) shows Δ for a number of initial conditions along the p-axis in Figure 11(b), with $\Delta p = 0.002$, clearly revealing the phase space islands near $p = 0.013$ and 0.02. The amount of complexity revealed by such a scan of course depends on the fineness of the mesh employed.

Next we consider the standard map for $K = 1.4$, which offers more complexity than Laskar's choice of $K = 1.3$. Figure 13(a) shows the separatrix layer surrounding the 6-fold Birkoff chain, and blown up in Figure 13(b). Several small islands are visible along the p-axis. Can the meander test find them all? Figure 14 depicts time histories of $\langle p^2 \rangle$ along the $q = 0$ line again illustrating the large differences between regular and chaotic orbits. The corresponding meander profile along this line reveals well-defined flat portions at the visible islands as well as several more which were not seen in Figure 13. These examples demonstrate that the meander method is an efficient "island finder," usually requiring fewer than 10000 iterations for each orbit. For strong chaos many fewer iterations are required, typically 100 steps to avoid the initial transient and about 500 steps to distinguish regular from chaotic orbits.

6. Discussion

We have revisited the classical virial theorem for bounded single-particle orbits generalized to natural Hamiltonian flows and natural symplectic maps of arbitrary dimension. For natural flows of the form $H = T + U$, where the kinetic energy T is positive definite, a strong and a weak form of the virial theorem were derived, which together with the conserved total energy yield an asymptotic invariant depending on the coordinates alone. Essential differences were found between regular and chaotic orbits. For chaotic orbits the time-averaged virial integrals converge extremely slowly, with large variations for moderate times (thousands of steps for discrete maps). Nevertheless, the two relevant averages vary together so that their differences converge to zero, behavior that might be described as co-asymptotic. We also found an asymptotically invariant function of the coordinates which converges equally rapidly for chaotic and regular orbits. The results were well verified numerically for the Hénon–Heiles system. Next we derived a strong and a weak virial theorem for natural symplectic maps, both of which closely

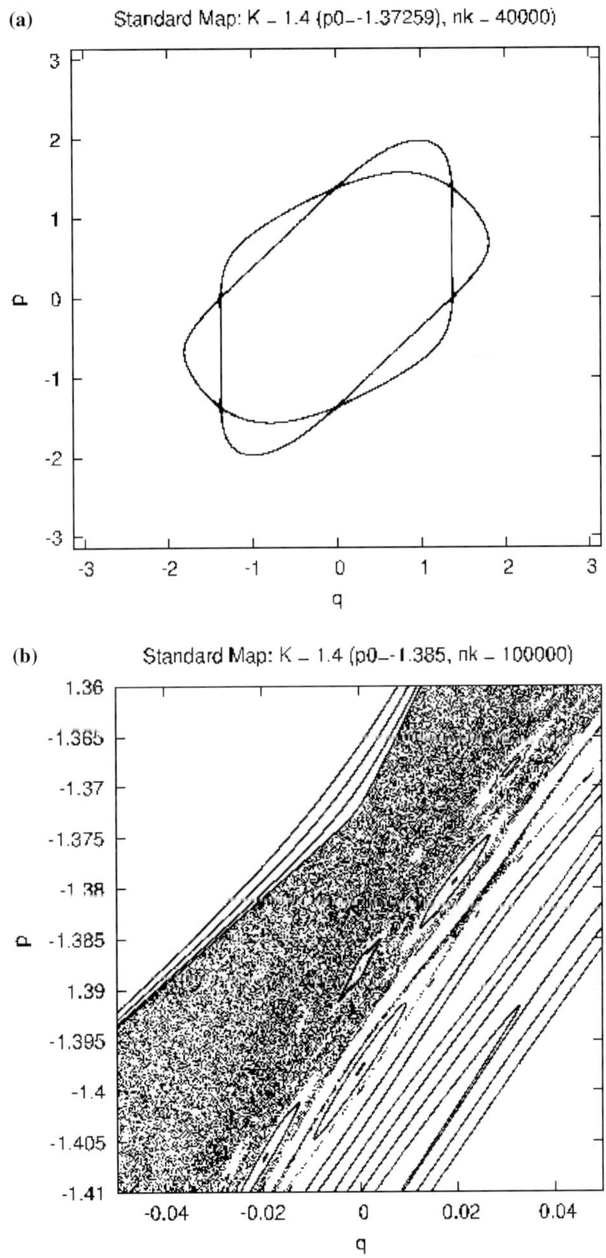

Figure 13. Separatrix layer for Birkhoff chain in standard map for $K = 1.4$ (a) full size and (b) enlarged.

resemble those for flows. However, since there is no conserved energy for symplectic maps, no analog of the asymptotic invariant was found. Again, significant differences were found between regular and chaotic orbits, the

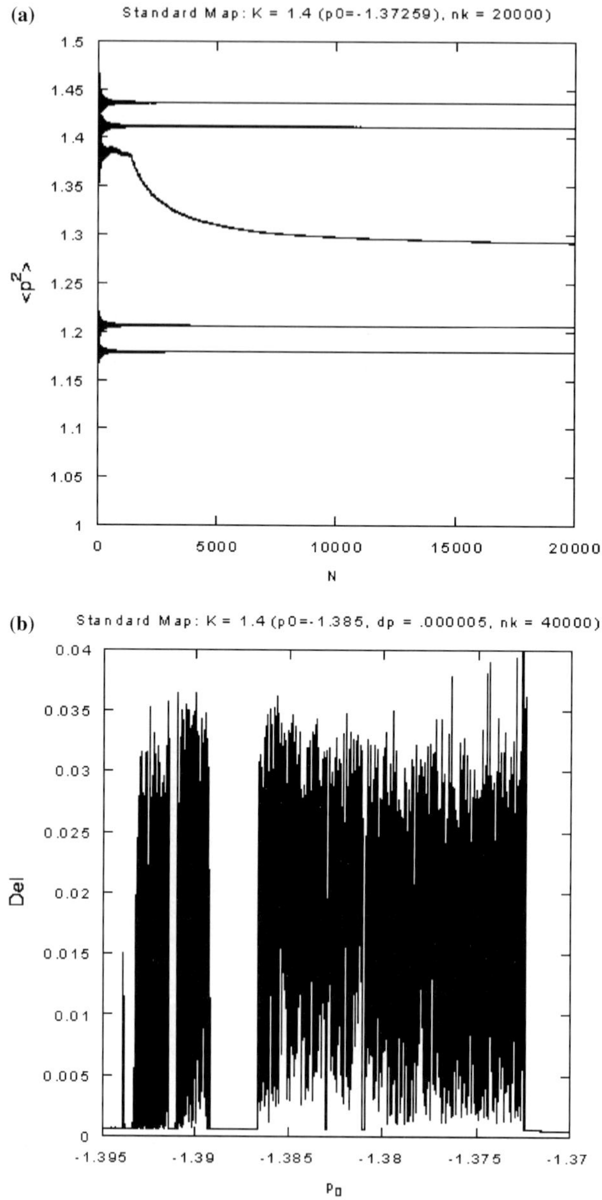

Figure 14. (a) Time histories with $K = 1.4$ of $\langle p^2 \rangle$ for several orbits in separatrix layer near the origin, (b) Meander scan along $q = 0$ line.

virial sums for chaotic orbits converging vastly slower than those for regular orbits. The resulting discrete virial theorem has been verified for several mappings of current interest, including a generalized Hénon map and weak chaos in the standard map. For periodic orbits of discrete maps the virial sums are finite and may be used to predict their locations in phase space.

The erratic behavior of the virial integrals and sums for chaotic orbits suggest a sensitive test for chaoticity, especially in thin chaotic layers where traditional tools such as Lyapunov exponents can converge very slowly. This behavior was quantified in a quantity we call meander, which measures the deviation of $\langle p^2 \rangle$ from its running time average. We have applied this technique to a typical instance of weak chaos in a separatrix layer of the standard map. The results show the $\Delta(N)$ method to be superior to the Lyapunov exponents and comparable to Laskar's frequency method and fast Lyapunov indicators. Whether our method is competitive with recent methods such as relative Lyapunov indicators remains to be shown. Perhaps the principal advantage of $\Delta(N)$ is that it is so simple and easy to program.

The results presented in this paper leave considerable scope for future work. More examples should be studied, both continuous and discrete. The convergence of the virial integrals should be determined quantitatively by some sort of variance function similar to the meander (29). Meiss' (1994) studies of long time behavior should be extended to other mappings having less symmetry than the standard map. Detailed comparisons should be made with the standard chaos detectors to determine under what circumstances the meander method would be suitable. For example, 4D mappings such as the Froeschlé map might be amenable to this approach. We also plan to try out more refined methods of gauging the variations in the virial sums.

Other types of virial theorems are also worth exploring in a canonical framework. For example, starting with the function $G_{ij} = q_i p_j$ will produce tensor virials similar to those of Parker (1954), for maps as well as flows. Magnetic virials for velocity dependent potentials could be constructed for axisymmetric systems via an effective potential of the form (Howard, 1999)

$$U^e = U(\rho, z) + \frac{(p_\phi - \Psi)^2}{2\rho^2}, \tag{30}$$

where $\Psi(\rho, z)$ is the magnetic stream function.

Further generalizations are possible. For example, virial theorems exist for nonnatural Hamiltonian flows and symplectic maps where kinetic and potential energies do not exist. So long as an invariant measure exists, convergence of the resulting virial sums is guaranteed by the Birkhoff Ergodic Theorem. Thus, we have derived and numerically verified a virial theorem for the symplectic version of the Hénon map $(x, y) \mapsto (y - k + x^2, -x)$ (Hénon, 1969). Beyond these examples, virial theorems have been derived for nonconservative flows, where an invariant measure does not necessarily exist, so that convergence of the virial averages is not rigorously known. We have also obtained an analogous theorem for nonsymplectic discrete maps (such as the generalized Hénon map), and are presently investigating the results for specific models.

Appendix A: Asymptotic Invariants and Hypervirial Theorems

The orbital time averages discussed in this paper are only valid for large time (with the exception of periodic orbits) and may be described as *asymptotic invariants*,

$$\langle f(p,q)\rangle = \text{const} + O(1/t). \tag{A.1}$$

In general an unlimited number of such invariants may be constructed for 1D natural flows by differentiating the function

$$G(p,q) = p^m q^n, \tag{A.2}$$

which leads to the *hypervirial theorem*

$$m\langle p^{m-1} q^n U_q\rangle = n\langle p^{m+1} q^{n-1}\rangle \tag{A.3}$$

provided of course that the two integrals converge. Using energy conservation then leads to the asymptotic invariant

$$\frac{\langle q^n U_q\rangle + 2n\langle q^{n-1} U\rangle}{2n\langle q^{n-1}\rangle} = E, \tag{A.4}$$

which holds for chaotic and regular orbits alike. More generally one may obtain (if one wishes) larger classes of invariants by working with an arbitrary differentiable function $G = f(p,q)$.

Acknowledgements

I am grateful to J. Laskar and J. M. Hawkins for helpful suggestions, and especially J. D. Meiss, who provided several insightful ideas. I am also grateful to l'Observatoire de Paris, where part of this work was carried out. This research was supported by NSF Grant DMS-0209856.

References

Arnold, V. I.: 1989, *Mathematical Methods of Classical Mechanics*, 2nd ed. Springer, New York.

Atkins, P. W.: 1983, *Molecular Quantum Mechanics*, 2nd ed. Oxford.

Binney, J. and Tremaine, S.: 1987, *Galactic Dynamics*, Princeton University Press, Princeton.

Birkhoff, G.:1931, 'Proof of the Ergodic Theorem', *Proc. Nat. Acad. Sci. USA* **17**, 656.

Blümel, R. and Reinhardt, W. P.: 1997, *Chaos in Atomic Physics*, Cambridge.

Clausius, R. J. E.: 1850,'Über die Bewegende Kraft der Wärme', *Ann. der Physik und Chemie* **79**, 368.

Contopoulos, G. and Voglis, N.:1977, 'A Fast Method for Distinquishing Between Order and Chaotic Orbits', *Astron. Astrophys.* **317**, 73.

Dragt, A. J.: 1979, 'A Method of Transfer Maps for Linear and Nonlinear Beam Elements',*IEEE Trans. Nucl. Sci.* **NS-26**, 601.

Dullin, H. R. and Meiss, J. D.: 2000, 'Generalized Hnon Maps: the Cubic Polynomial Diffeomorphisms of the Plane',*Physica* **D143**, 265.

Dumas, H. S. and Laskar, J.: 1993, 'Global Dynamics and Long-time Stability in Hamiltonian Systems via Numerical Frequency Analysis',*Phys. Rev. Lett.* **70**, 2975.

Froeschlé, C.: 1984, 'The Lyapunov Exponents – Applications to Celestial Mechanics', *Celest. Mech.* **34**, 95.

Goldstein, H., Poole, C. and Safko, J.: 1989, *Classical Mechanics*, 3rd ed. Addison-Wesley, New York.

Gottwald, G. A. and Melbourne, I.: 2004, 'A New Test for Chaos in Deterministic Systems', *Proc. Roy. Soc.* **460A**, 603.

Hénon, M. and Heiles, C.: 1964,'The Applicability of the Third Integral of Motion: Some Numerical Experiments',*Astron. J.* **69**, 73.

Hénon, M.: 1969, 'Numerical Study of Quadratic Area-Preserving Mappings', *Q. Appl. Math.* **27**, 291.

Howard, J. E., Lichtenberg, A. J. and Lieberman, M. A.: 1986, 'Four-Dimensional Mapping Model for Two-Frequency ECRH', *Physica* **20D**, 259.

Howard, J. E.: 1999, 'Stability of Relative Equilibria in Arbitrary Axisymmetric Gravitational and Magnetic Fields.', *Celest. Mech. Dyn. Astron.* **74**, 19.

Howard, J. E. and Dullin, H. R.: 2001, 'Spectral Stability of Natural Maps', *Phys. Lett.* **A88**, 225.

Laskar, J., Froeschlé, C. and Celletti, A.: 1992, 'The Measure of Chaos by the Numerical Analysis of the Fundamental Frequencies. Application to the Standard Mapping', *Physica* **D56**, 253.

Laskar, J.: 1993, 'Frequency Analysis for Multi-dimensional Systems. Global Dynamics and Diffusion', *Physica* **D67**, 257.

Lega, E. and Froeschlé, C.: 1997, 'Fast Lyapunov Indicators. Comparison with other Chaos Indicators. Application to Two and Four Dimensional Maps', In: R. Dvorak and J. Henrard, (eds.), *The Dynamical Behavior of our Planetary System*, Kluwer, Dordrecht.

Lichtenberg, A. J. and Lieberman, M. A.: 1992 *Regular and Chaotic Dynamics*, 2nd ed. Springer, New York.

Meiss, J. D.: 1994, 'Transient Measures in The Standard Map', *Physica* **D74**, 254.

Murray, C. D. and Dermott, S. F.: 1999, *Solar System Dynamics*, Cambridge.

Parker, E. N.: 1954, 'Tensor Virial Equations', *Phys. Rev.* **96**, 1686.

Pathria, R. K.: 1996, *Statistical Mechanics*, Butterworth and Heineman, Oxford.

Robinson, C.: 1998, *Dynamical Systems*, CRC Press, New York.

Sándor, Zs., Erdi, B. and Efthymiopoulos, C.: 2000, 'The Phase Space Structure Around L4 in the Restricted Three-Body Problem', *Celest. Mech. Dynam. Astron.* **78**, 113.

Shafranov, V. D.: 1966, *Reviews of Plasma Physics*, Consultants Bureau, New York.

Skokos, Ch.: 2001, 'Alignment Indices: A New, Simple Method for Determining the Ordered or Chaotic Nature of Orbits', *J. Phys.* **A34**, 10029.

Wolf, A., Swift, J. B., Swinney, H. L. and Vastano, J. A.: 1985, 'Determining Lyapunov Exponents from a Time Series', *Physica* **16D**, 285.

Zaslavskii, G. M., Sagdeev, R. Z., Usikov, D. A., Chernikov, A. A. and Sagdeev, A. R.: 1992, *Chaos and Quasi-Regular Patterns*, Cambridge University Press, Cambridge.

Celestial Mechanics and Dynamical Astronomy (2005) 92:243–255
DOI 10.1007/s10569-004-3834-6

LOCAL AND GLOBAL DIFFUSION ALONG RESONANT LINES IN DISCRETE QUASI-INTEGRABLE DYNAMICAL SYSTEMS

CLAUDE FROESCHLÉ[1], MASSIMILIANO GUZZO[2] and ELENA LEGA[1]
[1]*Observatoire de Nice, Bv. de l'Observatoire, B.P. 4229, 06304 Nice cedex 4, France, e-mail: elena@obs-nice.fr*
[2]*Dipartimento di Matematica Pura ed Applicata, Università degli Studi di Padova, via Belzoni 7, 35131 Padova, Italy.*

(Received: 5 July 2004; revised: 20 September 2004; accepted: 23 September 2004)

Abstract. We detect and measure diffusion along resonances in a quasi-integrable symplectic map for different values of the perturbation parameter. As in a previously studied Hamiltonian case (Lega et al., 2003) results agree with the prediction of the Nekhoroshev theorem. Moreover, for values of the perturbation parameter slightly below the critical value of the transition between Nekhoroshev and Chirikov regime we have also found a diffusion of some orbits along macroscopic portions of the phase space. Such a diffusion follows in a spectacular way the peculiar structure of resonant lines.

Key words: Global diffusion, Quasi-integrable Dynamical Systems

1. Introduction

In order to highlight the possibility of a drift of the actions along a resonance in an Hamiltonian quasi-integrable system. Arnold (1964) built an ad hoc model showing the existence of a very slow diffusion. This kind of diffusion is very difficult to detect numerically.

Within the framework of Nekhoroshev's (1977) theorem, one expects an exponentially slow drift of the actions along resonant lines. Moreover, the speed of the diffusion decreases as the order of the resonance increases (Morbidelli and Giorgilli, 1995; Giorgilli and Morbidelli, 1997).

In a previous work (Lega et al., 2003), using a time dependent Hamiltonian system with two degrees of freedom, we have provided numerical evidence for the existence of diffusive orbits along resonances as well as numerical estimate of the diffusion coefficient as a function of the perturbing parameter. The decreasing of the diffusion coefficient with the perturbing parameter is stronger than a power law, typical of Chirikov (1979) diffusion, and is compatible with an exponential law as expected in the Nekhoroshev regime.

In this paper, we perform the same kind of numerical experiments on a quasi-integrable symplectic mapping T of dimension four with a non integrable part analogous to that of the Hamiltonian used in Lega et al. (2003). Initial conditions were taken in the chaotic zone of a selected resonance. Such orbits were chosen using the fast Lyapunov indicator (FLI, Froeschlé and Lega, 2000), which allows to have a detailed knowledge of the topology of the resonances.

The use of the FLI allows to determine (Guzzo et al., 2002) both for slightly perturbed Hamiltonian systems and for weakly coupled symplectic mappings, the critical value of the perturbing parameter that corresponds to the transition from a Chirikov regime to the Nekhoroshev one. Therefore, chaotic initial conditions were taken for values of the coupling parameter still in the Nekhoroshev regime but relatively close to such a critical value.

The same procedure is applied in the present paper to the detection of the diffusion in the mapping T. The results obtained for the mapping are both qualitatively and quantitatively similar to those obtained for the Hamiltonian system in agreement with recent proofs of the Nekhoroshev theorem for nearly integrable symplectic maps (Kuksin and Pöschel, 1994; Guzzo, 2004). Moreover, thanks to the rapid computation of the mapping, which allows to explore much longer integration times with respect to the Hamiltonian system we have observed that Arnold's diffusion is relevant for global diffusion (Guzzo et al., 2005). More precisely, for a suitable choice of the perturbation parameter we show that a global diffusion occurs along the peculiar set of resonances forming the Arnold's web.

The paper is organized as follows. We recall in Section 2 the definition of the FLI and we give an application to the symplectic 4 dimensional map. We provide in Section 3 the evidence of the diffusion along resonant lines. The numerical estimate of the diffusion coefficient will be given in Section 4. Section 5 is devoted to the phenomenon of global Arnold's diffusion. Conclusions are provided in Section 6.

2. The FLI Revisited

When computing the Lyapunov characteristics indicators (LCI), the attention is focused on the length of time necessary to get a reliable value of their limit, but very little importance is given to the first part of the computation. In fact, this part is considered as a kind of transitory regime depending, among other factors, on the choice of an initial vector of the tangent manifold.

Already Froeschlé et al. (1997) have remarked that the intermediate value of the LCI (which was called FLI), taken at equal times for chaotic (even slow chaotic) and ordered motion, allows to distinguish between them. It turns out that the FLI allows also to distinguish among ordered motions of different origins, like resonant and non-resonant motions (Guzzo et al., 2002), despite the fact that in both cases the largest LCI tends to zero when time goes to infinity.

2.1. DEFINITION OF THE FLI

Given a mapping M from $\mathbb{R}^n$ to $\mathbb{R}^n$ an initial condition $\vec{x}(0) \in \mathbb{R}^n$, and an initial vector $\vec{v}(0) \in \mathbb{R}^n$ of norm 1 (we remark that in this particular case the manifold and the tangent manifold are both $\mathbb{R}^n$), let us define the FLI function $F(\vec{x}(0), \vec{v}(0), \tau)$, τ belonging to $\mathbb{Z}^+$, as:

$$F(\vec{x}(0), \vec{v}(0), \tau) = \sup_{0<t\leqslant\tau} \log \|\vec{v}(t)\| \tag{1}$$

where $v(t)$ is given by the system:

$$\begin{cases} \vec{x}(t+1) = M\vec{x}(t) \\ \vec{v}(t+1) = \frac{\partial M}{\partial \vec{x}}(\vec{x}(t))\vec{v}(t) \end{cases} \tag{2}$$

Let us remark that this definition has replaced, since Froeschlé et al.(2000), the first one given in Froeschlé et al. (1997). With the actual definition we could get rid of unnecessary complications and the introduction of the supremum of the norm has the same advantage of an averaging procedure easier to handle. Moreover the present definition is especially suited to the analysis of the neighborhood of a periodic orbit (Froeschlé and Lega, 2005).

2.2. A FOUR DIMENSIONAL MAP AS A MODEL PROBLEM

In previous papers (Froeschlé et al., 2000; Guzzo et al., 2002; Lega et al., 2003) we used the FLI to describe the geometry of the resonances of the Hamiltonian system:

$$H = \frac{I_1^2}{2} + \frac{I_2^2}{2} + I_3 + \epsilon \frac{1}{\cos(\phi_1) + \cos(\phi_2) + \cos(\phi_3) + 4} \tag{3}$$

where $I_1, I_2, I_3 \in \mathbb{R}$ and $\phi_1, \phi_2, \phi_3 \in \mathbb{T}$ are canonically conjugate and ϵ is a small parameter.

In this paper we use the FLI for the same kind of studies, for a symplectic mapping obtained through the leap-frog discretization (which ensures the symplecticity of the mapping) of the equations of motion of Equation (3):

$$T = \begin{cases} x_{j+1} = & x_j - \epsilon \frac{\sin(x_j+y_j)}{(\cos(x_j+y_j)+\cos(z_j+t_j)+4)^2} \\ y_{j+1} = & y_j + x_j \qquad (\text{mod} 2\pi) \\ z_{j+1} = & z_j - \epsilon \frac{\sin(z_j+t_j)}{(\cos(x_j+y_j)+\cos(z_j+t_j)+4)^2} \\ t_{j+1} = & z_j + t_j \qquad (\text{mod} 2\pi) \end{cases} \tag{4}$$

We still have a quasi-integrable system with the usual advantage of mappings: we can integrate for times longer than in the continuous case and this fact is crucial when dealing with the diffusion for small values of ϵ.

Figure 1 shows the variation of the FLI with time for three different kinds of orbits of T (indicated by arrows in Figure 2) with $\epsilon = 0.6$. The upper curve, with initial conditions in a chaotic zone in Figure 2: $(x = 2.07, y = 0, z = 2.1, t = 0)$ shows an exponential increase of the FLI with time. The intermediate curve corresponds to a regular invariant torus of initial conditions $(x = 1.8, y = 0, z = 1.2, t = 0)$ and the lowest one corresponds to a regular resonant curve of initial conditions $(x = 1.67, y = 0, z = 0.91, t = 0)$. Although the largest Lyapunov exponent of the two regular curves above is zero, the FLI, as shown in Figure 1, distinguishes between resonant and non resonant regular motions. More precisely, although the two curves exhibit essentially the same behavior they are parallel but distinct. At the origin of the different values of the FLI for regular non resonant and resonant motion there is the differential rotation which is not the same for the two dynamics. The FLI behavior has been extensively studied both numerically and analytically in (Guzzo et al., 2002).

Figure 2 shows, at $\tau = 1000$, the FLI for a grid of 500×500 initial conditions regularly spaced on x, z. The other initial conditions are $y = 0, t = 0$, and the initial vector is $\vec{v}(0) = (0.5(\sqrt{3} - 1), 1, 1, 1)$. The FLI is reported with a grey scale: the dark strips correspond to regular resonant

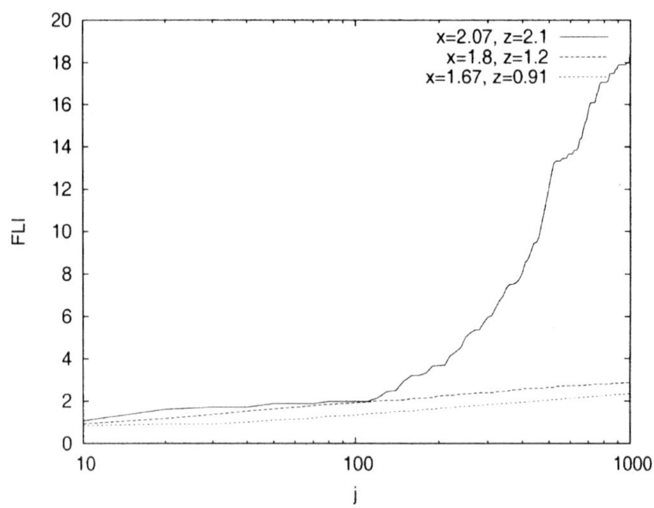

Figure 1. Variation of the FLI as a function of time for three orbits of the standard map T with $\epsilon = 0.6$. The upper curve is for a chaotic orbit with initial conditions $(x = 2.07, y = 0, z = 2.1, t = 0)$ the middle one is for a regular non resonant orbit with $(x = 1.8, y = 0, z = 1.2, t = 0)$ and the lowest one is for a regular resonant orbit with $(x = 1.67, y = 0, z = 0.91, t = 0)$.

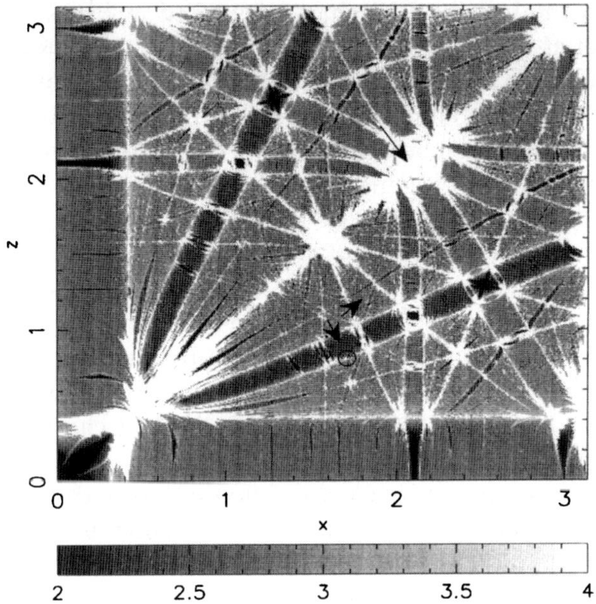

Figure 2. Geography of resonances in the plane $x - z$ for the mapping T with $\epsilon = 0.6$. The computation has been done for a set of 500 × 500 initial conditions regularly spaced on x, z. The other initial conditions are $y = 0, t = 0$, and the initial vector is $\vec{v} = (0.5(\sqrt{3} - 1), 1, 1, 1,)$. The grey scale ranges from black ($FLI \leqslant 2$) to white ($FLI \geqslant 4$).

motions while the white lines represent both chaotic resonant motions or the vicinity of a separatrix. The orbits having an FLI value of about $\log(\tau)$ constitute the background of KAM tori. Let us remark that for the Hamiltonian of Equation (3), when $\epsilon = 0$ the frequencies are strictly equal to the actions, therefore the FLI charts are nothing but frequencies charts, which display a very well detailed Arnold's web.

As far as the mapping is concerned, the same occurs when considering the variables $x(j)$ and $z(j)$ which are the frequencies of the unperturbed mapping $T(\epsilon = 0)$. Again, the FLI chart in the plane r–z (Figure 2) shows very clearly the Arnold's web.

3. Diffusion Along Resonances: Qualitative Aspects

For the four dimensional symplectic mapping T we have computed the FLI charts for different values of the perturbing parameter and we have selected a low order resonance. In order to compare the results obtained in the case of the Hamiltonian system of Equation 3 we have considered the same unperturbed resonance $x = 2z$.

Using the method of the FLI charts described in Froeschlé et al. (2000) and in Guzzo et al. (2002) we found that the critical perturbation parameter for the transition between the Nekhoroshev and the Chirikov regime is in the interval $0.8 < \epsilon < 1$.

Starting from an upper bound of $\epsilon = 0.7$, we have looked for diffusive orbits in the Nekhoroshev regime.

Figure 3 shows the FLI charts of the actions space for successive zoom around $x = 1.7$, $z = 0.8$ for different values of ϵ. In these pictures the region between the two white lines is the resonant strip approximated by $x - 2z < \sqrt{\epsilon}$, and the two white lines correspond to its hyperbolic border where diffusion is confined. These charts provide us the possibility of choosing initial conditions in the hyperbolic border. Then, we can also follow the evolution of the corresponding orbits on the FLI charts by plotting the points which intersect the double section $\sigma = |y| + |t| \leqslant 0.005$. Let us remark, that in such a way we minimize all projection effects and fast quasi-periodic movements. What remains is a very slow drift along the border of the resonance.

We have taken a set of 100 initial conditions corresponding to orbits of the FLI charts having FLI values larger than $1.5 \log(\tau)$, i.e., to chaotic orbits at the border of the resonance, far from the more stable crossing with other resonances. We have plotted on the FLI charts all the points in the double section described above. We remark that such points will appear on both side of the resonance (in fact the two white lines are connected by an hyperbolic region in the 4 dimensional phase space).

We show for $\epsilon = 0.6$, the evolution of the 100 orbits up to $j = 6 \times 10^6$ Figure 3 (top, left) and up to $j = 3 \times 10^8$ Figure 3(top, right). The diffusion along the resonance appears clearly, although the higher order resonances intersecting the main one become evident, and consequently the region of diffusion extends a little also on the direction transversal to the resonance.

When decreasing the perturbation, $\epsilon = 0.2$, the diffusion along the resonant line is more clear. In Figure 3 (middle) we have plotted again the intersections of the 100 orbits with the double section σ on the FLI chart after respectively $j = 3 \times 10^8$ (middle. left) and $j = 2 \times 10^{10}$ (middle, right). The phenomenon still appears very clearly even for $\epsilon = 0.07$, Figure 3 (bottom). i.e., an order of magnitude lower than the threshold of transition between the two regimes. We have observed the phenomenon down to $\epsilon = 0.03$.

4. Diffusion Along Resonances: Quantitative Aspects

In order to measure the diffusion coefficient we have considered the phenomenon as if it was a Brownian motion, since we do not have yet an analytic

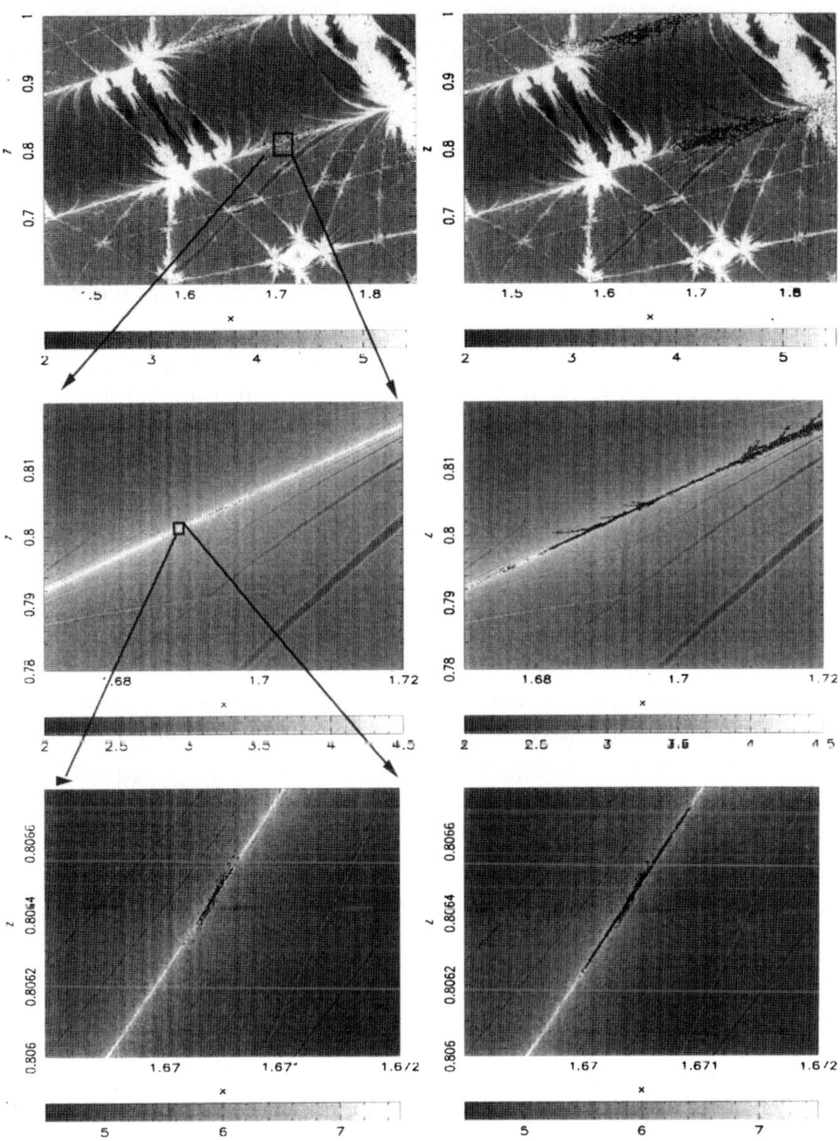

Figure 3. Diffusion along the resonant line $x = 2z$ of the mapping T for $\epsilon = 0.6$ (top), $\epsilon = 0.2$ (middle), $\epsilon = 0.07$ (bottom) for a set of 100 initial conditions taken in the chaotic border of the resonance. The figure in the middle and in the bottom correspond to the zone of the phase space contained in the square plotted respectively in the figure at top, left and in the figure at middle, left. The black points are the intersections of the orbits with the double section $|y| + |t| \leqslant 0.005$. The number of iterations on the set of orbits are respectively: $j = 6 \times 10^6$(top, left). $j = 3 \times 10^8$ (top, right). 3×10^8 (middle, left). 2×10^{10} (middle, right). 2×10^9 (bottom, left). 2.5×10^{10} (bottom, right). The grey scale ranges from black to white.

model for a diffusion like the one we observed. We look therefore for a linear increase with time of the mean square distance from the initial conditions. We are aware that this is a very crude assumption and that for systems like the standard map diffusion can be anomalous for very high values of the perturbation parameter Zaslavski and Edelman (2000), where no invariant curves remain. Instead, our computations concern quasi-integrable cases. This is a very interesting problem which goes behind the purpose of the present paper, mostly because we are submitted to computational limitations: we can't take a very large number of initial conditions which would require very long CPU times.

Notwithstanding these difficulties, we observed indeed an averaged linear increase with time of the mean squared distance from the initial conditions. Moreover, in order to reduce the contributions due to fast motion, we have only taken into account the points on the double section. More precisely, denoting with $x_i(0)$ and $z_i(0), i = 1, \dots, N$ the initial conditions of a set of N orbits, with $x_i(j)$ and $z_i(j)$ the corresponding values at time j, and choosing a fraction t^* of the total integration time, we considered the quantity:

$$S(nt^*) = \frac{1}{M_n} \sum_{i:(|y_i(j)|+|t_i(j)|)<0.005} [(z_i(j) - 2x_i(j)) - (z_i(0) - 2x_i(0))]^2 \qquad (5)$$

where M_n is the number of points on the double section for j in the interval $(n-1)t^* \leqslant j \leqslant nt^*$. We observe (Figure 4) a linear increase with time of S and we estimate the diffusion coefficient D as the slope of the regression line.

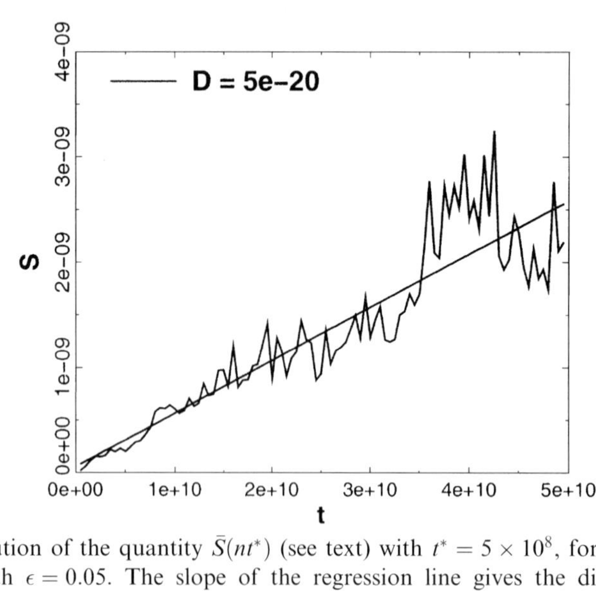

Figure 4. Evolution of the quantity $\bar{S}(nt^*)$ (see text) with $t^* = 5 \times 10^8$, for 100 orbits of the mapping T with $\epsilon = 0.05$. The slope of the regression line gives the diffusion coefficient $D = 5 \times 10^{-20}$.

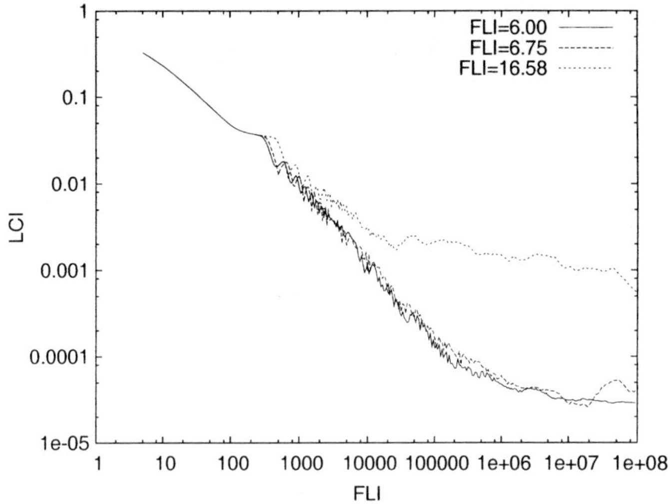

Figure 5. Evolution of the largest Lyapunov indicator with time for three orbits of the standard map T with $\epsilon = 0.05$. The initial conditions are $x = 1.67209489949749$ $y = 0.$ $z = 0.809297781072027$ $t = 0(FLI = 6.0)$, $x = 1.67192854271357$ $y = 0.$ $z = 0.809236608877722$ $t = 0.(FLI = 6.75), x = 1.67203944723618$ $y = 0.$ $z = 0.809286279229481$ $t = 0.(FLI = 16.58)$.

Following Zaslavski and Edelman (2001), we could have diffusion, more precisely anomalous diffusion, driven by orbits with zero Lyapunov exponent. Moreover, since the initial conditions were chosen using the FLI computed on a relatively short time we have checked the chaotic character of some of the selected orbits by computing the largest Lyapunov exponent. More precisely, we computed the Lyapunov indicators. i.e. the truncated values of the Lyapunov exponents which are defined by a limit for time going to infinity.

Figure 5 shows the evolution with time of the largest Lyapunov indicator of three orbits for $\epsilon = 0.05$. The orbits have been chosen considering the FLI distribution of the 100 initial conditions: the first one has the lowest FLI (6.00), the second has an FLI in the middle of the distribution (6.75) while the third has the largest FLI value (16.58). We recall that the FLI chart for $\epsilon = 0.05$ was computed on $\tau = 10^4$ iterations and that the orbits are considered chaotic when they have $FLI \geqslant 1.5 \log \tau$, i.e., $FLI \geqslant 6$ in the considered case. For the three-test orbits the largest Lyapunov indicator is small but positive (Figure 5) ensuring that we are following very weak chaotic motions. This kind of test comforts our confidence in using FLI method for distinguishing the dynamical character of the orbits on times which may be even some order of magnitude shorter than the time needed for the Lyapunov indicators to stop to decrease and to stabilize at a positive value.

The estimates of $ln(D)$ versus $ln(1/\epsilon)$ are reported in Figure 6. We have defined three sets of data, performing a local regression for each of them of the form $ln(D) = a + m\ln(1/\epsilon)$, and found three different slopes. The first set contains the values of D for $\ln(1/\epsilon) \leqslant 0.92$, the second for $1.17 \leqslant \ln(1/\epsilon) \leqslant 2.28$ and the third for $\ln(1/\epsilon) \geqslant 2.43$, and the corresponding slopes are respectively $m_1 = -4.2$, $m_2 = -8.5$ and $m_3 = -13.3$. Such changes of slope comfort the expected exponential decrease of D, as for the Hamiltonian case reported in (Lega et al., 2003).

It would be natural at this point to check if the exponential upper bound $D < \exp -(1/\epsilon)^b$ expected from Nekhoroshev theorem can be seen from our data, and in particular if one can obtain a numerical estimation of b. Indeed, an exponential fit of our data would give the value $b = 0.28$ (Figure 6) with apparently very good correlation coefficient of about 0.99. However, how much this computation is meaningful is a delicate matter: the apparently good correlation coefficient is due mainly to the small interval in ϵ used for the exponential fit (Figure 6): there are not theoretical predictions to compare to the detected value: numerical studies (Benettin and Fassó, 1999) have shown that the exponential upper bounds found by perturbation theories are indeed only upper bounds, while the true exchange of energy among the different degrees of freedom with respect to a perturbing parameter typically follows more complicated functional laws. For these reasons, we think it is necessary in the future to perform more numerical studies of the problem.

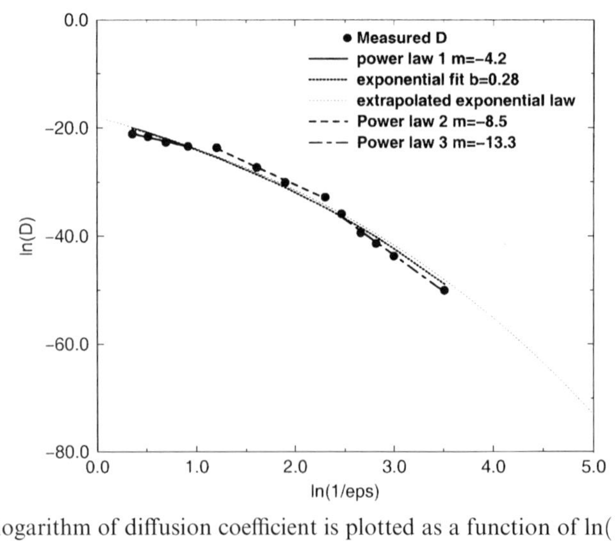

Figure 6. The logarithm of diffusion coefficient is plotted as a function of $\ln(1/\epsilon)$. The change of slope of the three power law fits is in agreement with the expected exponential decrease of D.

5. Global Diffusion

In the previous sections we have clearly showed the phenomenon of Arnold's diffusion occurring along a resonant line. It is clearly a phenomenon of local diffusion. When dealing with global diffusion we immediately think to the well known Chirikov's diffusion, due to the overlapping of resonances. Is it the only mechanism for global diffusion? At this purpose we have repeated the previous experiments for a given value of the perturbation parameter, close to the transition to the Chirikov regime, on a very long interval of time. We have considered twenty initial conditions in the vicinity of the point $(x, z) = (1.71,\ 0.81)$ for the symplectic map. Then, we computed numerically the map up to 10^{11} iterations.

The results are reported in Figure 7. Figure 7a shows only the location of initial conditions (inside the circles), on the FLI map of the action plane $(x,\ z)$, Figure 7a shows that we are in the Nekhoroshev regime since the resonances do not overlap and the majority of the orbits, i.e. dark grey and black zones in the FLI-chart, are regular. In Figure 7b we plotted as black dots all points of the orbits which have returned after some time on the section $|y| + |t| \leqslant 0.005$. We observe that the orbits filled a macroscopic region of the action plane whose structure is clearly that of the Arnold web (Guzzo et al., 2005). The orbits have moved along the single resonances, and avoided the center of the main resonance crossings, in agreement with the theoretical results which predict longer stability times for motions in these regions. The larger resonances (which correspond to the smaller orders) are practically all visited, while this is not the case for the thiner ones (which correspond to the higher orders). This is in agreement with the theoretical results of (Morbidelli and Giorgilli, 1995; Giorgilli and Morbidelli, 1997), which predict that the speed of diffusion on each resonance becomes smaller for resonances of high order. Therefore, the possibility of visiting all possible resonances is necessarily limited by finite computational time.

The described diffusion phenomenon is very different from Chirikov diffusion which is illustrated in Figure 7c, where the overlapping of resonances appears clearly since, except for the large strip of regular resonant zones corresponding to $x = 0$ and to $z = 0$, almost all regular orbits have disappeared. In the FLI-chart we observe a large white sea with small dark grey or black islands. In this case the diffusion for 20 initial conditions located in the circle of Figure 7c occurs in the same macroscopic region of Figure 7b of the phase space in a much shorter time scales (only 2×10^7 iterations) and without apparent peculiar topological properties of the stochastic region, i.e. the whole chaotic sea is densely visited.

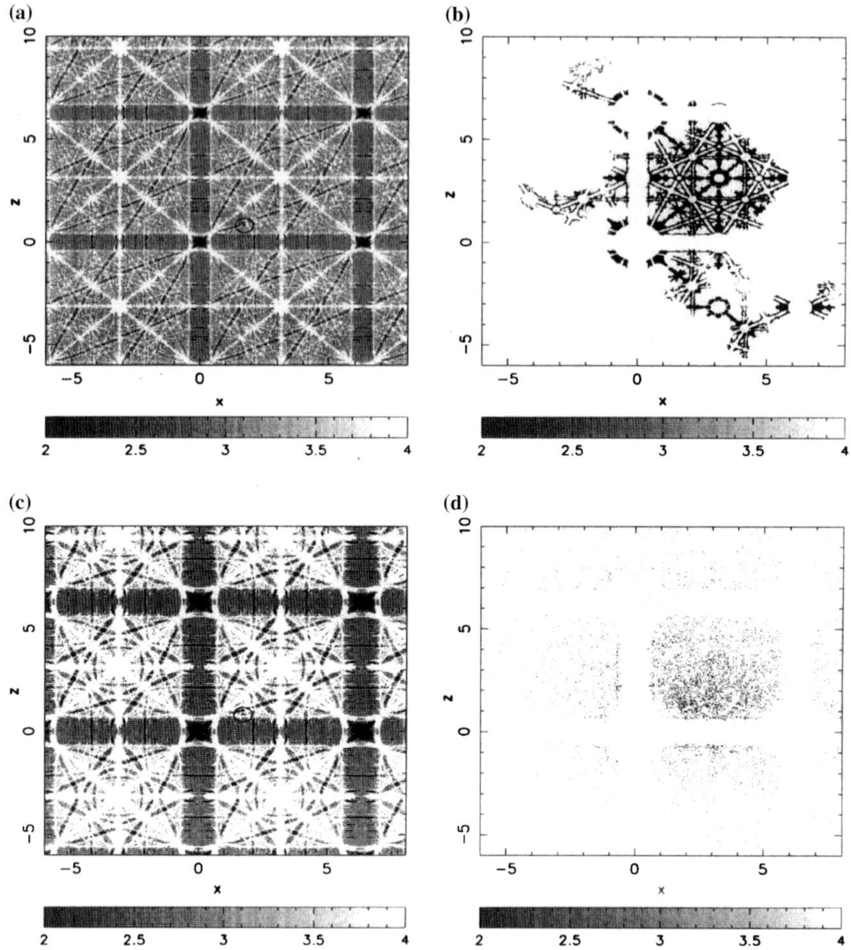

Figure 7. Panels (a) and (c) correspond to the FLI map of the action plane (x, z) for the map T for $\epsilon = 0.6$(a) and $\epsilon = 1.7$. The white region correspond to the chaotic part of the Arnold web. We marked with a circle the location of the twenty initial conditions chosen in the vicinity of the point $(x, z) = (1.71, 0.81)$. In panels (b) and (d) we marked with a black dot all points of the twenty orbits which have returned after some time on the section $|y| + |t| < 0.005$. We consider 10^{11} iterations for panel (b) and 2×10^7 iterations for panel (d).

6. Conclusion

By using a sensitive tool, the FLI, to detect the geography of resonances, we have been able to choose and follow orbits which exhibit diffusive behavior along resonant lines on a 4 dimensional symplectic map. We do not have an analytic model for such a diffusion, thus we make the hypothesis that it behaves as a Brownian motion. Under this hypothesis we measured a diffusion coefficient and showed that the dependence of the diffusion coefficient on the perturbing parameter does not follow a power law, and in the explored range is in agreement with the exponential decay predicted by Nekhoroshev theory. Results agree with those previously obtained for an Hamiltonian system (Lega et al., 2003). Moreover, we have shown that the phenomenon of Arnold's diffusion may also play an important role in the long-term evolution of quasi-integrable systems. More precisely the orbits can diffuse in a macroscopic portion of the phase space following in a spectacular way the peculiar structure of resonant lines.

References

Arnold, V. I.: 1964, 'Instability of dynamical systems with several degrees of freedom', *Sov. Math. Dokl.* **6**, 581–585.

Benettin, G. and Fassó, F.: 1999, 'From Hamiltonian perturbation theory to symplectic integrators and back', *App. Numer. Math.* **29**, 73–87.

Chirikov, B. V.: 1979, 'An universal instability of many dimensional oscillator system', *Phys. Rep.* **52**, 263–379.

Froeschlé, C., Lega, E. and Gonczi, R.: 1997, 'Fast Lyapunov indicators. Application to asteroidal motion', *Celest. Mech. Dyn. Astron.* **67,** 41–62.

Froeschlé, C., Guzzo, M. and Lega, E.: 2000, 'Graphical evolution of the Arnold's web: from order to chaos', *Science* **289(**N.5487), 2108–2110.

Froeschlé, C. and Lega, E.: 2000, 'On the structure of symplectic mappings. The fast Lyapunov indicator: a very sensitive tool', *Celest. Mech. Dyn. Astron.,* **78**, 167–195.

Froeschlé, C. and Lega, E.: 2005, 'The fast Lyapunov indicator. Applications to the study of the fine structure of Hamiltonian systems and to the detection of Arnold's diffusion', In: *Hamiltonian Systems and Fourier analysis*. New Prospectus for gravitational Dynamics, Cambridge Scientific Publishers.

Giorgilli, A. and Morbidelli, A.: 1997, *ZAMP*. **48**, 102–134.

Guzzo, M., Lega, E. and Froeschlé, C.: 2002,'On the numerical detection of the effective stability of chaotic motions in quasi-integrable systems', *Physica D* **163**, 1–25.

Guzzo, M.: 2004, 'A direct proof of the Nekhoroshev theorem for nearly integrable symplectic maps', *Ann. Henry poincaré*, **5**, 1013–1039.

Guzzo, M., Lega E. and Froeschlé, C.: 2005, 'First numerical evidence of global Arnold diffusion in quasi-integrable systems', DCDSB, in press.

Kuksin, S. and Pöschel, J.: 'On the inclusion of analytic symplectic maps in analytic Hamiltonian flows and its applications', Seminar on Dynamical Systems (St. Petersburg, 1991), 96–116. Progr. *Nonlinear Differential Equations Appl.* **12,** Birkhäuser, Basel.

Lega E., Guzzo, M. and Froeschlé, C.: 2003, 'Detection of Arnold diffusion in Hamiltonian systems', *Physica D* **182,** 179–187.

Morbidelli, A. and Giorgilli, A.: 1995, *Physica D* **86**, 514–516.

Nekhoroshev, N. N.: 1977, 'Exponential estimates of the stability time of near-integrable Hamiltonian systems', *Russ. Math. Surv.* **32,** 1–65.

Zaslavski, G. M. and Edelman, M.: 2000, 'Hierarchical structures in the phase space and fractional kinetics : I classical systems', *Chaos* **10**, 135–145.

Zaslavski, G. M. and Edelman, M.: 2001, 'Weak mixing and anomalous kinetics along filamented surfaces', Chaos **11**, 295–305.

Celestial Mechanics and Dynamical Astronomy (2005) 92:257–272
DOI 10.1007/s10569-005-2064-x

THE ROLE OF HYPERBOLIC INVARIANT SETS IN STICKINESS EFFECTS

YI-SUI SUN[1], LI-YONG ZHOU[1,2] and JI-LIN ZHOU[1]
[1]*Department of Astronomy, Nanjing University, Nanjing 210093, China,*
e-mail: sunys@nju.edu.cn
[2]*Tuorla Observatory, Väisäläntie 20, Piikkiö 21500, Finland*

(Received: 21 June 2004; revised: 28 December 2004; accepted: 10 February 2005)

Abstract. With the standard map model, we study the stickiness effect of invariant tori, particularly the role of hyperbolic sets in this effect. The diffusion of orbits originated from the neighborhoods of hyperbolic points, periodic islands and torus is studied. We find that they possess similar diffusion rules, but the diffusion of orbits originated from the neighborhood of a torus is faster than that originated near a hyperbolic set. The numerical results show that an orbit in the neighborhood of a torus spends most of time around hyperbolic invariant sets. We also calculate the areas of islands with different periods. The decay of areas with the periods obeys a power law, and the absolute values of the exponents increase monotonously with the perturbation parameter. According to the results obtained, we conclude that the stickiness effect of tori is caused mainly by the hyperbolic invariant sets near the tori, and the diffusion speed becomes larger when orbits diffuse away from the torus.

Key words: hyperbolic invariant sets, invariant tori, stickiness effect

1. Introduction

Hamiltonian systems are conservative dynamical systems which are encountered in various areas. The phase space of a nearly integrable Hamiltonian system typically consists of regular and chaotic regions. The study of orbital diffusion in the phase space is the basis of many topics in Hamiltonian dynamics. A chaotic orbit initialized close to a KAM torus will wander for a long time before it finally leaves the vicinity of the torus. Since Karney (1983) first uses the term "sticky" to describe such effect around islands, this phenomenon was then called the "stickiness effect" of invariant tori. The study of stickiness effect has now been extended to include all kinds of effects which slow down the diffusion. These effects may come from different invariant sets such as invariant tori (Lai et al., 1992; Perry and Wiggins, 1994; Sun and Fu, 1999), island-chains (Karney, 1983; Chirikov and Shepelyansky, 1984; Sun et al., 2002) and Cantori (Meiss and Ott, 1985; Contopoulos et al., 1997). We call this extended concept the "generalized stickiness effect". Moreover, it is also known that hyperbolic invariant sets possess the stickiness effect too (Froeschlé and Lega, 1998; Contopoulos et al., 1999; Zhou et al., 2002).

Zhou et al. (2002) suggested that the hyperbolic invariant sets is essential to the stickiness effect. In this paper we will clarify and confirm numerically this conclusion with the standard map model.

2. Model

We take the standard map T as the model,

$$T: \begin{cases} x_{n+1} = x_n + y_{n+1} \\ y_{n+1} = y_n - \frac{k}{2\pi}\sin(2\pi x_n) \end{cases} \pmod 1, \tag{1}$$

where k is the perturbation parameter. Figure 1 is the diagram of map T with $k = 1$.

We choose an island from the period-5 island-chain (indicated by the arrow in Figure 1) as the chief torus to analyze. The center of the chief torus is at $(x_0, y_0) = (0.2476544, 0.6638289)$. Figure 2 is the enlargement of the chief torus, around which come out islands and hyperbolic invariant sets with different orders due to the self-similarity. In the following sections, we will investigate the stickiness effect of the chief torus as well as of the periodic islands and the hyperbolic invariant sets near it. The role of the hyperbolic invariant sets in the generalized stickiness effect will be discussed in detail.

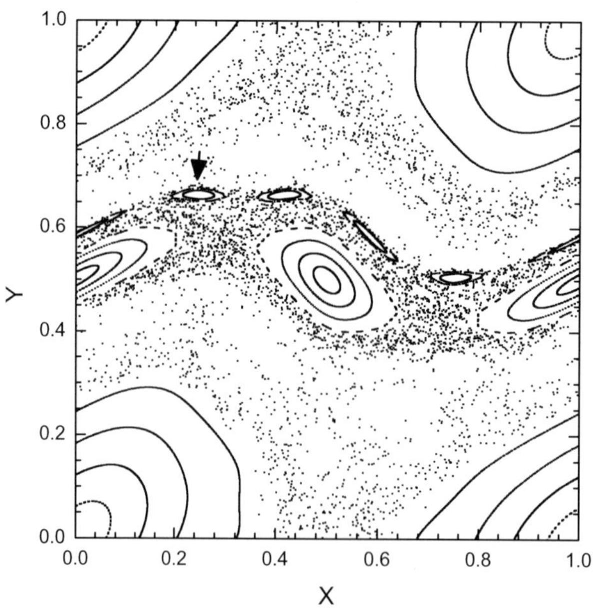

Figure 1. Phase diagram of map T with $k = 1.0$.

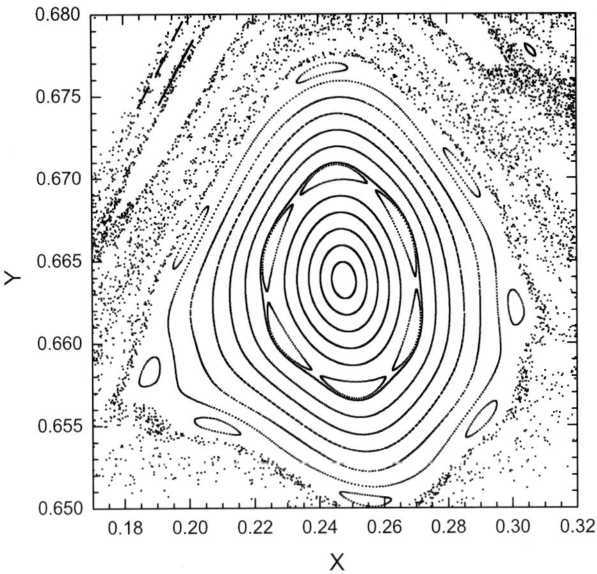

Figure 2. Enlargement of one island (called the chief torus in the paper) from the period-5 island-chain.

3. Numerical Results

3.1. STICKINESS EFFECT OF HYPERBOLIC INVARIANT SETS

From Figure 3 we can see that in the vicinity of the chief torus, there is an island-chain with 23 periodic islands and a hyperbolic invariant set consisting of 23 periodic hyperbolic fixed points as predicted by the Poincaré-Birkhoff fixed point theorem. We investigate the stickiness effect of this hyperbolic invariant set.

To study the stickiness effect of the hyperbolic invariant set, we trace an orbit with an initial point $(x, y) = (0.2972650, 0.6591000)$, which is very close to one of the period-23 hyperbolic fixed points $(x_h, y_h) = (0.2972704, 0.6590667)$. Figure 4 shows the diffusion process of the orbit. Because the orbit diffuses to a secondary island-chain (or a secondary hyperbolic invariant set) when the iterative number $n \sim 3 \times 10^5$, we trace the orbit up to $n = 2 \times 10^5$ iterations, before that the orbit diffuses around the hyperbolic invariant set. Since the orbit is very close to the "boundary" of the chief torus, we can calculate the distribution of points on the orbit along the "boundary", which can be approximatively defined as the possible outermost curve l of the chief torus. Every point near l can be projected to l, so that gets a reference coordinate from the projection on l. In such a way we can see where the orbit is "stuck" during its diffusion.

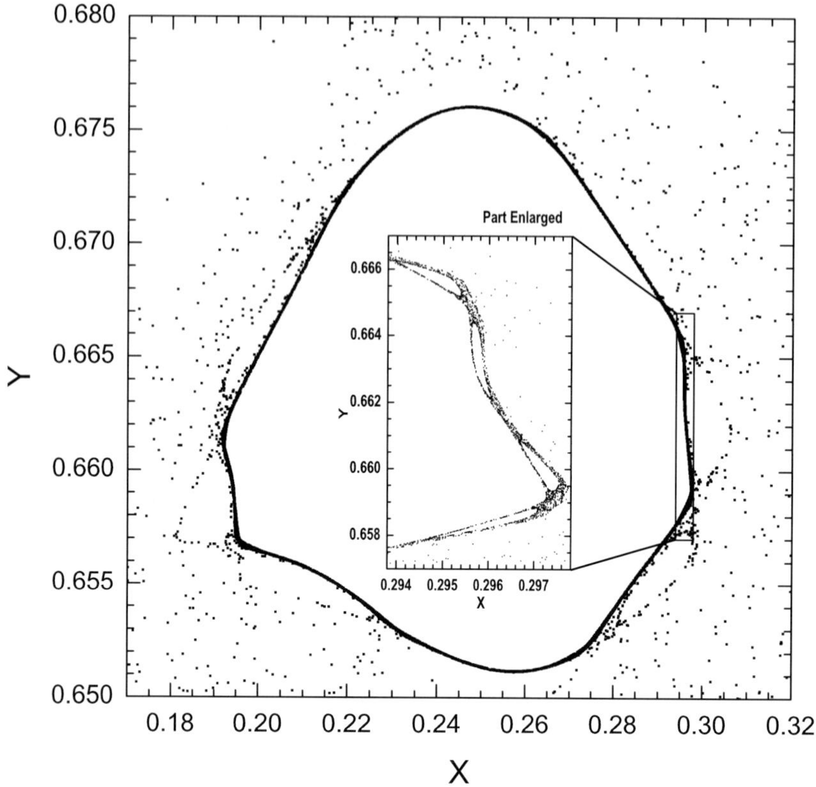

Figure 3. Diagram of the period-23 secondary islands around the chief torus.

In Figure 5 we show the distribution of orbit points with respect to the length of *l*, and the positions of the 23 hyperbolic fixed points are marked too. The outstanding feature of Figure 5 is that every peak evidently corresponds just to one of the positions of the period-23 hyperbolic fixed points (hyperbolic invariant set). The distribution of points on other orbits starting from points not very close to the hyperbolic fixed point, but lying between two neighboring hyperbolic fixed points in the same hyperbolic invariant set, are found to have very similar distribution to the one in Figure 5. We repeated the same calculation around a higher-order island (one of the 23 islands around the chief torus), and obtained the same results as above.

Now it should be stressed that an orbit spreads in fact in a 2-dimensional zone, so that the density of orbit points in the sticky zone should be counted in definite areas rather than along a curve. However, from a practical view, there is no straightforward method to define such areas, in which higher-order islands are embedded as holes and around them there are secondary hyperbolic invariant sets related with the stickiness effects of the

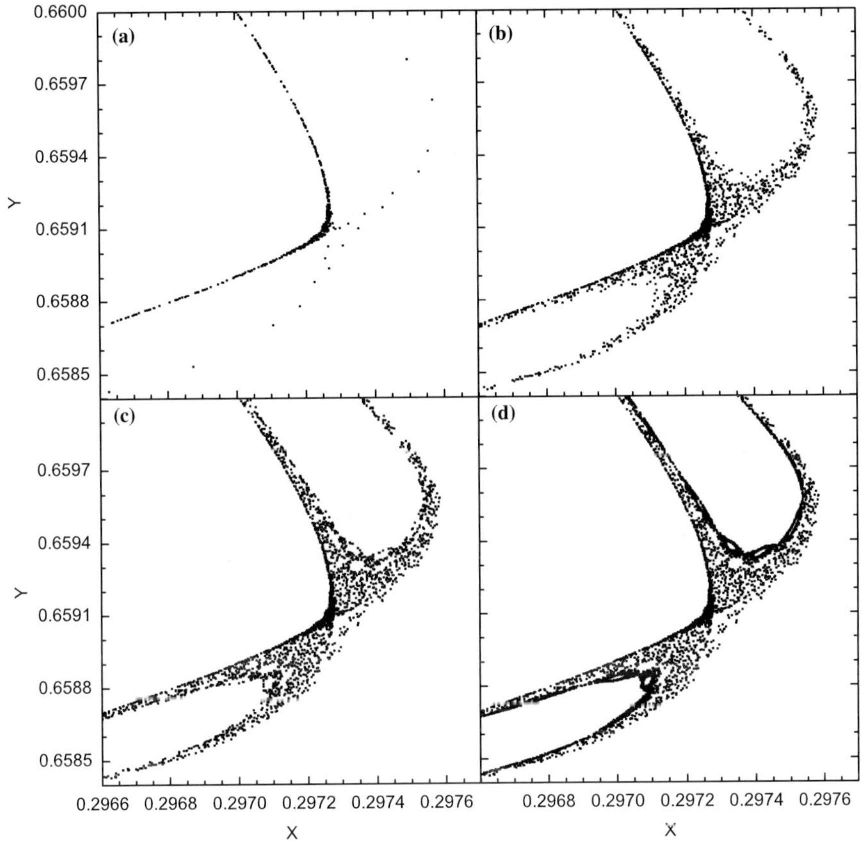

Figure 4. Diffusion of an orbit around a hyperbolic invariant fixed point. a, b, c and d are the snapshots of the orbit up to 5×10^4, 2×10^5, 3×10^5 and 6×10^5 iterations, respectively.

higher-order islands. On the other hand, the sticky zone occupied by an orbit shown in Figure 4 as an example shows a little thinner width around the hyperbolic points than around the rest of the zone, and in fact, from Figure 4 we can see that the orbit points near the hyperbolic fixed point are denser than in the zones far from it. Therefore, even if the sticky zone is considered roughly as a uniformly wide strip around the chief torus, the distribution along a curve in Figure 5 would reflect approximately the distribution of orbit points in the sticky area near the chief torus.

It is well known that there are many different unstable periodic orbits in a chaotic region. In Figure 5 the match between the positions of the outstanding peaks and the hyperbolic fixed points reveals that this hyperbolic invariant set is much more important than others in this region. It also implies that such orbits spend more time around this hyperbolic invariant set than elsewhere, therefore the hyperbolic invariant set plays an important role

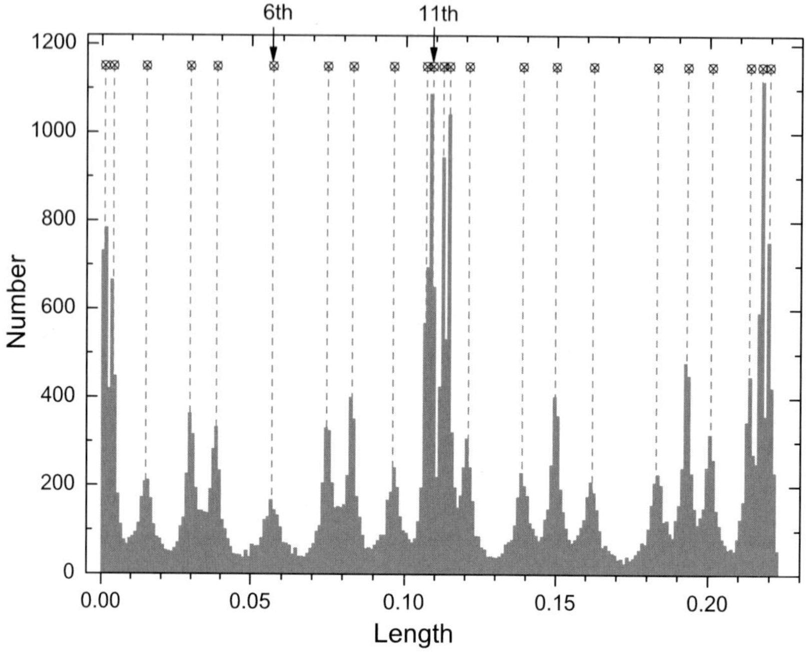

Figure 5. Distribution of points on orbit. The crosses inside circles correspond to the positions of the 23 hyperbolic fixed points.

in the stickiness effects. Moreover, an orbit on the "outermost" invariant curve also has a similar accumulation of points (the distribution is very similar to Figure 5). Since this invariant curve is very close to the hyperbolic set located among the island-chain, this accumulation can be explained by the continuous dependence of the density of orbits on the initial conditions.

We note the peaks in Figure 5 have different heights. With the eigenvalues and eigenvectors of the mapping on the hyperbolic points, it may be given an explanation. By iterating 115 ($= 5 \times 23$) times of the tangent map of (1), we get the matrix of the corresponding tangent map and then calculate the eigenvalues and eigenvectors of this matrix. The results show that for a hyperbolic point corresponding to a higher peak, the angle between eigenvectors (standing for directions of stable and unstable manifold at this point) is relatively bigger, and the eigenvalue corresponding to the unstable direction has a smaller value. While on a lower peak, the angle is smaller and the eigenvalue is bigger. For instance, the eigenvalues and eigenvectors of the sixth (counting from left to right, indicated by an arrow in Figure 5) hyperbolic points, which has a lower peak, are $\lambda_1 = 20.1326, \mathbf{v}_1 = (1.00000, -1.23649 \times 10^{-4})^T$; $\lambda_2 = 4.96707 \times 10^{-2}, \mathbf{v}_2 = (0.998415, 5.62761 \times 10^{-2})^T$. The angle between $\mathbf{v}_1$ and $\mathbf{v}_2$ is $\alpha = 3^\circ.23345$. In comparison, for the 11th point in Figure 5, which possesses a higher peak, we have $\lambda'_1 = 3.84879, \mathbf{v}'_1 = (-0.687897, -0.725808)^T$;

$\lambda_2' = 0.259822, \mathbf{v}_2' = (-0.841136,\ 0.540823)^T$; and $\alpha' = 79°.2758$. The same phenomena appear in the case of other hyperbolic points. In the case of a lower peak, the orbit spreads quickly due to the bigger eigenvalue λ_1 in the unstable direction and occupies a wide region along the invariant curve. While in the case with a higher peak, orbit diffuses along the unstable direction in a relatively smaller speed ($\lambda_1' < \lambda_1$), and the points of an orbit focus in a relatively narrower region ($\alpha' > \alpha$). Consequently, the peaks have different heights.

3.2. STICKINESS EFFECT OF TORI

In order to clarify further the role of hyperbolic invariant sets in the stickiness effect of tori, we discuss the diffusion of orbits in the vicinity of tori. We define a "neighboring zone" of the torus, and if an orbit diffuses away from this zone, we regard it as "escaped". To do this, we first calculate the length L of the boundary curve l of the chief torus and the area A surrounded by it. Then we define the neighboring zone around the chief torus by expanding the curve l outwards a width $D = (A/10L)$, i.e., the area of the neighboring zone is taken as $A/10$, where $A = 1.8341447 \times 10^{-3}$, $L = 0.2221069$. As soon as the distance of an orbit to the chief torus is larger than $2D$, we regard it as escaped.

3.2.1. *Diffusion of orbits near the chief torus*

We take 20,000 initial points spread uniformly in the neighboring zone of the chief torus, P_0^{ij}: $(r_0^i + (j/20)D, \theta_0^i = (2\pi/1000)i)$, $i = 1, 2, \ldots, 1000, j = 1, 2, ..., 20$, where (r_0^i, θ_0^i) are the polar coordinates of the points P_0^i on l. Then we choose 2000 points $P_c^k = (r_c^k, \theta_c^k)$, $k = 1, 2, \ldots, 2000$ spread uniformly on l as the standard points. After n iterations of the map T, a point P_m^{ij} ($m = (n/5)$, noting the period 5) on the orbit initialized from point P_0^{ij} is regarded as escaped, if the distance $d = \min(P_c^k, P_m^{ij}) > 2D$, where $1 \leqslant k \leqslant 2000$. We regard an orbit as never escaped if it is not escaped before $n = 1.15 \times 10^7$. In the same way we take another 20,000 initial points near one of the 23 periodic hyperbolic fixed points $(x_h, y_h) = (0.2972704, 0.6590667)$. The selected initial points are all in a segment of the above mentioned neighboring zone with angle coordinates from $\theta_1 = 6.1851$ to $\theta_2 = 6.1903$. Interpretatively, this fixed point and its two neighboring fixed points respectively have angle coordinates of $\theta = 6.1875$, 6.1478 and 6.2337. With the above definitions, we illustrate in Figure 6 the variations of the surviving orbit number with time (iteration number) for both cases. From Figure 6 we see the two curves are very similar, but for the orbits starting

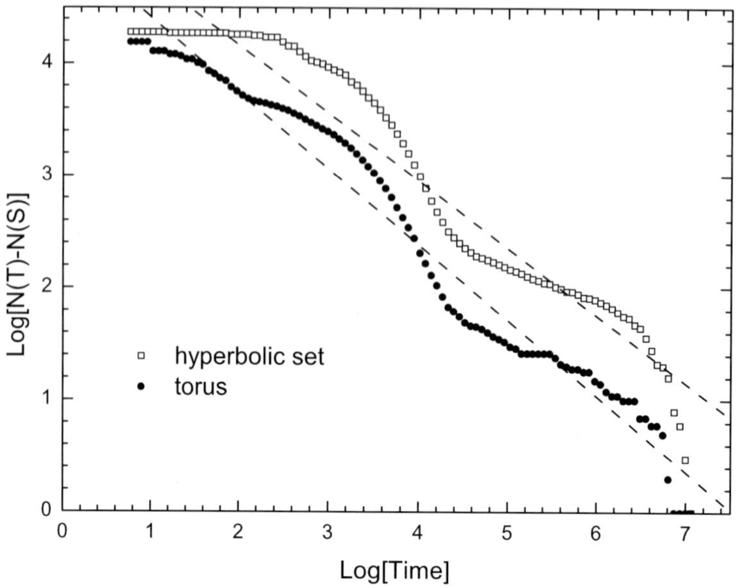

Figure 6. Diffusion of orbits started closed to the chief torus and close to the hyperbolic fixed point. The total number of points that have not escaped at time T is denoted by $N(T)$ and the number of points that will never escape is denoted by $N(S)$. Dots and squares stand for the cases close to the chief torus and close to the hyperbolic fixed points, respectively. The dashed lines are the linear fittings.

near the hyperbolic fixed point, the diffusion is slower than those started from the vicinity of the chief torus.

Two different quantities can quantify the stickiness effect: the exponent of the power law of orbit diffusion and the average time of orbit escape. According to the results shown by Ding et al. (1990), the surviving number N of orbits starting from a mixed region in the phase space of a Hamiltonian system will decrease in a power law with respect to time T: $N \sim T^{-z}$, where z is a positive number depending on the dimension D of the system. When $D = 2$, $z = 1.2 \sim 1.5$. In our cases, we find that each curve in Figure 6 can be divided into several segments, which can be linearly fitted with different slopes. Particularly, when $\log[\text{Time}] \in [3.6, 4.4]$ both of the two curves can be well fitted by lines with slopes of -1.5, that is, the exponent here is $z = 1.5$. Besides, there are segments with smaller z, and we also find $z = 3$ when $\log[\text{Time}] > 6.4$, which is consistent with the result in Chirikov and Shepelyansky (1999). Despite these interesting details, in this paper we focus mainly on the similar profiles of the two varying curves, therefore here we report the average exponents over the whole time range from 0 to 10^7, which are $z_1 = 0.6756$ and $z_2 = 0.6028$ respectively for both cases with starting points near the chief torus and the hyperbolic sets as mentioned above. The small

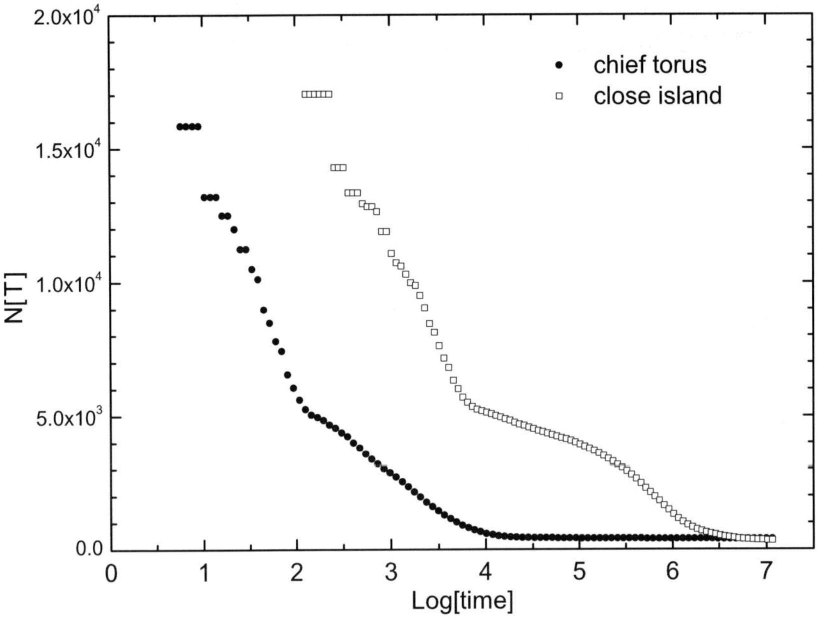

Figure 7. Diffusion of orbits started close to the chief torus and close to the island in the vicinity of the chief torus. Dots denote the case close to the chief torus, and squares denote the case close to the island.

values of z_1, z_2 are due to the existence of segments with flat slope. The average times of escape $\bar{T} = (1/N_0)\int_0^{N_0} T\,\mathrm{d}N$ are also numerically calculated and they are $\bar{T}_1 = 4815$ and $\bar{T}_2 = 18677$ (iteration times) for the cases of torus and hyperbolic set, respectively.

Both the exponent z and the average time of escape $\bar{T}$ imply that, the orbits with initial points in the vicinity of the chief torus, will diffuse faster than those started near the hyperbolic invariant set. These results also imply the hyperbolic invariant sets will play a major role in slowing down the diffusion of orbits near the torus, or in the stickiness effect of tori.

3.2.2. *Diffusion of orbits near the islands*

The phase space of map T possesses the self-similarity structure and an orbit may diffuse around islands of different orders on its way of escaping. In this section, we study the orbital diffusion near one of the above mentioned 23 periodic islands around the chief torus with center at $(x, y) = (0.2972144, 0.6599696)$. The neighboring zone and the escape criteria are the same as that in last subsection, but the A and L are now $A = 2.8499899 \times 10^{-7}$, $L = 3.0598743 \times 10^{-3}$. With the same way of taking initial points, we follow the evolution of 20,000 orbits starting from the neighboring zone of this island. Figure 7 displays the diffusion of orbits near the island and near the chief torus as a comparison. The similarity between the profiles of the two curves in Figure 7 implies the similar sources of the stickiness effects in these two cases. Combined with Figure 6 we may argue once again that the stickiness effect of island is also caused mainly by the hyperbolic invariant set near it. By the way, Figure 7 does not mean that the diffusion speed of orbits near island is slower than that near the chief torus, because the definitions of the escaping zone of chief torus and island are only similar, but not the same.

3.3. SIZE OF ISLANDS

Froeschlé and Lega (1998) studied the variations of sizes of Fibonacci islands with the distance to the chief torus. Efthymiopoulos et al. (1997) discussed the variations of islands sizes with perturbation parameter. In this paper we measure the size of different islands including the chief torus in the phase space, then investigate the variations of islands' area with respect to the periods of island-chains and to the system perturbation parameter k. With these results we can understand further the orbital diffusion in phase space and the role of hyperbolic invariant sets in the stickiness effect of tori. Here the area of an island is approximated by the summation of areas of 2000 triangles inside the "outermost" invariant curve surrounding the island.

We compute the areas of islands of different periods (the chief torus is a period-5 island) in the main (island) sequences (island-chain sequence surrounding the central island) for $k = 0.90, 0.95, 1.00, 1.05$ and 1.10, respectively. The results are shown in Figure 8, in which each square denotes the total area of an island-chain of a definite period. From Figure 8 we can find that the decay of island area with respect to the period obeys roughly a power law, and the absolute values of the slopes of fitting lines increase monotonically with k. This means that the decay rate of the area for island sequences increases with the perturbation parameter k (Figure 9).

As well known, the self-similarity property indicates that there are secondary islands around an island in the main sequence, and then higher order islands around this secondary islands, and so on. To study the area of such "islands around islands", we select the most outstanding secondary island chain around an island, and successively repeat such selecting to higher and higher order. We call such an island sequence "hierarchical (island) sequence". Taking $k = 1.0$, we compute the areas of islands in several hierarchical sequences starting respectively from islands with periods of 4, 16 and 29 in the main sequence. The numerical results displayed in Figure 10 show

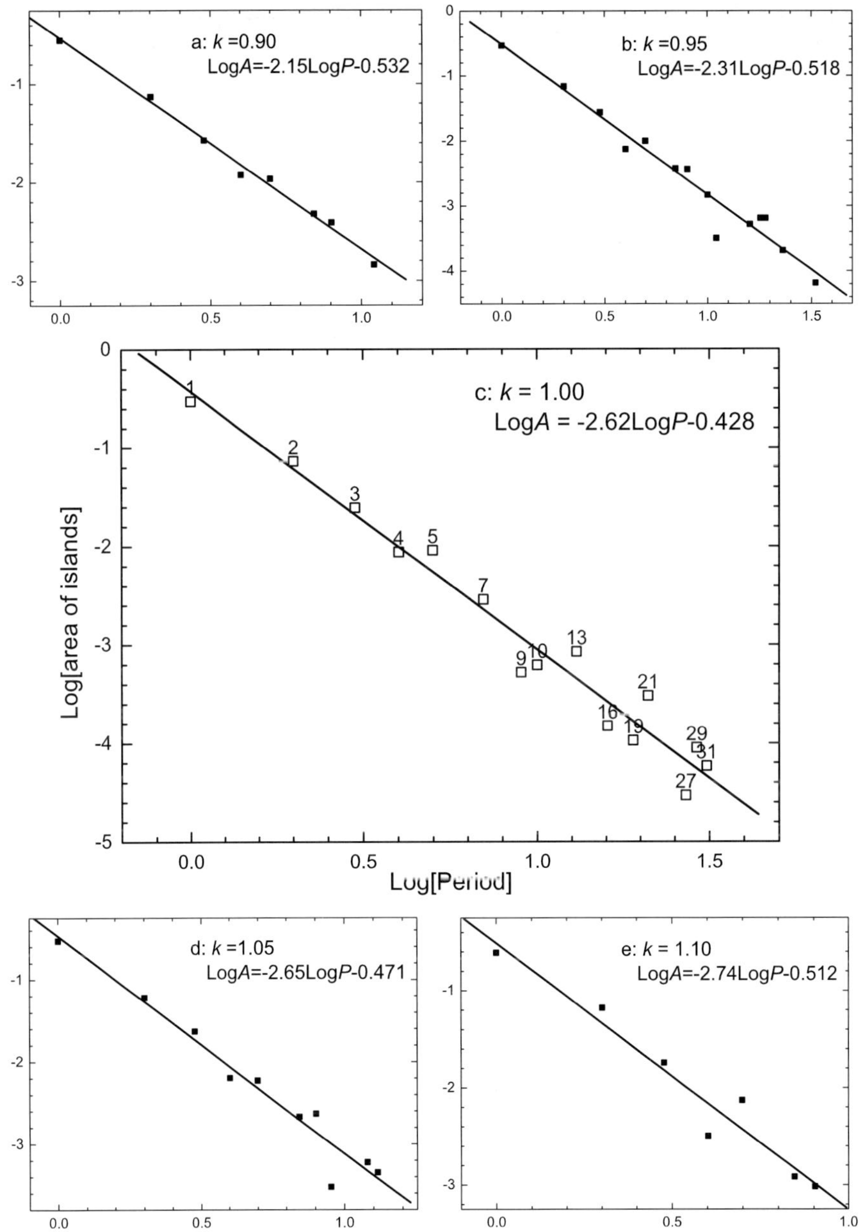

Figure 8. Variations of area of islands in the main island sequence with respect to the period of island. The linear fit of the data is indicated by both a line and a linear function. (a), (b), (c), (d) and (e) are the situations of $k = 0.90, 0.95, 1.00, 1.05$ and 1.10, respectively. The numbers above the squares in (c) denote the periods of corresponding islands.

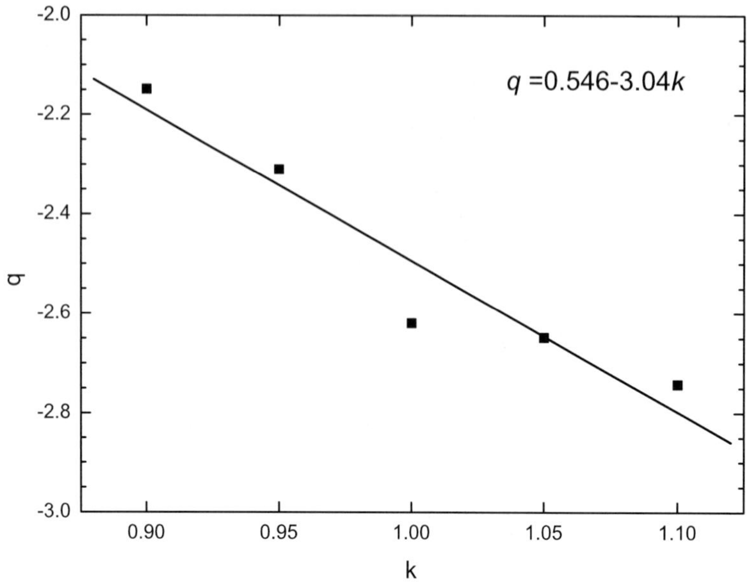

Figure 9. Variation of slopes q of fitting lines in Figure 9 with the perturbation parameter k.

the decay of the area of islands in the hierarchical sequence possesses a power law too, but compared with the case for main sequence, the absolute values of the slopes q of fitting lines are smaller. This implies the decay of areas in the hierarchical sequences is slower than that in the main sequence.

Efthymiopoulos et al. (1999) found that the sizes A of islands with odd and even "multiplicities" (periods) P in a sequence approaching to a cantorus decrease following the same power law $A = CP^{-2.75}$ but with different constants C. In this paper, we investigate the relations between the areas and the periods of islands in the main and the hierarchical island sequences, respectively. They are all proved to be power laws with different exponents. The variations of such exponents with respect to parameter k are also studied and we found that the exponents are the same for island sequences with even and odd periods provided the parameters k are the same.

Now we try to estimate the upper bound of the total area of islands in phase space. As indicated above, the variation of island area with period possesses roughly a power law, i.e. $\log A_P(k) = q(k) \log P + C(k)$, where $A_P(k)$ is the total area of period-P islands for a given k, $q(k)$ the slope and $A_1(k) = 10^{C(k)}$ the area of the central island (period-1 island). According to the result in Figure 9, we have $q(k) = 0.546 - 3.04k$ and $A_P(k) = P^{0.546-3.04k} A_1(k)$, $P \in \{N\}$, while $\{N\}$ is the set of period of islands in the main sequence for $k \in [0.90, 1.10]$. Based on these we obtain the total area of islands in the main island sequence

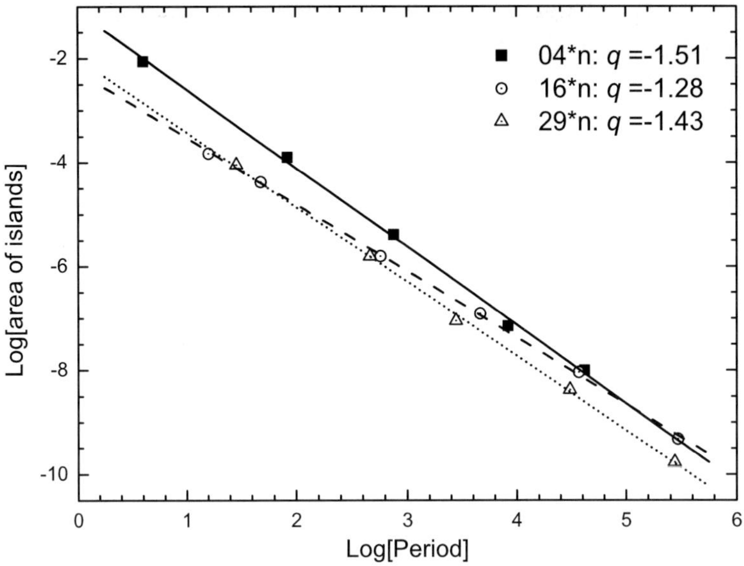

Figure 10. Variations of area of island in the hierarchical island sequences with the period of islands, and the corresponding fitting lines. The filled squares, circles and triangles represent the situations in the sequence starting from the period-4, 16, and 29 islands in the main sequence. q is the slope of the fitting line.

$$A(k) = \sum_{P\in\{N\}} A_P(k) \leqslant \sum_{P\in[1,\infty)} A_P(k) = A_1(k) \sum_{P\in[1,\infty)} P^{q(k)}. \quad (2)$$

We know that series $\sum_{n=1}^{\infty}(1/n^s)$ is convergent if $s > 1$, and the sum is called Riemann function $\zeta(s)$. After elementary calculus, we get

$$\zeta(s) < \frac{1 - s - 2^{-s}(1+s)}{1-s} = 1 - \frac{2^{-s}(1+s)}{1-s}. \quad (3)$$

Thus we have

$$A(k) < A_1(k)[1 - \frac{2^{q(k)}(1-q(k))}{1+q(k)}] = U(k), k \in [0.90, 1.10]. \quad (4)$$

Generally the area of central island decreases as k increases in a long term for large variation of k, but the varying is not smooth. It can even increase temporarily (Efthymiopoulos et al., 1997), as happening to occur in the present case. However we take the average $\overline{C}(k)$ for the five values of k, $\overline{C}(k) = -0.502$.

Substituting $q(k) = 0.546 - 3.04k$ and $A_1(k) = 10^{-0.502}$, we finally get an upper bound $U(k)$ of the total area of islands in the main island sequence. Because the total area of higher-order islands in the hierarchical island sequences is much smaller than that in the main island sequence, $U(k)$ is roughly an upper bound of the total area of islands in phase space. Calculating of Equation (4) shows $U(k) = 0.500, 0.469, 0.445, 0.426,$ and 0.410 when $k = 0.90, 0.95, 1.00, 1.05,$ and 1.10, respectively.

If there were more values of island area for large variation of k, and if we could get a better expressions on $q(k)$ and $C(k)$, we would have a better estimation of the upper bound of the total area of islands in phase space for given k.

On the other hand, the whole phase space in our model consists of islands and other invariant sets. If the total area of islands is smaller than 1.0, the residue of them must have a positive measure. This may also imply the importance of hyperbolic sets in the orbital diffusion. Although the above calculation is only a coarse estimate, we hope it can give some valuable hints.

4. Conclusions

From the above results, we conclude that:

(1) An orbit started close to tori will spend most of time near the hyperbolic invariant sets surrounding the tori, before escaping from the vicinity. This is reasonable because of the following fact: according to the character of the hyperbolic invariant set, an orbit would spend very long time in approaching to (leaving from) a hyperbolic fixed point along the stable (unstable) manifold.

(2) The orbits started close to tori and those started close to hyperbolic invariant sets obey the same diffusion rules, but in the latter case the orbits diffuse slower than that in the former case.

(3) For the orbits initially close to the chief torus and orbits initially close to the secondary island, respectively, they possess the similar diffusion rule. This is consistent with the character of self-similarity structure of phase space.

(4) The decay of island area with period in both the case of main island sequence and hierarchical island sequence obeys roughly a power law, and in the former case the absolute values of the slope of fitting line increase monotonously with k. From these results it seems that when an orbit diffuses outwards from the chief torus more and more, the probability of its encounter with the islands should be smaller and smaller, and this could be a reason for the gradually faster diffusion of an orbits farther from the chief torus. Moreover, the speed of diffusion increases with k, coinciding with the above conclusions.

Finally, we conclude that the stickiness effect of tori is indeed caused mainly by the hyperbolic invariant sets in the vicinity of tori. According to the Poincaré-Birkhoff fixed point theorem, between the islands in an island-chain, there exist the hyperbolic invariant sets. And cantori are known to consist of hyperbolic invariant sets too. Therefore, when the orbits diffuse through island-chain and cantori, the corresponding hyperbolic invariant sets would slow down the speed of diffusion. So we can conclude also that the generalized stickiness effect is caused mainly by the hyperbolic invariant sets.

Acknowledgements

We are grateful to Drs. C. Efthymiopoulos, C. Froeschlé, E. Lega, R. Dvorak and H. Varvoglis for very helpful discussions and suggestions during the preparation of the manuscript. The first author thanks the IMS of CUHK in Hong Kong for the hospitality during his visit. This work is supported by the Natural Science Foundation of China (No. 10233020), the Special Funds for Major State Basic Research Project (G200077303), and a grant from the Department of Education of China for Ph.D candidate training (20020284011). The second author is also supported by the Academy of Finland (No. 44011) and NSFC (No. 10403004).

References

Chirikov B. and Shepelyansky D. L.: 1984, 'Correlation properties of dynamical chaos in Hamiltonian systems', *Physica D* **13**, 395–400.

Chirikov B. and Shepelyansky D. L.: 1999, 'Asymptotics statistics of Poincaré recurrences in Hamiltonian systems with divided phase space', *Phys. Rev. Lett.* **82**, 528–531.

Contopoulos G., Harsoula M., Voglis N. and Dvorak R.. 1999, 'Destruction of islands of stability', *J. Phys. A: Math. Gen.* **32**, 5213–5232.

Contopoulos G., Voglis N., Efthymiopoulos C., Froeschlé Cl., Gonczi R., Lega E., Dvorak R. and Lohinger E.: 1997, 'Transition spectra of dynamical systems', *Celest. Mech. Dyn. Astron.* **67**, 293–317.

Ding M., Bountis T. and Ott E.: 1990, 'Algebraic escape in higher dimensional Hamiltonian systems', *Phys. Lett. A* **151**, 395–400.

Efthymiopoulos C., Contopoulos G. and Voglis N.:, 1999, 'Cantori, islands and asympototic curves in the stickiniess region', *Celest. Mech. Dyn. Astron.* **73**, 221–230.

Efthymiopoulos C., Contopoulos G., Voglis N. and Dvorak R.: 1997, 'Stickiniess and cantori', *J. Phys. A: Math. Gen.* **30**, 8167–8186.

Froeschlé Cl. and Lega E.: 1998, 'Modelling mappings: an aim and a tool for the study of dynamical systems', In: D. Benest & Cl. Froeschlé (eds.), *Analysis and Modelling of Discrete Dynamical Systems*, pp. 3–54. The Netherlands, Gordon and Breach Science Publishers.

Karney C. F.: 1983, 'Long-time correlations in the stochastic regime', *Physica D* **8**, 360–380.

Lai Y. C., Ding M. Z., Grebogi C. and Blumel R.: 1992, 'Algebraic decay and fluctuations of the decay exponent in Hamiltonian systems', *Phys. Rev. A* **46**, 4661–4669.

Meiss J. D. and Ott E.: 1985, 'Markov-tree model of intrinsic transport in Hamiltonian systems', *Phys. Rev. Lett.* **55**, 2741–2744.

Perry A. D. and Wiggins S.: 1994, 'KAM tori are very sticky: rigorous lower bounds on the time to move away from an invariant Lagrangian torus with linear flow', *Physica D* **71**, 102–121.

Sun Y. S. and Fu Y. N.: 1999, 'Diffusion character in four-dimensional volume preserving map', *Celest. Mech. Dyn. Astron.* **73**, 249–258.

Sun Y. S., Zhou J. L., Zheng J. Q. and Valtonen M.: 2002, 'Diffusion in comet motion', *Contemporary Mathematics* **292**, 229–238.

Zhou J. L., Zhou L. Y. and Sun Y. S.: 2002, 'Hyperbolic structure and stickiness effect', *Chin. Phys. Lett.* **19**, 1254–1256.

R. Dvorak and S. Ferraz-Mello (eds.), A Comparison of the Dynamical Evolution of Planetary Systems (2005) 273–285
DOI 10.1007/s10569-005-0004-4

STELLAR-PLANETARY RELATIONS: ATMOSPHERIC STABILITY AS A PREREQUISITE FOR PLANETARY HABITABILITY

H. LAMMER[1], YU. N. KULIKOV[2], T. PENZ[1], M. LEITNER[1], H. K. BIERNAT[1] and N. V. ERKAEV[3]

[1] *Space Research Institute, Austrian Academy of Sciences, Schmiedlstr. 6, A-8042 Graz, Austria, e-mails: {helmut.lammer, thomas.penz, helfried.biernat, martin.leitner}@oeaw.ac.at*

[2] *Polar Geophysical Institute, Russian Academy of Sciences, Khalturina Str. 15, Murmansk, 183010, Russian Federation, e-mail: kulikov@pgi.ru*

[3] *Institute for Computational Modelling, Russian Academy of Sciences, 660036 Krasnoyarsk 36, Russian Federation, e-mail: erkaev@icm.krasn.ru*

Abstract. The region around a star where a life-supporting biosphere can evolve is the so-called Habitable Zone (HZ). The current definition of the HZ is based only on the mass-luminosity relation of the star and climatological and meteorological considerations of Earth-like planets, but neglects atmospheric loss processes due to the interaction with the stellar radiation and particle environment. From the knowledge of the planets in the Solar System, we know that planets can only evolve into a habitable world if they have a stable orbit around its host star and if they keep the atmosphere and water inventory during: (i) the period of heavy bombardment by asteroids and comets and (ii) during the host stars' active X-ray and extreme ultraviolet (XUV) and stellar wind periods. Impacts play a minor role for planets with the size and mass like Earth, while high XUV fluxes and strong stellar winds during the active periods of the young host star can destroy the atmospheres and water inventories. We show that XUV produced temperatures in the upper atmospheres of Earth-like planets can lead to hydrodynamic "blow off", resulting in the total loss of the planets water inventory and atmosphere, even if their orbits lie inside the HZ. Further, our study indicates that Earth-like planets inside the HZ of low mass stars may not develop an atmosphere, because at orbital distances closer than 0.3 AU, their atmospheres are highly affected by strong stellar winds and coronal mass ejections (CME's). Our study suggests that planetary magnetospheres will not protect the atmosphere of such planets, because the strong stellar wind of the young star can compress the magnetopause to the atmospheric obstacle. Moreover, planets inside close-in HZ's are tidally locked, therefore, their magnetic moments are weaker than those of an Earth-like planet at 1 AU. Our results indicate that Earth-like planets in orbits of low mass stars may not develop stable biospheres. From this point of view, a HZ, where higher life forms like on Earth may evolve is possibly restricted to higher mass K stars and G stars.

Key words: Atmospheric erosion, Atmospheric XUV heating, Stellar radiation, Stellar wind, Water loss

1. Introduction

An important requirement for the evolution of a biosphere is the orbital long-term stability of a terrestrial planet moving in the Habitable Zone (HZ) of its host star (Dvorak et al., 2003; Menou and Tabachnik, 2003). The circumstellar HZ was defined and delimited by Kasting et al. (1993) based on the consideration of a planet with the size and mass comparable to Earth, containing large H_2O and CO_2 reservoirs, at an orbital distance where the atmospheric CO_2 is able to sustain stable liquid H_2O at the planetary surface. If the planet orbit is below a critical distance, the planet would experience a so-called *runaway greenhouse effect*: the vaporized H_2O starts to increase the surface warming and thus enhance the evaporation in a positive feedback (Kasting, 1988).

This results in a H_2O-rich atmosphere that keeps the surface temperature at high level, in which H_2O vapor reaches the high altitudes of the atmosphere, where it is photolyzed. Within a period of the order of 10–100 Myr, the planet loses the hydrogen to space and becomes dry. At the outer border of the HZ, an increase of the CO_2 pressure results in a surface cooling more than heating, due to enhanced Rayleigh back-scattering of the incoming stellar radiation to space. With this definition, the HZ only depends on the luminosity of the star and the surface temperature which is assumed to be stabilized above 0° Celsius through the carbon-silicate cycle.

The range of the HZ is slightly larger for planets that are larger than Earth and for exoplanets, which have higher N_2 partial pressures. The HZ moves to greater orbital distances with time because the star's luminosity increases as it ages. The HZ of F-type stars is larger and may reach orbital distances between 1 and 2.5 AU, while the HZ of K and M dwarf stars is smaller and occurs closer to the star.

Recently, Menou and Tabachnik (2003) investigated 85 detected exoplanetary systems concerning the possibility of harboring "hypothetical" terrestrial planets in the HZ of their host stars. For the global statistics they classified these systems according to the remaining habitable test bodies after an integration time of one Myr and found that about 25% of the investigated systems have a high probability of hosting terrestrial planets in their HZ. These systems are mostly those with relatively close giant planets moving on nearly circular orbits. Another 25% may allow additional terrestrial planets, but the probability is not very high. The remaining 50% are very unlikely for hosting additional planets, due to the strong perturbations of the giant planets in these systems.

The current simplified definition of the HZ and the estimation of its inner and outer boundaries (Kasting et al., 1993) are useful but several uncertainties still need to be discussed. Indeed, if a planet has a stable orbit within the HZ, this condition is far from being sufficient that an Earth-like biosphere may develop. The HZ of low mass K stars and M stars are in orbital distances, where the planets are tidally locked, which can result in hostile climate effects, where the whole atmosphere can snow out on the planets nightside (Joshi, 2003). Furthermore, due to the slow rotation tidally locked Earth-like planets shall have weak magnetic dynamos resulting in weak intrinsic magnetic moments and magnetospheres which can not protect the planet from atmospheric erosion processes by the stellar wind (Griessmeier et al., 2004). Because planets inside the HZ of M and the majority of K stars are tidally locked, plate tectonics may not develop and super-volcanoes like on Venus (big hot spot volcanoes) may frustrate life periodically or destroy the long-time habitability of the planet (Courtillot, 2002).

However, we will focus in this work only on problems related to aeronomy and long-time stability of planetary atmospheres, which can affect the atmospheres of terrestrial planets inside stable orbits of close-in HZ's of low mass stars. The HZ of low mass stars (late K and M: the large majority of stars) are exposed to strong X-ray and extreme ultraviolet (XUV) irradiation (λ: 1–1000 Å) and with a high probability, to strong stellar winds as well (e.g. Wood et al., 2002; Lammer et al., 2003a,b). We show that this radiation and particle exposure can make the atmosphere of an Earth-size planet unstable, can destroy the planets water inventory and may also limit the range of the currently defined HZ.

We review the latest results on observations of the evolution of the radiation environment of solar-like stars in Section 1. In Section 2, we present the effects of thermospheric heating on an Earth-like planet due to XUV radiation and discuss the implications for an evolving biosphere. In Section 3, we describe observational evidence of strong mass loss of young solar-like G and K-type stars and use empirical correlations of stellar mass loss rates with X-ray surface flux values to estimate the stellar winds of young stars. The obtained stellar wind fluxes are used in Section 4 for the estimation of atmospheric erosion on Earth-size exoplanets in orbits of close-in HZ's. Further, we discuss the effects of coronal mass ejection[1] (CME) from the host stars on the atmospheric environments of hypothetical Earth-like planets in orbital distances $\leqslant 0.1$ AU and discuss the implications for the search of habitable terrestrial exoplanets.

[1]Coronal mass ejections or CME's are huge bubbles of gas threaded with magnetic field lines that are ejected from the Sun over the course of several hours.

2. The XUV Radiation Environment of Solar-like Stars

Because the escape of atmospheric constituents in planetary atmospheres depends on the evolution of the stellar XUV radiation ($\lambda \leqslant 1000$ Å), which affects the thermosphere[2] and exosphere[3] temperature, the evolution of planetary atmospheres must be understood within the context of the evolving stellar energy and particle fluxes.

The relevant wavelengths for the heating of upper atmospheres of planets are the ionizing ones less than 1000 Å (e.g. Bauer and Lammer, 2004; Lammer et al., 2004), which contain only a small fraction of the stellar spectral power. The wavelength range with an energy flux of $\geqslant 2\,\mathrm{erg\,cm^{-2}\,s^{-1}}$ represents the predominant XUV heat source for the upper atmosphere of an Earth-like planet.

Astrophysical observations, obtained with the ASCA, ROSAT, EUVE, FUSE and IUE satellites, show that coronal XUV emissions of young main-sequence G-type stars are about 100–1000 times stronger than those of the 4.5 Gyr old Sun. The resulting relative XUV fluxes yield an excellent correlation between the emitted flux and stellar age. In the 1000–1 Å interval, the fluxes follow a power-law (Lammer et al., 2003a,b; Ribas et al., 2005), which is valid for solar like G-type stars of ages between 0.1–7 Gyr. One finds fluxes of about six times the present XUV flux about 3.5 Gyr ago, and about 100 the present XUV flux 100 Myr after a G-type star arrived on the Zero-Age-Main-Sequence (ZAMS). The total X-ray flux of stars with different age and rotation period shows also a decreasing behavior with time (Guedel et al., 1997).

For an initial estimation of the evolution of XUV irradiances on stars with lower masses like K and M stars one can use as a proxy indicator, the ratio of the X-ray luminosity L_{X} to the bolometric luminosity L_{bol}. This ratio is highest for the more active stars with the shortest rotation periods and decrease monotonically with decreasing level of chromospheric activity (Pizzolato et al., 2003).

With the same underlying physical mechanism responsible for XUV emissions and a supposedly similar spectral energy distribution, it is reasonable to assume that stars with similar values of $\log(L_{\mathrm{X}}/L_{\mathrm{bol}})$ will also have similar $\log(L_{\mathrm{XUV}}/L_{\mathrm{bol}})$ (i.e., $1\,\text{Å} < \lambda < 1000\,\text{Å}$).

[2]Thermosphere: the region in the upper atmosphere where the stellar XUV radiation is absorbed and in part used for atmospheric heating (Bauer and Lammer, 2004).

[3]Exosphere: the outermost atmospheric layer where no collisions between atmospheric particles take place and light atoms with energies higher than the escape energy can be lost thermally from the planet (Bauer and Lammer, 2004).

Recent studies of K type stars show that they stay at saturated emission levels of about 100 times the present solar XUV flux for a little longer time and then also decrease following a power–law relationship of a very similar slope. Interestingly, M0–M5 stars seem to have saturated emission levels up to 1 Gyr and possibly longer and then decrease in an analogous way to G and K stars (Ribas et al., 2005). These preliminary results indicate that early K stars and early M-type stars may have XUV irradiances that are about 3–4 times and about 10–100 times higher, respectively, than solar-type stars of the same age. More accurate investigations are currently being carried out with an extended sample and a large variety of observational data (Ribas et al., 2005).

3. Atmospheric Blow off Due to XUV Heating

The critical phase if a water-bearing terrestrial planet, at a dynamically stable orbit inside the HZ, can evolve a biosphere, is its survival during the early period of the high XUV flux of the young host star. As discussed before, the energy budget of the upper atmosphere of Earth-like planets is primarily governed by the heating of the gas due to the absorbtion of the XUV radiation by the atmospheric species, by heat transport due to conduction and convection and by heat loss due to emissions in the infrared (IR) (e.g. Bauer and Lammer, 2004).

One can summarize the important heating and cooling processes in the upper atmosphere of the Earth due to the N_2, O_2, and O photo-ionization, photo-dissociation of O_2 and O_3 molecules, chemical heating in exothermic reactions with O and O_3 and neutral gas heat conduction (e.g. Gordiets et al., 1982). Radiative loss by IR occurs when atmospheric constituents are present which have transition levels in the IR. This is the case for atomic oxygen in the terrestrial atmosphere and NO, CO, CO_2, OH, O_3, etc. in other planetary atmospheres.

The effective heat production Q_{XUV} in the upper atmosphere of an Earth-like planet is balanced by the divergence of a conductive heat flux in the thermosphere due to the incoming XUV radiation and the heat energy L_{IR} lost by emitted IR radiation per unit volume

$$\rho c_v \left[\frac{\partial T}{\partial t} + \vec{v}_n \cdot \vec{\nabla} T \right] + p \vec{\nabla} \cdot \vec{v}_n - \vec{\nabla} \cdot (K_n \vec{\nabla} T) = Q_{XUV} - L_{IR}, \quad (1)$$

where ρ is the atmospheric mass density, c_v the corresponding specific heat at constant volume, $\vec{v}_n$ the velocity and p the pressure of the neutral atmosphere and K_n is the thermal conductivity. A simplification in the one-dimensional case (vertical variability z only) can be applied in the following form

$$\rho c_p \left(\frac{\partial T}{\partial t} + v_{nz} \frac{\partial T}{\partial z} \right) - \frac{\partial}{\partial z} \left(K_n(T) \frac{\partial T}{\partial z} \right) = Q_{\mathrm{XUV}} - L_{\mathrm{IR}}. \tag{2}$$

For the calculation of temperature profiles as function of various stellar XUV radiation fluxes, we use the thermospheric model of Gordiets et al. (1982) and solve the time-dependent 1D equations of continuity, hydrostatic and heat balance simultaneously in the height region above the base of the thermosphere up to the exosphere level.

When a large amount of XUV energy is deposited at the top of an atmosphere, heated atoms can overcome the planetary gravity field and a planetary wind consisting of the heated atoms develops. By assuming a progressive increase of stellar XUV flux from present features (Sun = 1 XUV) to 10 or even 100 times the present XUV flux, the following conditions shown in Figure 1 occur:

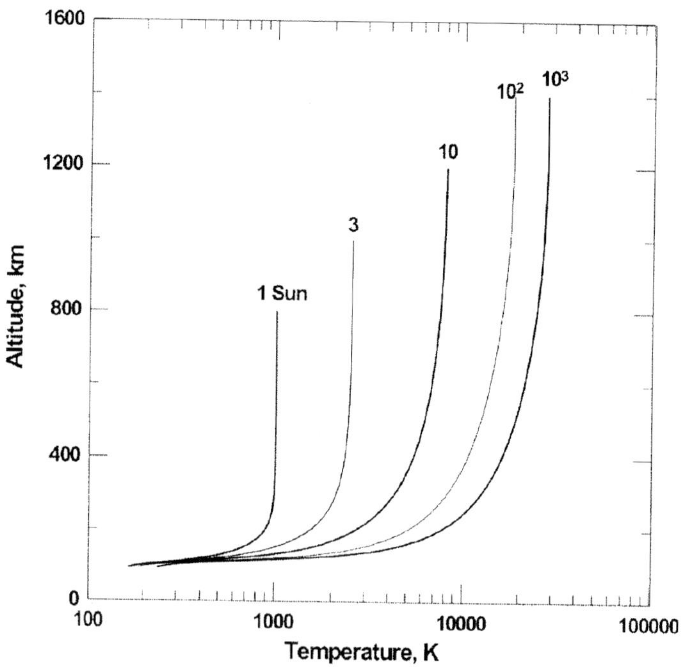

Figure 1. Illustration of the rise of exobase level and temperature in Earth's present thermosphere and exosphere for various XUV levels times the present Sun-value (1 Sun). If the exosphere temperature reaches about 5000 K (6 XUV: 3.5 Gyr ago), diffusive or even energy-limited escape for hydrogen atoms originates. 100 time higher XUV fluxes can even remove heavy gases like O, C and N atoms with a very high efficiency from an Earth-like planet.

First step: the upper atmospheric temperature increases, with a subsequent increase of the altitude of the exobase[4] (see Figure 1).

Second step: at a certain level of the XUV flux, there is no (quasi) static solution anymore and atmospheric blow off occurs.

If IR cooling of the thermosphere can not balance the incoming stellar XUV energy anymore, the excess thermal energy is directly converted into kinetic energy and hydrodynamic escape occurs. Jeans escape occurs when only a very small fraction of atoms, in the energetic wing of the velocity Maxwellian distribution, are lost to space (in quasi steady state), hydrodynamic diffusion-limited or energy-limited escape[5] results in a rapid depletion of the full distribution, which cannot be re-populated over sufficiently short time-scales. The critical temperature for H atoms on an Earth-mass planet is about 5000 K. One can see from Figure 1 that the temperature would have been overcome during the first Gyr, indicating that the early Earth had a different atmosphere than today. Our calculation indicates that only dense Venus-like CO_2 atmospheres can protect the atmospheres of young Earth-like exoplanets from atmospheric evaporation and survival of their water inventories during active XUV periods of their host stars because of strong IR cooling. However, long-time XUV radiation fluxes in the order of about 50–100 times the present value can dramatically affect the water inventory of a terrestrial exoplanet and possible even the stability of its whole atmosphere.

The latter problem may occur on a terrestrial H_2O-rich planet inside the HZ of a low mass M or K star, where there is observational evidence that their XUV energy fluxes stay active over longer time periods. First, the CO_2 may prevent extreme hydrodynamic escape conditions like on early Earth due to its heavy mass and good IR cooling capabilities, but after the CO_2 is removed from the atmosphere like on Earth due to chemical weathering in a humid wet environment (Franck et al., 2002), N_2 like on present Earth may become the dominant constituent in the atmosphere, so that the upper atmosphere is heated like in Figure 1 and large escape

[4]Exobase: atmospheric level where the mean free path of the main atmospheric species is similar to the scale height $H = (kT/mg)$ of the gas, with k the Boltzmann constant, T the exospheric temperature, g the gravitational acceleration and m is the mass of the atmospheric species.

[5]Hydrodynamic escape: diffusion-limited escape means that all atoms especially the light ones, which diffuse through the surrounding heavy gas up to the upper atmosphere can escape from the planet. Energy-limited escape occurs if only one species (i.e., hydrogen) is available in large amounts in the upper atmosphere so that it can escape as a planetary wind as observed at the giant hydrogen-rich exoplanet HD 209458 b (e.g. Lammer et al., 2003b).

rates develop, which evaporate the planets H_2O inventory (e.g. Chassefiere, 1997a; Lammer et al., 2003b; Bauer and Lammer, 2004).

Parallel to chemical weathering and removal of CO_2 from the planets atmosphere a CO_2 atmosphere can also be eroded by the interaction of the stellar wind plasma of the young star.

4. Stellar Wind Evolution of Solar-like Stars

The temporal evolution of the stellar wind velocity can be derived from lunar and meteorite fossil records (Newkirk, 1980). For the estimation of the early stellar wind density, one can use the mass loss estimations provided by Wood et al. (2002), which are IR from Hubble Space Telescope high–resolution spectroscopic observations of the H Lyman-α feature of several near–by main–sequence G and K stars.

Wood et al. (2002) found from their observations that the mass loss $\mathrm{d}M/\mathrm{d}t$ of solar-like G and K stars is proportional to their observed X-ray surface flux, which is correlated to the rotation periods P_{rot} of the stars. The product of the mass loss and the stellar wind velocity v_{sw} can be determined by (Griessmeier et al., 2004; Lammer et al., 2004)

$$\frac{\mathrm{d}M}{\mathrm{d}t} v_{\mathrm{sw}} \propto P_{\mathrm{rot}}^{-3.34\pm0.67}. \tag{3}$$

By using the temporal behavior of the stellar wind from Newkirk (1980) as $v_{\mathrm{sw}} = v_0(1 + t/t_c)^{-0.4}$, the time behavior of the stellar wind density can be found as $n_{\mathrm{sw}} = n_0(1 + t/t_c)^{-1.54\pm0.47}$. The time constant is $t_c = 2.56 \times 10^7$ year. The proportionality constants n_0 and v_0 can be derived for present–day conditions at 1 AU. For distances other than 1 AU, the constants can be evaluated with a r^{-2} dependence.

Nevertheless it should be noted that Newkirk (1980) pointed out that a lower initial rotation rate in this model would give slightly lower values for the early solar/stellar wind velocities and vice versa, which is not considered in this work. Therefore, more active young solar-like G and K stars with X-ray surface fluxes larger than $10^6\,\mathrm{erg\,cm^{-2}\,s^{-1}}$ must be studied during the near future (Linsky and Wood, 2004).

5. Atmospheric Loss Induced by Stellar Wind and CME Events

The boundary between the stellar wind and the planetary magnetosphere, which protects an atmosphere from stellar wind erosion processes is called the magnetopause. The precise location and shape of the magnetopause are determined mainly by the stellar wind parameters, which depend on the orbital distance from the host star and the planetary magnetic field strength.

Planets in orbits close to their host star are subject to strong tidal dissipation, leading to tidal locking on a very short timescale. For tidally locked planets the rotation period is equal to the orbital period, therefore, a fast rotation of the planet is not possible. All common scaling laws for the planetary magnetic moment yield a magnetic moment rapidly decreasing with decreasing rotation rate (Griessmeier et al., 2004). Due to the reduced internal magnetic moment the pressure balance is shifted closer to the planet until the planetary atmosphere acts as an obstacle (Figure 2). In this case, the atmosphere is subject to various non-thermal loss processes. Moreover, the magnetic and flare activity of young stars is much higher than at a 4.5 Gyr old star like our Sun, therefore, it is reasonable to assume that more CME's occur at these stars. These CME's reach close-in exoplanets at orbital distances < 0.1 and affect strongly their magnetospheres and atmospheres. Scaling laws from the observation of CME's for the spatial evolution of the maximum and minimum density $n_{\max} = 7.1 \times r^{-2.99}\ \mathrm{cm}^{-3}$

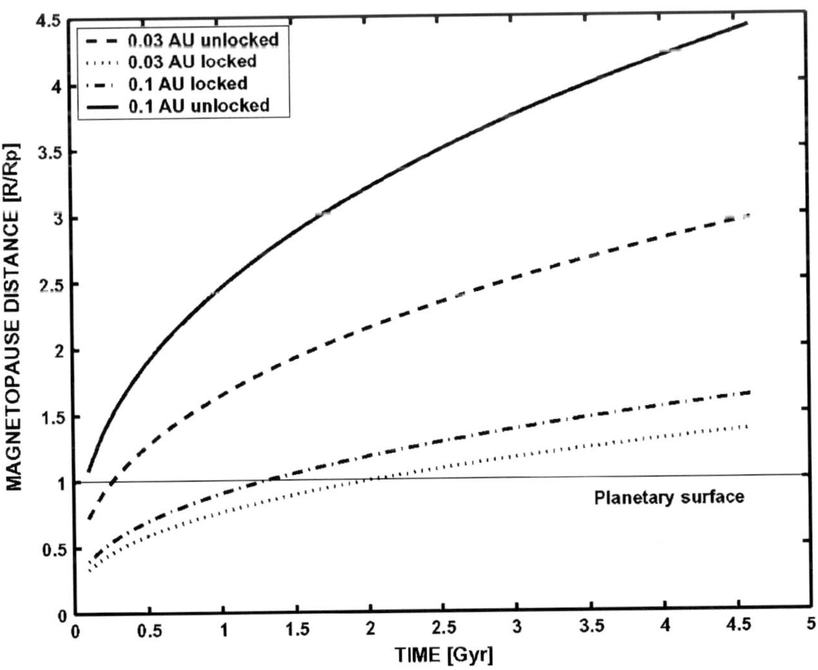

Figure 2. Compression of the magnetopause with and without tidal locking (Magnetic field strength is similar than on Earth) at orbital distances of 0.03 and 0.1 AU in units of the planetary radii R_p.

and $n_{\min} = 4.88 \times r^{-2.31}\ \mathrm{cm}^{-3}$, respectively, and the bulk velocity is about 500 km/s (Kodachenko et al., 2005). For example the planetary magnetic field compression of a Jupiter-class exoplanet due to the interaction with a CME is shown in Figure 3. We use a rough estimation for the calculation of the temporal evolution of an unprotected planetary atmosphere due to mass loss estimations caused by photo-ionization, electron impact ionization and charge exchange from the upper atmosphere into the stellar wind plasma flow around the planetary obstacle for the changing solar wind parameters (Michel, 1971; Bauer, 1983).

For preliminary mass loss estimations we use the model of Michel (1971) and Bauer (1983), which assumed that the solar wind interaction is confined to the scale height H of the atmospheric gas and the mass loss is produced by the ionization of the neutral gas above the planetary obstacle corresponding to the compressed magnetopause/ionopause distance. By applying momentum balance considerations between the stellar wind and

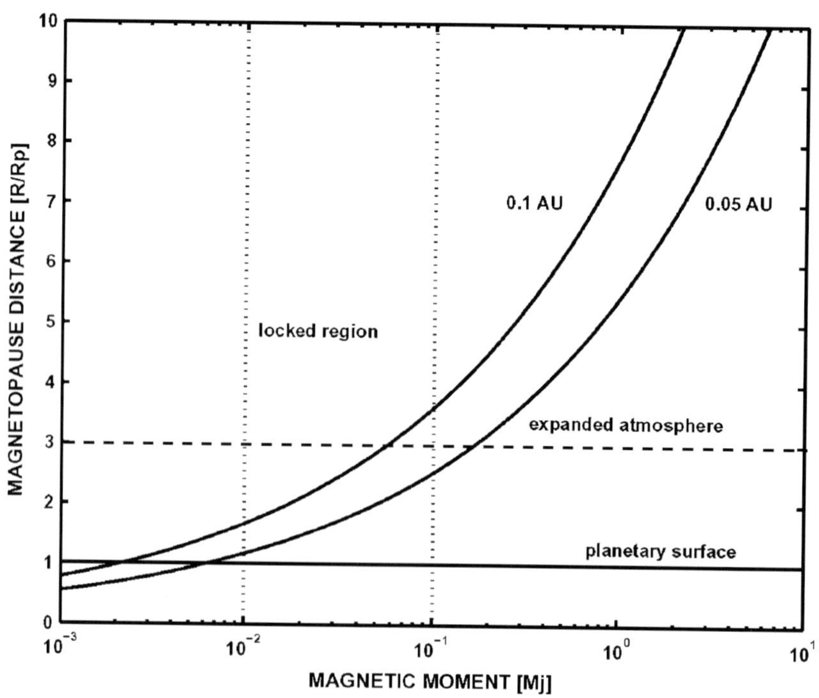

Figure 3. Compression of the magnetopause of a Jupiter-class exoplanet due to a CME for orbital distances of 0.05 and 0.1 AU. Note that the Earth magnetic moment is less than 10^{-4} times the Jupiter magnetic moment, and therefore the atmosphere of terrestrial planets, which are hit by a CME at this orbital distances will be eroded or destroyed.

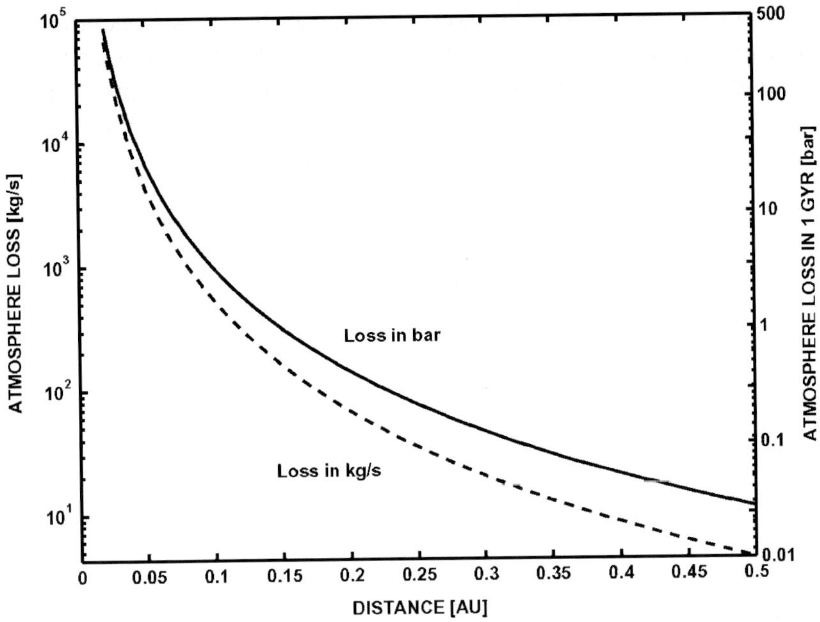

Figure 4. Atmospheric loss rates for Earth-like exoplanets with low magnetic moments due to tidal locking as a function of the orbital distance in bar during 1 Gyr (solid line, right axis) and in kg/s (dashed line, left axis).

the photo-ions and steady-state conditions (Bauer, 1983) the atmospheric mass loss rate $\mathrm{d}M_a/\mathrm{d}t$ due to the solar wind interaction is given as

$$\frac{\mathrm{d}M_a}{\mathrm{d}t} \approx -2K\rho_{\mathrm{sw}} v_{\mathrm{sw}} R_{ip}^2, \tag{4}$$

where $K \approx 0.3$ is a so-called mass loading limit[6] (Michel, 1971; Bauer, 1983), and the subscript "sw" indicates stellar wind conditions. The estimated loss rates for Earth-like planets as function of orbital distance are shown in Figure 4 for low magnetic moments due to tidal locking. One can see that atmospheres of more than about 500 bar can be removed within 1 Gyr for close-in orbits of 0.03 AU. At orbits of 0.1 AU, the atmospheric loss is in the range of several bars during 1 Gyr. The loss rate per second is about 500 kg/s for planets orbiting at 0.1 AU, more than 5000 kg/s at 0.05 AU and less than 5 kg/s at 0.5 AU. For comparison, the present loss rate at Venus is about 1 kg/s (Bauer, 1983). A recent study by Lammer et al. (2005) applied a more complex numerical test particle model to

[6]Mass loading limit: Only a fraction of about 1/3 of the atmospheric species, which are affected by the stellar wind plasma flow will be ionized and picked up.

Earth-like exoplanets-CME interaction in close-in HZ's. They found that a combination of weak magnetic moments, high XUV fluxes and extended upper atmospheres can result in loss rates up to 100 of bars even at 0.2 AU, which is much higher than the estimates obtained from equation (4) and shown in Figure 4.

6. Conclusion

Our study suggests that exoplanets orbiting inside close-in HZ's of M and low mass K stars may not develop Earth-like biospheres, because these planets may lose their atmospheres and water inventories due to the long-time activity in high XUV radiation, strong stellar winds and CME's of their host stars. Moreover, exoplanets inside the HZ of these stars are tidally locked so that their magnetic dynamos and the resulting magnetospheres will be weaker than the magnetosphere of Earth at 1 AU. From this point of view we suggest that higher mass K and G stars should be considered as good primary stellar candidates for the search of "hypothetically" habitable planets in future terrestrial planet finding missions like Darwin (ESA) and TPF-C (NASA). Thus, dynamical stability with further restrictions on the orbit of a planet is the requirement of a stable atmosphere, the result of our investigation has also an impact on studies, which are underway to produce catalogues of hypothetical planetary systems where terrestrial exoplanets can have stable dynamical orbits but can also keep their atmospheres and surface water inventories.

Acknowledgements

H. Lammer, Yu. N. Kulikov, H. K. Biernat, and T. Penz thank the "Österreichischer Austauschdienst", which supported this work by the projects I.12/04 and I.2/08. The authors thank also the Austrian "Fonds zur Förderung der wissenschaftlichen Forschung" which supported this study under projects P17100-N08 and P17099-N08 and the support by the Austrian Academy of Sciences, "Verwaltungsstelle für Auslandsbeziehungen" and by the Russian Academy of Sciences. Yu. N. Kulikov also thanks the "Russian Fund for Basic Research" of the Russian Academy of Sciences which supported this study as the Russian-Austrian project RFBR No. 03-05-20003 BWTZ a. This study was supported by International Space Science Institute (ISSI) and benifited from the ISSI team "Evolution of Habitable Planets".

References

Bauer, S. J.: 1983, *Ann. Geophysics* **1**, 477.

Bauer, S. J. and Lammer, H.: 2004, *Planetary Aeronomy: Atmosphere Environments in Planetary Systems*, Springer-Verlag, New York.

Chassefiere, E.: 1997a, *Icarus* **124**, 537.

Chassefiere, E.: 1997b, *Icarus* **126**, 229.

Courtillot, V.: 2002, *Evolutionary Catastrophes: The Science of Mass Extinctions*, Cambridge University Press, Cambridge.

Dvorak, R., Pilat-Lohinger, E., Funk, B. and Freistetter, F.: 2003, *Astron. Astrophys.*, **398**, L1.

Franck, S., Kossacki, K. J., v. Bloh, W. and Bounama, C.: 2002, *Tellus* **54B**, 325.

Gordiets, B. F., Kulikov, Yu., Markov, N., Marov, M. N., Ya M.: 1982, *J. Geophys. Res.* **87**, 4504.

Griessmeier, J.-M., Stadelmann, A., Penz, T., Lammer, H., Selsis, F., Ribas, I., Guinan, E. F., Motschmann, U., Biernat, H. K. and Weiss, W. W.: 2004, *Astron. Astrophys.* **425**, 753.

Guedel, M., Guinan, E. F. and Skinner, S. L.:1997, *Astrophys. J.* **483**, 947.

Joshi, M. M.: 2003, *Astrobiology* **3**, 415.

Kasting, J. F.: 1988, *Icarus* **74**, 472.

Kasting, J. F., Whitmire, D. P. and Reynolds, R. T.: 1993, *Icarus* **101**, 108.

Kodachenko, M. L., Ribas, I., Lammer, H., Griessmeier, J.–M., Selsis, F., Leitner, M., Penz, T., Eirora, C., Hanslmeier, A., Biernat, H. K., Farrugia, C. J. and Rucker, H. O.: 2005, *Astrobiology*, submitted.

Lammer, H., Lichtenegger, H. I. M., Kolb, C., Ribas, I., Guinan, E. F. and Bauer, S. J.: 2003a, *Icarus* **165**, 9.

Lammer, H., Selsis, F., Ribas, I., Guinan, E. F., Bauer, S. J. and Weiss, W. W.: 2003b, *Astrophys J. Lett.* **598**, L121.

Lammer, H., Ribas, I., Griessmeier, J.-M., Penz, T., Hanslmeier, A. and Biernat, H. K.: 2004, *Hvar. Obs. Bull.* **28**, 139.

Lammer, H., Lichtenegger, H. I. M., Yu, N., Kulikov, Yu, N., Grießmeier, J.-M., Erkaev, N. V., Biernat, H. K., Khodachenko, M. L., Ribas, I., Penz, T. and Selsis, F.: 2005, *Astrobiol.* submitted.

Linsky, J. L. and Wood, B. E.: 2004, in *The Sun and the Heliosphere as an Integrated System*, Kluwer Academic Publ., lordrecht in press.

Menou K. and Tabachnik, S.: 2003, *Astrophys. J.* **583**, 473.

Michel, C. F.: 1971, *Planet. Space Sci.* **19**, 1580.

Newkirk, Jr., G.: 1980. *Geochi. Cosmochi. Acta Suppl.* **13**, 293.

Pizzolato, N., Maggio, A., Micela, G., Sciortino, S. and Ventura, P.: 2003, *Astron. Astrophys.* **397**, 147.

Ribas, I., Guinan, E. F., Güdel, M. and Audard M.: 2005, *Astrophys. J.* **622**, 680.

Stauffer, J. R., Caillault, J.-P., Gagne, M., Prosser, C. F. and Hartmann, L. W.: 1994, *Astrophys. J. Suppl.* **91**, 625.

Stelzer, B. and Neuhäuser, R.: 2001, *Astron. Astrophys.* **377**, 538.

Wood, B. E., Müller, H.-R., Zank, G. and Linsky, J. L.: 2002, *Astrophys. J.* **574**, 412.

R. Dvorak and S. Ferraz-Mello (eds.), A Comparison of the Dynamical
Evolution of Planetary Systems (2005) 287–300
DOI 10.1007/s10569-005-0005-3

DYNAMIC HABITABILITY OF EXTRASOLAR PLANETARY SYSTEMS

W. VON BLOH, C. BOUNAMA and S. FRANCK
Potsdam Institute for Climate Impact Research (PIK), P.O. Box 601203, 14412 Potsdam, Germany, e-mail: bloh@pik-potsdam.de

Abstract. In this paper, we estimate the likelihood to find habitable Earth-like planets on stable orbits for the extrasolar planetary systems ε Eridani, 55 Cancri, 47 Ursae Majoris, and ρ Coronae Borealis and provide a new tool to assess habitability of extrasolar planetary systems. For determining the habitable zone in these systems an integrated system approach is used taking into account a variety of climatological, biogeochemical, and geodynamical processes. Habitability is linked to the photosynthetic activity on the planetary surface. We find that habitability strongly depends on the age of the stellar system and the characteristics of the Earth-like planet. In particular the portion of land/ocean coverages plays an important role. In general, a high percentage of ocean area significantly increases the probability for habitability of planets on stable orbits in extrasolar planetary systems. We show that the systems ε Eridani and 55 Cancri are most promising to find dynamic habitable Earth-like planets, while the system ρ Coronae Borealis is most unpromising.

Key words: extrasolar planets, geodynamics, habitable zone, orbital stability, planetary climate.

1. Introduction

The search for extrasolar Earth-like planets is one of the main goals of present research. More than 150 extrasolar giant planets are known to orbit around Sun-like stars including several multiple-planet systems. These giant planets, with hydrogen and helium as the main constituents, have atmospheres too turbulent to permit the emergence of life and have no underlying solid surfaces or oceans that could support a biosphere. The distribution of masses of all known exoplanets lets scientists suppose that there must be a multitude of planets with lower masses (Marcy et al., 2003). The existence of Earth-type planets around stars other than the Sun is strongly implied by various observational findings including (1) the steep rise of the mass distribution of planets with decreasing mass, which implies that more small planets form than giant ones; (2) the detection of protoplanetary disks (with masses between 10 and 100 times that of Jupiter) around many solar-type stars younger than ~3 Myr; and (3) the discovery of "debris disks" around middle-aged stars, the presumed analogs of the Kuiper Belt and zodiacal dust (Marcy and Butler, 2000 and references therein).

Even if it seems today beyond the technical feasibility to detect Earth-mass planets we can apply computer models to investigate known exoplanetary systems to determine whether they could host Earth-like planets with surface conditions sufficient for the emergence and maintenance of life on a stable orbit. Such a configuration is described as dynamically habitable. Jones et al. (2001) have investigated the dynamical habitability of known exoplanetary systems. They used the boundaries of the habitable zone (HZ) originating from Kasting et al. (1993). The inner boundary is defined as the maximum distance from the star where a runaway greenhouse effect would lead to the evaporation of all the surface water, and the outer boundary as the maximum distance at which a cloud-free CO_2 atmosphere could maintain a surface temperature above 0°C. To test the intersection of stable orbits and the HZ, putative Earth-mass planets were launched into various orbits in the HZ and a symplectic integrator was used to calculate the celestial evolution of the extrasolar planetary system. In this paper, we adopt a somewhat different definition of HZ already used by Franck et al. (1999, 2000a, b). Here habitability (i.e., presence of liquid water at all times) does not just depend on the parameters of the central star, but also on the properties of the planet itself. In particular, habitability is linked to the photosynthetic activity of the planet, which in turn depends on the planetary atmospheric CO_2 concentration, and is thus strongly influenced by planetary geodynamics. This leads to additional spatial and temporal limitations of habitability, as the stellar HZ (defined for a specific type of planet) becomes narrower with time due to the persistent decrease of the planetary atmospheric CO_2 concentration.

2. Habitability of Extrasolar Planetary Systems

Hart (1978,1979) calculated the evolution of the atmosphere of a terrestrial planet over geologic time by varying its distance from the Sun. In his approach, the HZ is the region within which an Earth-like planet might enjoy moderate surface temperatures needed for advanced life forms. He found that the HZ ($[R_{inner}, R_{outer}]$) between runaway greenhouse and runaway glaciation is surprisingly narrow for G2 stars like our Sun: $R_{inner} = 0.958$ AU, $R_{outer} = 1.004$ AU. A main drawback of these calculations is the neglect of the negative feedback between atmospheric CO_2 content and mean global surface temperature discovered later by Walker et al. (1981).

The implementation of this feedback by Kasting et al. (1988) provided an almost constant inner boundary ($R_{inner} = 0.95$ AU) but a remarkable extension of the outer boundary near Martian distance ($R_{outer} = 1.5$ AU). Later Kasting et al. (1993) and Kasting (1997) recalculated the present HZ boundaries as $R_{inner} = 0.95$ AU and $R_{outer} = 1.37$ AU. Kasting et al. (1993) defined the HZ of an Earth-like planet as the region where liquid water is present at the surface. According to this definition the inner boundary of the HZ is determined by the loss of water via photolysis and hydrogen escape. The outer boundary of the HZ is determined by the condensation of CO_2 crystals out of the atmosphere that attenuate the incident sunlight by Rayleigh scattering. The critical CO_2 partial pressure for the onset of this effect is about 5–6 bar (Kasting, personal communication, 1999). On the other hand, the effects of CO_2 clouds have been challenged by Forget and Pierrehumbert (1997). The CO_2 clouds have the additional effect of reflecting the outgoing thermal radiation back to the surface. The result for the HZ as a function of central star mass for a zero-age main sequence star is shown in Figure 1.

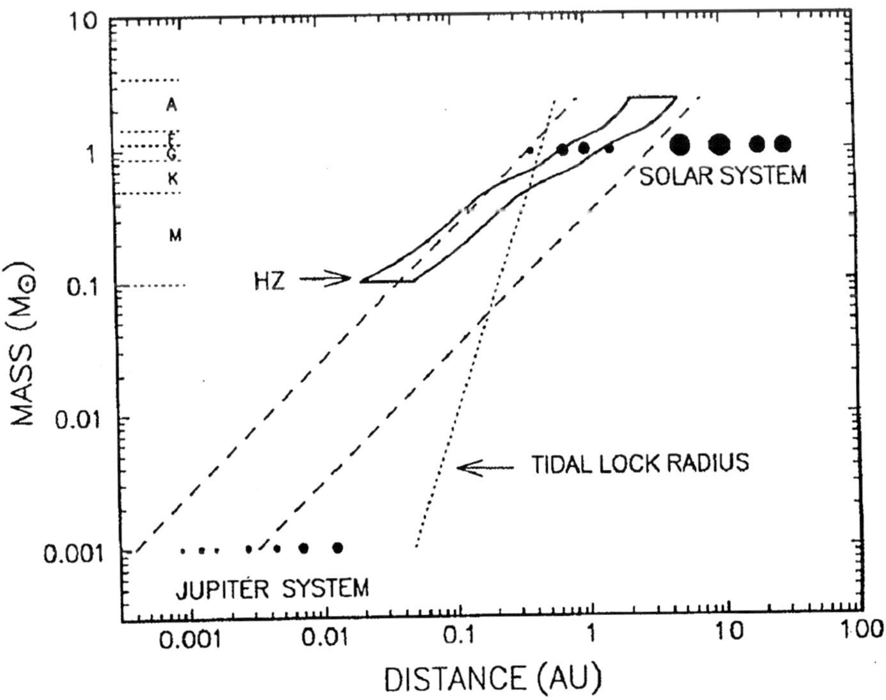

Figure 1. The habitable zone of a zero age main sequence star as a function of the stellar mass from Kasting et al., (1993).

In our calculation of the HZ we are following an integrated system approach. On Earth, the carbonate-silicate cycle is the crucial element for a long-term homeostasis under increasing solar luminosity. In most studies (see, e.g., Caldeira and Kasting, 1992), the cycling of carbon is related to the present tectonic activities and to the present continental area as a snapshot of the Earth's evolution. On the other hand, on geological timescales the deeper parts of the Earth are considerable sinks and sources for carbon. In addition, the tectonic activity and the continental area change noticeably. Therefore, we favour the so-called geodynamical models that take into account both the growth of continental area and the decline in the spreading rate (Franck et al., 2000a). Our numerical model couples the stellar luminosity, L, the silicate-rock weathering rate, F_{wr}, and the global energy balance to allow estimates of the partial pressure of atmospheric and soil carbon dioxide, P_{atm} and P_{soil}, respectively, the mean global surface temperature, T_{surf}, and the biological productivity, Π, as a function of time, t (Figure 2). The main point is the persistent balance between the CO_2 sink in the atmosphere-ocean system and the metamorphic (plate-tectonic) sources. This is expressed with the help of dimensionless quantities (Berner et al., 1983, Kasting, 1984)

$$f_{\mathrm{wr}} \cdot f_A = f_{\mathrm{sr}}, \tag{1}$$

where $f_{\mathrm{wr}} \equiv F_{\mathrm{wr}}/F_{\mathrm{wr},0}$ is the weathering rate normalised by the present value, $f_A \equiv A_c/A_{c,0}$ is the continental area normalised by the present value, and $f_{\mathrm{sr}} \equiv S/S_0$ is the spreading rate normalised by the present value. Equation (1) can be rearranged by introducing the geophysical forcing ratio GFR (Volk, 1987):

$$f_{\mathrm{wr}} = \frac{f_{\mathrm{sr}}}{f_A} =: \mathrm{GFR}. \tag{2}$$

With the help of Equation (2), we can calculate the normalised weathering rate from geodynamics based on the continental growth model and the spreading rate (Franck et al., 2000a). The continental area can be prescribed by several scenarios, e.g. delayed growth, linear growth or constant area. The spreading rate is determined with the help of the boundary layer theory of whole mantle convection (Turcotte and Schubert, 1982). On the other hand, the weathering rate, f_{wr}, depends directly on the surface temperature and the CO_2 partial pressure (Walker et al., 1981):

$$f_{\mathrm{wr}} = f_{\mathrm{wr}}(T_{\mathrm{surf}}, P_{\mathrm{atm}}). \tag{3}$$

The connection between the stellar parameters and the planetary climate can be formulated by using a radiation balance equation (Williams, 1998)

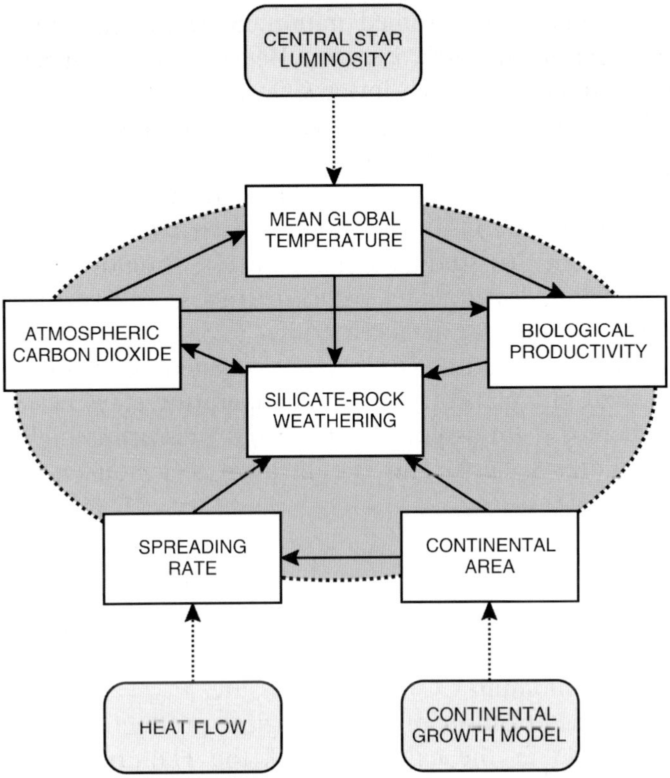

Figure 2. Box model of the integrated system approach (Franck et al., 2000a). The arrows indicate the different forcings (dotted lines) and feedback mechanisms (solid lines).

$$\frac{L}{4\pi R^2}[1-\alpha(T_{\text{surf}}, P_{\text{atm}})] = 4I_R(T_{\text{surf}}, P_{\text{atm}}). \tag{4}$$

Here α denotes the planetary albedo, I_R the outgoing infrared flux, and R the distance from the central star. The evolution of the surface temperature and the CO_2 partial pressure can be derived by solving Equations (2)–(4) simultaneously.

In our model, biological productivity is considered to be solely a function of the surface temperature and the CO_2 partial pressure in the atmosphere

$$\frac{\Pi}{\Pi_{\text{max}}} = \left(1-\left(\frac{T_{\text{surf}}-50^\circ\text{C}}{50^\circ\text{C}}\right)^2\right)\left(\frac{P_{\text{atm}}-P_{\text{min}}}{P_{1/2}+(P_{\text{atm}}-P_{\text{min}})}\right). \tag{5}$$

Here Π_{max} denotes the maximum biological productivity, which is assumed to amount to twice the present value Π_0 (Volk, 1987). $P_{1/2}+P_{\text{min}}$ is the value at which the pressure-dependent factor is equal to 1/2, and P_{min}

is fixed at 10^{-5} bar, the presumed minimum value for C_4-photosynthesis (Pearcy and Ehleringer, 1984; Larcher 1995). The evolution of the biosphere and its adaption to even lower CO_2 partial pressures are not taken into account in our model. For a given P_{atm}, Equation (5) yields maximum productivity at $T_{surf} = 50°C$ and zero productivity for $T_{surf} \leqslant 0°C$ and $T_{surf} \geqslant 100°C$. There exist hyperthermophilic life forms with a temperature tolerance well above 100°C. In general, these are chemoautotrophic organisms not included in this study. At this point we should emphasise that all calculations are done for a planet with Earth mass and size, and an Earth-like radioactive heating rate in its interior.

The HZ around an extrasolar planetary system is defined as the spatial domain where the planetary surface temperature stays between 0°C and 100°C and where the atmospheric CO_2 partial pressure is higher than 10^{-5} bar to allow photosynthesis. This is equivalent to a non-vanishing biological productivity, $\Pi > 0$, i.e.,

$$HZ := \{R \mid \Pi(P_{atm}(R,t), T_{surf}(R,t)) > 0\}. \tag{6}$$

According to the definition in Equation (6) the boundaries of the HZ are determined by the surface temperature extrema, $T_{surf} = 0°C \vee T_{surf} = 100°C$, or by the minimum CO_2 partial pressure, $P_{atm} = 10^{-5}$ bar. Therefore, the specific parameterisation of the biological productivity (Equation (5)) plays a minor role in the calculation of the HZ. In the approach by Kasting et al. (1993) the HZ is limited only by climatic constraints invoked by the luminosity of the central star, while our method relies on additional constraints. First, habitability is linked to the photosynthetic activity of the planet (Equation (6)) and second, habitability is strongly affected by the planetary geodynamics. In principle, this leads to additional spatial and temporal limitations of habitability. To present the results of our modelling approach we have delineated the HZ for an Earth-like extrasolar planet at a given but arbitrary distance R in the stellar mass-time plane (Figure 3). For this particular case a linear continental growth model was applied, i.e. $f_A \equiv f_A(t) = t/(4.6\,\text{Gyr})$. A detailed discussion about the influence of continental growth models can be found in Franck et al. (2000a). The qualitative behaviour does not depend on the choice of a specific scenario. In general, the HZ is limited by the following effects:

1. Stellar lifetime on the main sequence τ_H decreases strongly with mass. Using simple scaling laws (Kippenhahn and Weigert, 1990), we estimated the central hydrogen burning period and got $\tau_H < 0.8\,\text{Gyr}$ for $M > 2.2\,M_\odot$. Therefore there is no point in considering central stars with masses larger than $2.2\,M_\odot$ because an Earth-like planet may need approximately 0.8 Gyr of habitable conditions for the development of

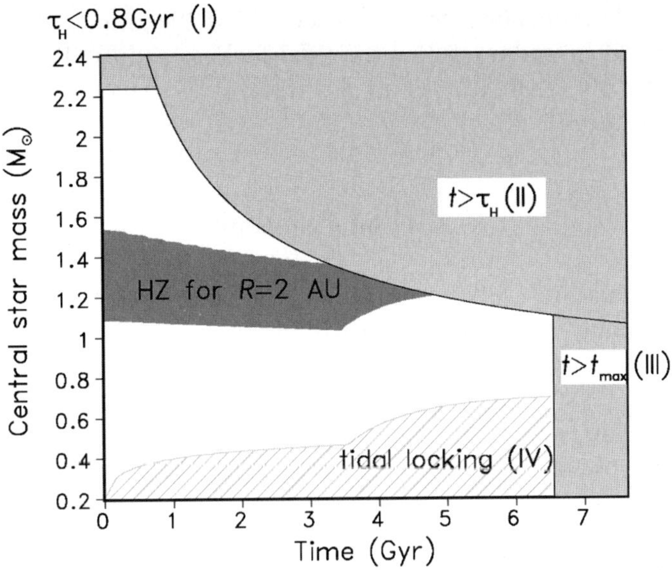

Figure 3. Shape of the HZ (grey shaded) in the mass-time plane for an Earth-like planet at distance $R = 2$ AU from the central star. The potential overall domain for accommodating the HZ for planets at some arbitrary distance is limited by a number of factors that are independent of R: (I) Minimum time for biosphere development ($\tau_H < 0.8$ Gyr excluded); (II) Central-star life time on the main sequence ($t > \tau_H$ excluded); (III) Geodynamics of the Earth-like planet ($t > t_{\max}$ excluded); (IV) Tidal locking of the planet (non-trivial sub-domain excluded). The excluded realms are marked by light grey shading in case of the first three factors, and by grey hatching for the tidal-locking effect. Picture is taken from Franck et al., (2000b).

life (Hart, 1978, 1979). Quite recently, evidence for liquid water at the Earth's surface already at 0.2 Gyr after formation has been found (Wilde et al., 2002). Because liquid water is a necessary condition for life, this is the lower limit for the origin of life (Bada, 2004). If we perform calculations with $\tau_H < 0.2$ Gyr, we obtain qualitatively similar results, but the upper bound of central star masses is shifted to 3.4 $M_\odot$.

2. When a star leaves the main sequence to turn into a red giant, there clearly remains no HZ for an Earth-like planet. This limitation is relevant for stellar masses in the range between 1.1 and 2.2 $M_\odot$.
3. In the stellar mass range between 0.6 and 1.1 $M_\odot$ the maximum life span of the biosphere is determined exclusively by planetary geodynamics, which is independent (in a first approximation, but see limiting effect 4) of R. So we obtain the limitation $t < t_{\max}$.
4. There have been discussions about the habitability of tidally locked planets. We take this complication into account and indicate the domain where an Earth-like planet on a circular orbit experiences tidal lock-

ing. That domain consists of the set of (M, t) couples which generate an outer HZ boundary below the tidal-locking radius. This limitation is relevant for $M < 0.6\,M_\odot$. As an illustration we depict the HZ for $R = 2$ AU in Figure 3.

3. Orbital Stability

Planetary habitability requires orbital stability of the Earth-type planet over a biologically significant length of times in the HZ. The analysis of orbital stability of (hypothetical) terrestrial planets in extrasolar planetary systems has to take into account the effects of the giant planet(s) in those systems. In many cases the giant planets restrict the orbital stability of the terrestrial planet to a small or very small orbital domain or prevent orbital stability completely. There exists a variety of papers discussing the orbital stability of (hypothetical) terrestrial planets in extrasolar planetary systems, which is strongly influenced by the masses, orbital positions and eccentricities of Jupiter-size planets in such systems. Jones et al., (2001) analysed the stability of orbits of terrestrial planets in several known extra-solar planetary systems. They used a mixed-variable symplectic integrator by Chambers (1999) over a time scale of $\approx 10^9$ years, while Pilat-Lohinger and Dvorak (2002) used a Lie-series method (e.g., Hanslmeier and Dvorak, 1984) for the calculation of the orbits of an Earth-like planet. The stability can be determined by checking for close encounter or using the maximum eccentricity method. If the eccentricity of the Earth-like planet reaches a critical threshold the orbit is temporarily leaving the habitable zone. Nevertheless, Williams and Pollard (2002) investigated Earth-like worlds on high eccentric orbits with excursions beyond the HZ. Such planets can remain habitable only in the case of sufficient high atmospheric density. In a first approximation planets with such orbits are excluded from habitability. If the Earth-like planet approaches three Hill radii of the giant planet severe orbital perturbations of the terrestrial planet occur. The Hill radius R_H is defined as

$$R_H = \left(\frac{m}{3M}\right)^{1/3} a, \tag{7}$$

where m is the mass of the giant planet, M the central star mass and a is the semimajor axis.

Another method to assess the stability of planetary systems is the indicator Mean Exponential Growth of Nearby Orbits (MEGNO) applied by Goździewski (2002) to extrasolar planetary systems. The stability is determined by calculating the Lyapunov exponents of nearby orbits. It is based on the same ideas as the method of Lyapunov Characteristic Numbers (LCN). In contrast to LCN which requires integration over long evolutionary time the MEGNO method is very fast.

In our study we are using results taken from Jones and Sleep (2003). They calculated extensively the stability of orbits for the systems 47 Ursae Majoris (47 UMa), ε Eridani (ε Eri), ρ Coronae Borealis (ρ CrB), Gliese 876 and ν Andromeda and found that the first three systems can in principle host habitable Earth-like planets on stable orbits. The HZ has been determined by using results from Kasting et al. (1993). Additionally we are using our results from a previous paper (von Bloh et al., 2003) for the system 55 Cancri (55 Cnc). In Table I the luminosity, stellar age and range of stable orbits for terrestrial planets of the four extrasolar planetary systems under investigation (ε Eri, 55 Cnc, 47 UMa, and ρ CrB) are provided. For the stellar age of ρ CrB we use a value from Noyes et al., (1997), and for 47 UMa from Cuntz et al. (2003).

4. Results and Discussion

The HZ for a fixed central star luminosity of $L \equiv 1\,L_\odot$ is calculated as a function of the age of the stellar system. For the investigation of an Earth-like planet under the external forcing, we adopt a model planet with a prescribed continental area. The fraction of continental area to the total planetary surface is varied between 0.1 and 0.9. According to Franck et al. (2000a, Table I) a constant continental area yields the maximum life span of the biosphere. Therefore, we have chosen this scenario in order to get the

TABLE I

Stellar parameters (luminosity, age) and range of orbital stability of an Earth-like planet in the extrasolar planetary systems ε Eri, 55 Cnc, 47 UMa, and ρ CrB. The sources of data are denoted by footnotes.

System	Luminosity (solar units)	Age (Gyr)	Orbital stability (AU)
ε Eri	0.28[b]	1[b]	$R < 0.59$[b]
55 Cnc	0.69[a]	4.5[a]	$R > 0.72$[a]
47 UMa	1.55[c]	6.3[c]	$R < 1.25$[c]
ρ CrB	1.61[b]	10[d]	$R > 0.80$[b]

[a] von Bloh et al. (2003)
[b] Jones and Sleep (2003)
[c] Cuntz et al. (2003) and Franck et al. (2003)
[d] Noyes et al. (1996)

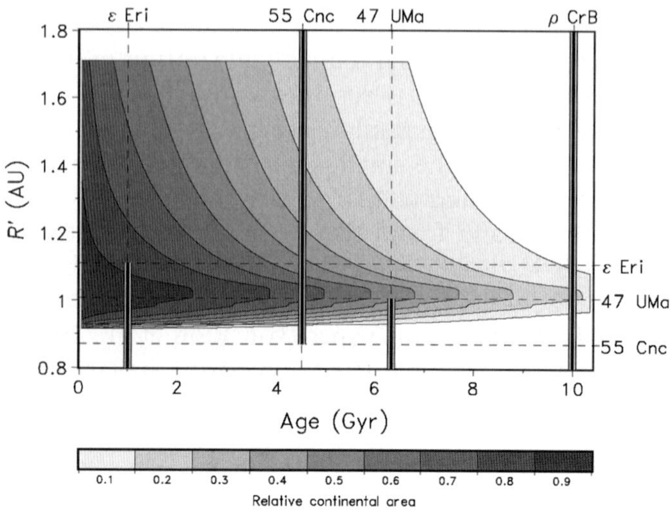

Figure 4. The HZ around an central star for a luminosity of $L = 1L_{\odot}$. The grey shaded areas indicate the extent of the HZ for different relative continental areas. The vertical bars indicate the permissible orbital range as constraint by the stellar age and the orbital stability limit for the systems ε Eri, 55 Cnc, 47 UMa, and ρ CrB. The limits R' are rescaled to $1L_{\odot}$, $R' = R\sqrt{L_{\odot}/L}$.

most optimistic estimation of habitability of an extrasolar planetary system. The HZ is calculated by combining the Equations (5) and (6) with the solutions for T_{surf} and P_{atm} from Equations (2) to (4).

In Figure 4 the grey shaded areas indicate the HZs for different constant continental areas of the Earth-like planet. While the inner limit of the HZ does not change significantly with age, the outer limit shows a non trivial behaviour. Up to a critical age the outer limit is constant and is determined by the maximum CO_2 atmospheric pressure (5 bar). Beyond this critical point the outer boundary moves inward due to geodynamic effects. At this point the source of carbon released into the atmosphere is too low to prevent the planet from freezing. An ultimate life span t_{max} of the Earth-like planet is determined by the coincidence of the outer and inner boundary. For a planet older than this ultimate life span, $t > t_{max}$, no habitability can be found. The critical age and the ultimate time span is an decreasing function of the relative continental area of the Earth-like planet. It is obvious that an almost completely ocean-covered planet ("water world") has the highest likelihood of being habitable (Franck et al., 2003).

The intervals of orbital stability (see Table I) for the four extrasolar planetary systems are plotted as vertical bars in Figure 4. For comparison they are rescaled to a luminosity of $1\,L_{\odot}$, $R' = R\sqrt{L_{\odot}/L}$. The range of dynamical habitability can be easily derived by the intersection of the vertical bars with the HZ for a certain relative continental area. The likelihood that an Earth-like planet is both on a stable orbit and also within the HZ can be quantitatively estimated from the width of the intersection of the HZ and the range of stable orbits (O_{stable}):

$$\Delta R = \max(HZ \cap O_{\text{stable}}) - \min(HZ \cap O_{\text{stable}}). \tag{8}$$

The widths ΔR of these intersections are shown in Figure 5 for the four extrasolar planetary systems as a function of the relative continental area. We find that the system 55 Cnc and ε Eri are most promising for finding Earth-like planets in the HZ on stable orbits. While for the system ρ CrB the habitability is strongly reduced by its high age, dynamical constraints are limiting the width for the system 47 UMa. The likelihood of finding Earth-like planets in 55 Cnc is only slightly lower than in our solar system (dashed line in Figure 5). In particular the likelihood of finding dynamic habitable planets in the system ε Eri is almost independent of the continental area. In general, we can state that finding an Earth-like habitable extrasolar planet is the more promising the younger the system and the lower its land coverage on its surface. Younger systems like ε Eri tend to be more geodynamically active and therefore contain more carbon dioxide in the planetary atmosphere. This leads to a stronger greenhouse effect and a broader HZ. As a consequence, habitability is maintained at larger distances from the stars, i.e., regions of lower stellar flux densities, as also pointed out by Forget and Pierrehumbert (1997), Mischna et al.,

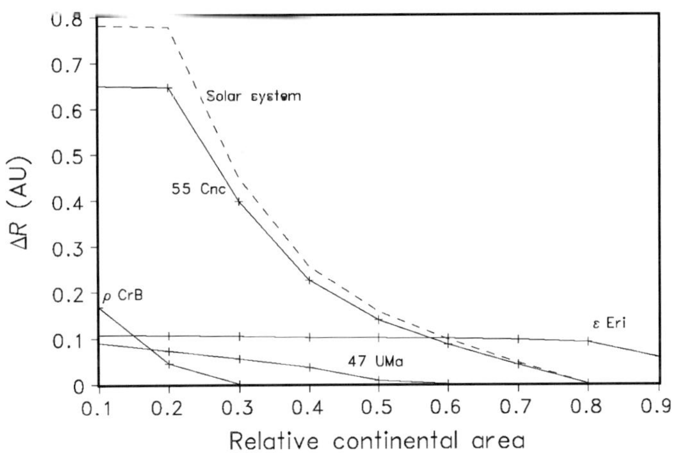

Figure 5. Widths ΔR (Equation (8)) of the orbital range both warranting habitability and orbital stability for the four extrasolar planetary systems (ε Eri, 55 Cnc, 47 UMa, ρ CrB) as a function of relative continental area. The dashed line denotes the result for the solar system.

(2000) and others in their studies on the early Martian climate. Planets with a relative high percentage of land coverage show a stronger weathering and therefore an enhanced removal of carbon dioxide from the atmosphere. This leads to a weaker greenhouse effect and habitability ceases at smaller ages.

5. Conclusions

We studied the principle possibility of Earth-like habitable planets in the systems ε Eri, 55 Cnc, 47 UMa, and ρ CrB under the condition that these planets have successfully formed and are orbitally stable. In particular, we considered Earth-like planets with different ratios of land/ocean coverages. This study is based on the integrated system approach, which describes the photosynthetic biomass production under geodynamic conditions. We show that the likelihood to find a habitable Earth-like planet on a stable orbit around a central star in general depends critically on the percentage of the planetary land/ocean coverage. This is explicitly true for the system 55 Cnc where an almost completely ocean-covered planet (“water world") has the highest likelihood of being both habitable and orbitally stable. On the other hand, for relatively young systems like ε Eri, the habitability is only limited by dynamical constraints. The reason is that planets with increased land coverage show a stronger weathering and therefore increased removal of carbon dioxide from the atmosphere thus limiting habitability. Relatively young planets have a high geodynamic activity increasing the amount of carbon dioxide in the atmosphere by emissions at mid-ocean ridges. In contrast to Jones and Sleep (2003) the likelihood to find a habitable Earth-like planet in ρ CrB is rather low. The high age of $\approx 10\,\mathrm{Gyr}$ strongly reduces the width of the HZ. For an Earth-like planet covered with more than 30% continental area habitability vanishes.

Our integrated system approach provides an easy to use method to determine the dynamic habitability for virtual Earth-like planets of newly discovered extrasolar planetary systems. Once the luminosity, the stellar age and the range of stability are known one can directly derive the likelihood of finding a habitable Earth twin using the HZ diagram (Figure 4). In this way we offer a new tool to choose potential targets for future space missions (e.g., TPF, Darwin).

Acknowledgements

We thank Manfred Cuntz from the University of Texas at Arlington for helpful and stimulating discussions.

References

Bada, J. L.: 2004, How life began on Earth: a status report', *Earth Planet. Sci. Lett.* **226**, 1–15.

Berner, R. A., Lasaga, A. C. and Garrels, R. M.: 1983, 'The carbonate-silicate geochemical cycle and its effect on atmospheric carbon dioxide over the past 100 million years', *Am. J. Sci.* **283**, 641–683.

Caldeira, K. and Kasting, J. F.: 1992, 'The life span of the biosphere revisited', *Nature* **360**, 721–723.

Chambers, J. E.: 1999, 'A hybrid symplectic integrator that permits close encounters between massive bodies', *MNRAS* **305**, 793–799.

Cuntz, M., von Bloh, W., Bounama, C. and Franck, S.: 2003, On the possibility of Earth-type habitable planets around 47 UMa', *Icarus* **162**, 214–221.

Forget, F. and Pierrehumbert, R. T.: 1997, 'Warming early Mars with carbon dioxide clouds that scatter infrared radiation', *Science* **278**, 1273–1276.

Franck, S., Kossacki, K. and Bounama, C.: 1999, 'Modelling the global carbon cycle for the past and future evolution of the earth system', *Chem. Geol.* **159**, 305–317.

Franck, S., Block, A., von Bloh, W., Bounama, C., Schellnhuber, H.-J. and Svirezhev, Y.: 2000a. 'Reduction of biosphere life span as a consequence of geodynamics', *Tellus* **52**B, 94–107.

Franck, S., von Bloh, W., Bounama, C., Steffen, M., Schönberner, D. and Schellnhuber, H.-J.: 2000b, 'Determination of habitable zones in extrasolar planetary systems: where are Gaia's sisters?', *J. Geophys. Res.* **105**(E1), 1651–1658.

Franck, S., Cuntz, M., von Bloh. W. and Bounama C.: 2003, 'The habitable zone of Earth-mass planets around 47 UMa: results for land and water worlds', *Int. J. Astrobiol.* **2**(1), 35–39.

Goździewski, K.: 2002, 'Stability of the 47 UMa planetary system', *Astron. Astrophys.* **393**, 997–1013.

Hanslmeier, A. and Dvorak, R.: 1984, 'Numerical integration with Lie series', *Astron. Astrophys.* **132**, 203–207.

Hart, M. H.: 1978, 'The evolution of the atmosphere of the Earth', *Icarus* **33**, 23–39.

Hart, M. H.: 1979, 'Habitable zones about main sequence stars', *Icarus* **37**, 351–357.

Jones, B. W. and Sleep, P. N.: 2003, 'The orbits of terrestrial planets in the habitable zones of known extrasolar planetary systems', In: D. Deming and S. Saeger (eds.), *Scientific Frontiers in Research on Extrasolar Planets*, ASP Confer. Ser. CS-294, pp. 225–230.

Jones, B. W., Sleep, P. N. and Chambers, J. E.: 2001, 'The stability of the orbits of terrestrial planets in the habitable zones of known exoplanetary systems', *Astron. Astrophys.* **366**, 254–262.

Kasting, J. F.: 1984, 'Comments on the BLAG model: the carbonate-silicate geochemical cycle and its effect on atmospheric carbon dioxide over the past 100 million years', *Am. J. Sci.* **284**, 1175–1182.

Kasting, J. F.: 1997, 'Habitable zones around low mass stars and the search for extraterrestrial life', *Origins Life* **27**, 291–307.

Kasting, J. F., Toon, O. B. and Pollack, J. B.: 1988, 'How climate evolved on the terrestrial planets', *Sci. Am.* **256**, 90–97.

Kasting, J. F., Whitmire, D. P. and Reynolds, R. T.: 1993, 'Habitable zones around main sequence stars', *Icarus* **101**, 108–128.

Kippenhahn, R. and Weigert, A.: 1990, *Stellar Structure and Evolution*, Springer-Verlag, Berlin, Heidelberg.

Larcher, W.: 1995, *Physiological Plant Ecology: Ecophysiology of Functional Groups*, Springer-Verlag, New York.

Marcy, G. W. and Butler, R. P.: 2000, 'Planets orbiting other suns', *Astron. Soc. Pac.* **112**, 137–140.

Marcy, G. W., Butler, R. P., Fischer, D. A. and Vogt, S. S.: 2003, 'Properties of extrasolar planets', In: D. Deming and S. Saeger (eds.), *Scientific Frontiers in Research on Extrasolar Planets*, ASP Confer. Ser. CS-294, 1–16.

Mischna, M. A., Kasting, J. F., Pavlov, A. and Freedman, R.: 2000, 'Influence of carbon dioxide clouds on early Martian climate', *Icarus* **145**, 546–554.

Noyes, R. W., Jha, S., Korzennik, S. G., Krockenberger, M., Nisenson, P., Brown, T. M., Kennelly, E. J. and Horner, S. D.: 1997, 'A planet orbiting the star ρ Coronae Borealis', *Astrophys. J.* **483**, L111–L114.

Pearcy, R. W. and Ehleringer, J.: 1984, 'Comparative ecophysiology of C_3 and C_4 plants', *Plant Cell Environ.* **7**, 1–13.

Pilat-Lohinger, E. and Dvorak, R.: 2002, 'Stability of S-type orbits in binaries', *Celest. Mech. Dyn. Astron.* **82**, 143–153.

Turcotte, D. C. and Schubert, G.: 1982, *Geodynamics*, John Wiley, New York.

Volk, T.: 1987, 'Feedbacks between weathering and atmospheric CO_2 over the last 100 million years', *Am. J. Sci.* **287**, 763–779.

Von Bloh, W., Cuntz, M., Franck, S. and Bounama, C.: 2003, 'On the possibility of Earth-type habitable planets in the 55 Cancri system, *Astrobiology* **3**(4), 681–688.

Walker, J. C. G., Hays, P. B. and Kasting, J. F.: 1981, 'A negative feedback mechanism for the long-term stabilization of Earth's surface temperature', *J. Geophys. Res.* **86**, 9776–9782.

Wilde, S. A., Valley, J. W., Peck, W. H. and Graham, C. M.: 2002, 'Evidence from detrital zircons for the existence of continental crust and oceans on the Earth 4.4 Gyr ago', *Nature* **409**, 175–178.

Williams, D. M.: 1998, *The Stability of Habitable Planetary Environments*, PhD Dissertation, Pennsylvania State University, University Park.